Reflecting on Practice for STEM Educators

Reflecting on Practice for STEM Educators is a guidebook to lead a professional learning program for educators working in STEM learning environments.

Making research on the science of human learning accessible to educational professionals around the world, this book shows educators how to relate this research to their own practice. Educators' collective work broadens the scope of an organization's reach, and through this effort, the organization grows its social capital in its local community and beyond. This book offers opportunities to engage in processes that lead toward organizational learning by attending to the professional growth of the educators. Tran and Halversen show how learning together can shape the language and meanings by which educators do and talk about their work to support visitors' experiences. The book provides guidance on how teams of educators can build community as they engage in reflective practice.

Reflecting on Practice for STEM Educators will be essential reading for leaders of any organization that aims to educate and engage the public in science, technology, engineering, and mathematics. It will be particularly useful to educators who work in museums, zoos, aquariums, botanical gardens, youth organizations, after-school programs, and nature, science, and conservation centers.

Lynn Uyen Tran is Research Director in the Learning and Teaching Group at the Lawrence Hall of Science, University of California Berkeley, USA. Her work focuses on relating the science of learning to teaching practices of informal educators and university faculty, and cultivating reflective practice and professional learning among these communities.

Catherine Halversen is Senior Program Director in the Learning and Teaching Group at the Lawrence Hall of Science, University of California Berkeley, USA. Her work focuses on developing and disseminating in-person and online professional learning programs and instructional materials for STEM university/college faculty, informal educators, and K–12 teachers.

Reflecting on Practice for STEM Educators

A Guide for Museums, Out-of-School, and Other Informal Settings

Lynn Uyen Tran

Catherine Halversen

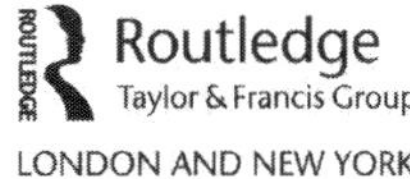

Routledge
Taylor & Francis Group

LONDON AND NEW YORK

First published 2021
by Routledge
2 Park Square, Milton Park,
Abingdon, Oxon OX14 4RN

and by Routledge
52 Vanderbilt Avenue,
New York, NY 10017

Routledge is an imprint of the Taylor & Francis Group, an informa business

© 2021 Regents of the University of California

The right of Lynn Uyen Tran and Catherine Halversen to be identified as authors of this work has been asserted by them in accordance with sections 77 and 78 of the Copyright, Designs and Patents Act 1988.

All rights reserved. Handouts and other sheets intended to be reproduced during the activities and sessions may be duplicated for educational use and are marked as such. No other part of this book may be reprinted or reproduced or utilized in any form or by any electronic, mechanical, or other means, now known or hereafter invented, including photocopying and recording, or in any information storage or retrieval system, without permission in writing from the publishers.

Trademark notice: Product or corporate names may be trademarks or registered trademarks, and are used only for identification and explanation without intent to infringe.

British Library Cataloguing-in-Publication Data
A catalogue record for this book is available from the British Library

Library of Congress Cataloging-in-Publication Data
A catalog record has been requested for this book

ISBN: 978-0-367-48826-0 (hbk)

ISBN: 978-0-367-48828-4 (pbk)

ISBN: 978-1-003-04299-0 (ebk)

Typeset in Minion, Myriad Pro and Whitney

Access the companion website: www.routledge.com/CW/Tran

Publisher's Note
This book has been prepared from camera-ready copy provided by the authors.

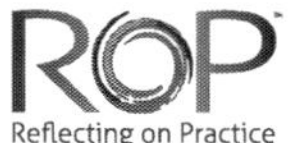

CONTENTS

Preface vii
Acknowledgements & Credits viii
Introduction 1

Module 1: Learning, Reflections, and Science

Introduces the routines and foundations of the *Reflecting on Practice* program.

Overview 9
Session 1: Learning Beliefs, Behaviors, and Goals 10

Initiates exploration of foundational ideas on learning.

Session 2: Effective Reflective Practice 36

Examines professional learning and reflective practice, including participating in a Video
Reflection in whole community.

Session 3: Nature and Practices of Science 68

Focuses on what is science, how science works, and science as a way of knowing.

Module 2: How People Learn

Articulates how learning builds on prior knowledge.

Overview 89
Session 1: How People Learn 90

Delves further into Five Foundational Ideas on Learning, and introduces the Learning Cycle.

Session 2: Learners' Prior Knowledge 122

Scrutinizes the connections between prior knowledge and conceptual change.

Session 3: Whole-Community Video Reflection on Prior Knowledge 156

Community engages in Video Reflections, first in whole community then small groups,
focusing on prior knowledge.

Module 3: Learning Conversations

Emphasizes that talking is learning.

Overview 171
Session 1: Talking to Learn 172

Examines how talking is learning, and introduces Discussion Map and Teaching Purposes
for facilitating learning conversations.

Session 2: Facilitating Conversations 208

Explores use of Facilitation Approaches to alleviate tension between everyday and
disciplinary views and language, leverage power responsibly, and foster discourse.

Session 3: Whole-Community Video Reflection on Learning Conversations 240

Community engages in Video Reflections, in whole community and then small groups,
focusing on facilitating conversations.

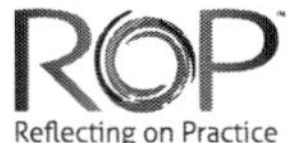

CONTENTS

Module 4: Objects and Design
Spotlights how we use objects in our learning designs.

Overview 253
Session 1: Teaching with Objects 254
Determines limitations and potentials for learning from different types of objects.
Session 2: Designing Experiences for Learning 284
Revisits research and uses Rapid Design Feedback to revise the design of learning
experiences.
Session 3: Whole-Community Video Reflection on Objects 300
Community engages in Video Reflections, and uses Critiquing Objects
to reflect on their use of objects in learning experiences.

References 311

Appendix A: Tools for Reflective Practice 315
Protocol, guides, checklists, and worksheets for making, observing, and discussing videos
and other work products during reflection sessions.

Appendix B: Design and Teaching Tools 333
Models, frameworks, and worksheets for designing and teaching learning experiences used
and examined in the *Reflecting on Practice* program.

Index 344

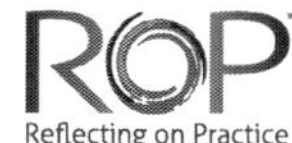

Preface

This guidebook puts into practice the *Reflecting on Practice*™ program that was developed at the Lawrence Hall of Science, University of California, Berkeley by the authors. It was grounded in Tran's areas of scholarship in the work, knowledge, and identities of informal science educators. It was inspired by Halversen's work teaching scientists, formal and informal educators, and undergraduate science students how to communicate science based on how people learn. The authors garnered multiple federal grants over the years to develop, test, refine, and disseminate the program and written curriculum. Hundreds of educators from informal science education organizations (predominantly across the US) have used this curriculum to implement *Reflecting on Practice* with their colleagues, transforming the ways they engage with the many thousands of learners they reach annually. This guidebook enables the authors to make the materials and program more broadly available to the informal science education community globally.

As our field recovers from the coronavirus pandemic of 2020, rebuilding our workforce must include dismantling structural racism. That work requires constant reflections and learning, both individually and collectively. We hope this guidebook can be a useful resource towards that effort.

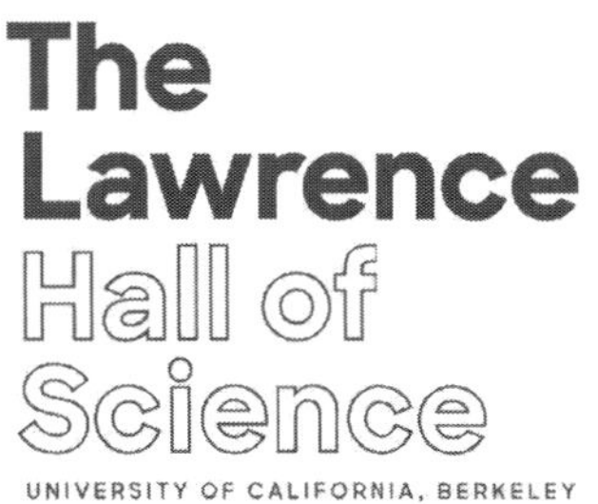

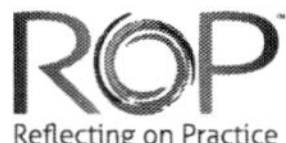

Acknowledgements

We're thankful to our incredible colleagues for their unwavering leadership and commitment in helping to make this guidebook a contribution to advance our field (Naina Abowd, Dave Bader, Lindzy Bivings, Marilyn Brink, Karen Burns, Kristin Evans, Preeti Gupta, Agnes Kovacs, Alie Lebeau, Katherine Miller, Sarah Pedemonte, Sara Pelleteri, Lindsay Thomas, Emily Yam, Daniel Zeiger).

Much gratitude to our curriculum editor (Nicole Parizeau) and designer (Jean Butterfield) for their expertise, creativity, professionalism, and patience.

We thank our colleagues at the Lawrence Hall of Science for their support.

Development of this material was supported by grants and awards from the National Science Foundation (Grant No. AISL-1612515 and ESI-0540417) and the Institute for Museums and Library Services (Grant No. MP-09-0002-09). Any opinions, findings, conclusions, or recommendations expressed in this publication do not necessarily reflect the views of the funders.

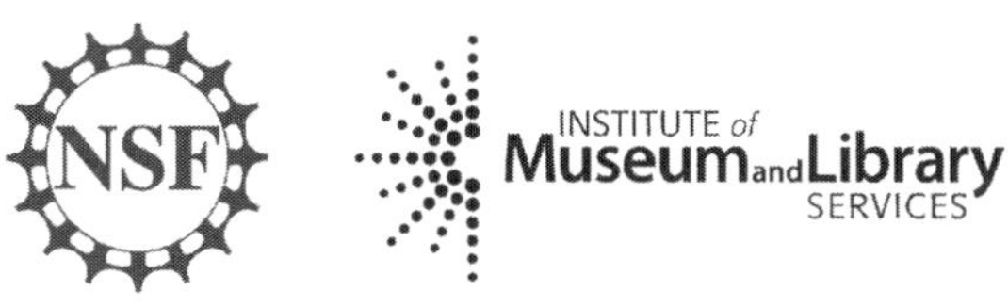

Credits

Lynn Uyen Tran, Author

Catherine Halversen, Author

Nicole Parizeau, Curriculum Editor

Jean Butterfield, Designer

Jeff Nesmith, Jr., Logo Creator

Introduction

Reflecting on Practice is a professional learning program designed for educators in informal science learning environments. It's intended for organizations to adopt and implement themselves. This book guides educational leaders within an organization to facilitate experiences with colleagues to engage in ongoing learning about their practice.

Experiences described in this guidebook make research on learning and teaching science available for educational professionals to talk about, reflect on, and apply to their practice. Staff at the organization learn together and from one another, and in the process negotiate meanings and develop shared language by which they talk about and do their work to support learning. This program requires commitment from both the organization and individuals. Management in the organization offers time, resources, and freedom needed for professionals to learn, while individual participants remain open-minded and willing to scrutinize their practice and change as they learn. Over time, new leaders within the community emerge to lead colleagues at the organization through *Reflecting on Practice*, and the cycle continues. Thus, cultivating leadership capacity from within is inherently built into this program.

The goal of *Reflecting on Practice* is to **advance the informal science education field by cultivating communities of learners among its professionals**. In support of this goal are three key objectives:

- Build shared language and understanding among professionals by delving into research on learning and teaching.

- Engage practitioners in habits of reflection through observing their own teaching, to develop their practice and make it public.

- Nurture a tradition of continued professional learning among participants, and thereby build a professional learning community.

Learning as Professionals and Why it Matters

Few would disagree that workforce learning is a necessity for organizations to stay competitive, especially in knowledge-based work like education. What remains elusive is how learning is positioned and enacted within the organization. In workplaces, the key objective is not learning, but rather successful delivery of their goods and/or services (Unwin, 2004). In education-focused organizations, like museums and universities, those goods and services are learning experiences, but for their patrons, not their employees. Learning happens for employees because humans are social organisms who learn as we participate in social activities (Brown, Collins, & Duguid, 1989; Lave & Wenger, 1991). Work is a prominent social activity throughout our lifetime, especially in adulthood. Thus, places of work are places of learning, regardless of whether workers and managers consider the workplace as such. As we participate in our social activities, humans learn by imitation, collaboration, and instruction (Bransford et al., 2006; Meltzoff & Decety, 2003; Vygotsky, 1978; Wood, Bruner, & Ross, 1976); in work environments, the ways that learning happens have been described as informal, incidental, and formal (Elkjaer & Wahlgren, 2005; Marsick & Watkins, 2001).

While efforts to identify direct correlation between staff development and organizational performance have fallen out of favor (Unwin, 2004), the desire for improved performance from investments in staff remains. Understandably, it's valuable for learning acquired by individual staff to transfer to the organization. What's inappropriate, however, is how learning for staff is often viewed and carried out (Tran, Gupta, & Bader, 2019). Entrenched in traditions of workforce development and organizational management, staff development has common features that are counterproductive to learning or transforming practice. Learning is situated as separate from working—*if they're learning, they're not doing their work*; thus, staff development is typically treated as a burdensome transaction that's undertaken to remain competitive. Staff learning is approached from a deficit perspective—*if they know this, they'll be better at their job*; thus, discrete trainings to "fix people" are offered. Learning experiences are usually one-way transmission of information— *let me tell you what you need to know*; thus, demonstrating the flawed association that telling is teaching. Learning opportunities are available hierarchically—*managers first*; thus, reserving opportunities to those in positions of power. Education-focused organizations aren't immune to these approaches towards workplace learning.

We consider learning at work from another mindset: the workplace is a learning environment and learning occurs through participation in the social practices of work (Billett, 2004). This broader viewpoint on learning is foundational to educational work in informal science learning environments but is rarely extended to the staff in the organization. Figure 1 illustrates how different areas of scholarship in learning and teaching are structured in the *Reflecting on Practice* program. This guidebook is a curriculum for workplace learning (Billett, 2002); staff learn as they participate in doing their work and talking about what they're doing. We position the educational professionals who participate in the program as learners, learning about their own work. They're

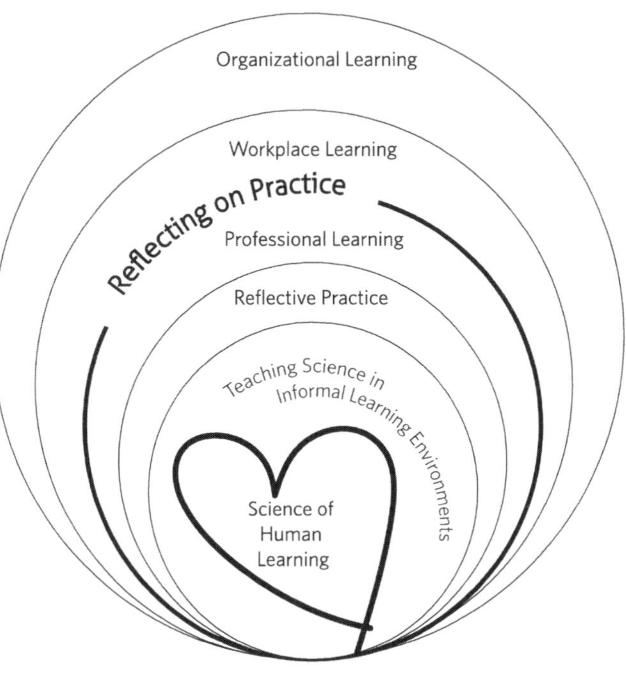

Figure 1. Areas of scholarship in the *Reflecting on Practice* program.

partaking in professional learning, which is fundamental to developing expertise in any profession (Webster-Wright, 2009). The **science of human learning** is at the heart of the program (and guidebook). Participants explore this content on learning and consider how to use it in their practice as they teach science in informal learning environments; we use this content to inform design of the experiences. Facilitators of this program must also use this content to support their staff's learning. Learning about their practice entails engaging in reflective practice, individually and with colleagues. Participants negotiate meanings and language to talk and think about their work, and make connections across experiences. Facilitators who use this guidebook with their staff, learn alongside their colleagues. As the community learns together, the learning gains and knowledge cycle between the individuals and the organization (Crossan, Lane, & White, 1999).

It's convenient to treat this guidebook as a well-designed training program for new or temporary staff. Indeed, it can be sensible to take such a restrictive approach towards workplace learning for seasonal staff hired to do discrete work (Fuller & Unwin, 2004). However, limiting usage of this guidebook as training material for onboarding sells short the potential. Stories from places that have adopted the program demonstrate the long-term potential (Tran et al., 2019; Tran, Werner-Avidon, & Newton, 2013). This program and guidebook are designed to cultivate communities of learners and habits towards mastery among professionals within an organization (Chadwick & Raver, 2015), and facilitate learning and knowledge creation to move from individuals into the organization and back again (Crossan et al., 1999). Ultimately, the community and habits you cultivate, and the learning gains your organization achieve through this program, are yours to keep.

Learning from Individual to Organizational, and the Structure of this Guidebook

An important issue in organizational learning that needs to be explored pertains to transfer of learning from the individual to the organization. While calculating whether *this amount of investment* in staff will yield *that amount of return* in their work is a futile exercise (humans aren't robots!), it's necessary to put our energy into ensuring pathways for learning to move from individuals to the organization. Learning is done by humans, not organizations. The organization benefits from the learning gains only when people can apply what they learned into how they do their work, and those gains endure when they're shaped by the community and become artifacts of the community. The organization "learns" when people enact changes from what they've learned and those new ways of doing and thinking become systematized, taken-for-granted patterns. Consequently, organizational learning falls short if the individual's learning isn't attended to appropriately.

Valuable learning occurs incidentally and informally at work (Marsick & Watkins, 2001), but we shouldn't rely exclusively on happenstance (Billett, 2004). If there are learning outcomes desired in the workplace, then intentional pathways can be designed to support them. Crossan, Lane, and White (1999) offer a useful, multi-level framework to conceptualize organizational learning that links the three levels within organizations (individuals, groups, and the organization) based on four social and psychological processes (intuiting, interpreting, integrating, and institutionalizing). These processes inform each other to lead to organizational learning. **Intuiting** is "the preconscious recognition of the pattern and/or possibilities inherent in a personal stream of experience" (p. 525) that happens individually; it affects others if they interact with the individual. **Interpreting** is the process of explaining a new idea to oneself and others through words or actions, which can occur at both the individual and group levels. **Integrating** is the process through which a shared understanding is developed among individuals and coordinated action is taken through mutual adjustment, and this process takes place at the group and organizational levels. **Institutionalizing** occurs at the organizational level when routinized actions transpire or become part of the taken-for-granted patterns.

Figure 2 illustrates how we envisage this framework in the *Reflecting on Practice* program; the three circular objects are each a level, bolded text are the processes, solid arrows represent experiences in the guidebook that facilitate flow within and between the processes, and the broken arrows offer the potential for feeding back from the institution to other processes and levels. The tasks, exercises, and tools (structures) in this guidebook offer opportunities to engage in these processes in a deliberate way. Table 1 lists which type of processing each of the structures generally support. Every session includes a combination of these structures, described briefly in the suggested session agenda.

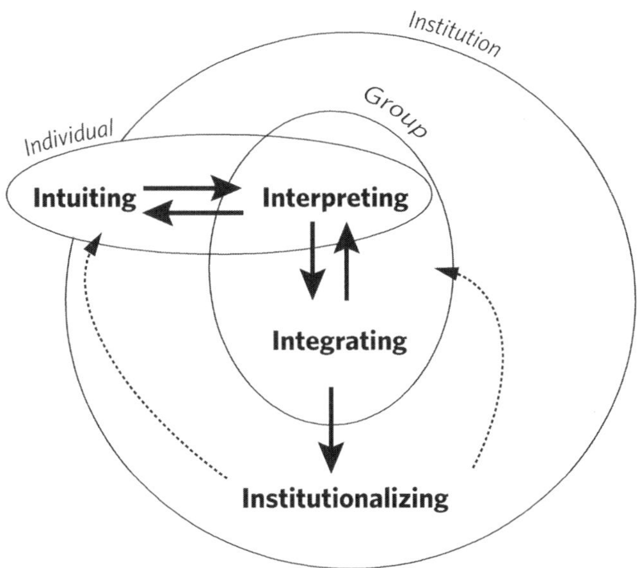

Figure 2. Application of Crossan, Lane and White's (1999) multi-level framework on organizational learning in *Reflecting on Practice*.

Table 1. Type of social and psychological processing generally supported in the tasks, exercises, and tools in *Reflecting on Practice*.

Reflecting on Practice			
Level	**Process**	**Tasks**	**Exercises & Tools**
Individual	Intuiting	• Hands-on • Continue the Learning	
Group	Interpreting	• Retrieve & Connect • From Learners' Perspective • Research Discussions	• Learning Journals • Reflection Worksheets • Video Reflections—presenter
	Integrating	• Let's Talk Practice	• Educator Moves • Video Reflections—peers • Institutional Practices
Organization	Institutionalizing		• Design and Teaching Tools • Tools for Reflective Practice

There are 12 sessions (~2.5 hours each) detailed in this guidebook, organized into four modules that examine topics (learning, prior knowledge, conversations, objects and design of learning experiences) relevant to educational practice in informal science learning environments. There are two **interactive** sessions and one **reflective** session in each module. The tasks and tools in **interactive sessions** engage participants in reflections and discussions on learning and teaching— both drawing on their personal and professional experiences and integrating research from many disciplines, including psychology, education, and sociology. The tasks are enacted through various Thinking Routines that make learners' thinking recognizable and visible to themselves and others. Participants express their beliefs and understanding, review their habits and assumptions, consider applicability of ideas from research, and determine ways to change their practice.

Reflection sessions are occasions when the community meets to present samples of their work and provide feedback on that work, in both small and whole group. Video is the primary type of work sample reviewed in this guidebook. By using video clips of their real-world interactions to reflect on their practice and draw on facilitated observations and feedback from colleagues, participants experience profound change in their teaching habits and mindsets. ***Reflecting on Practice* Tools for Reflective Practice** are designed to make the process, roles, and responsibilities known to all. This transparency is intended to let everyone know what to expect and what is expected of them.

Learning Content in Service of Equity and Justice

Informal science learning environments have gained recognition as important cultural institutions—places to learn about science, develop career aspirations in science, and cultivate interest and enjoyment in science with family and friends (Ellenbogen, Luke, & Dierking, 2004; Falk & Dierking, 2010; Lehr et al., 2007). There is also growing acknowledgement that these same environments aren't equally beneficial and inviting to all. Typically, the staff is predominately white and middle class, as are the usual visitors (Dawson, 2014a). Learning experiences often favor English language practices and Western representations of science and people in exhibits, signage, etc., excluding members of ethnic groups and competing ways of knowing (Ash, 2004; Garibay, 2009). The physical spaces are structured in ways that tend to exclude the needs of people with disabilities (Sandell, Dodd, & Garland-Thomson, 2010). Ways to explore and navigate the space may reinforce social disadvantages for some visitors (Dawson, 2014b). Collectively, this research shows how the education sector isn't impervious to society's inequities and injustices, nor is it entirely innocent from perpetrating acts that inflict undue harm on marginalized communities.

Education, whether in formal or informal environments, exists within the sociopolitical context. It's part of society's system, which means it affects and is affected by what happens in society. With this awareness comes the need for educators to understand oppression in individual and systemic terms, and realize the implications on how they do their work; simply treating everyone the same won't make limits to equal opportunity disappear. It's also important for educators to understand power in personal and social terms, and how to exercise their power responsibly. Without socially conscious interrogation, there is risk of reducing equity pedagogies to a list of strategies to check off or focus on celebrating culture over transforming curricula (Sleeter, 2012). These familiar actions reveal the flawed assumption that valuing learners' cultural diversity will solve the problems of systemic inequities. In short, despite their good intentions, educators can't achieve the positive impacts they desire if they aren't engaging in critical conversations. To engage in such conversations requires educators' willingness to reflect upon themselves and to sometimes be uncomfortable.

Certainly, informal learning environments are just one component in the larger system of society, and educators are just one group in the larger organizational system. However, we're also not entirely powerless. As organizations and individuals, we can **choose to be critically conscious** of how injustices are perpetuated, both through individual and institutional actions. We can **choose to enact change** where we can. Taking action involves hard choices concerning organizational structures that perpetuate inequities and exclusion, such as governance, funding, hiring practices, leadership development, etc. In educators' daily practices those choices include, but aren't limited to, how they: design learning experiences, interact with learners, and treat one another. Those choices rely on the strength and mindset of the community because the work towards equity and justice is difficult and ongoing.

The content featured in this guidebook is focused on how people learn and how to support their learning, which is of value for all learners. However, the critical conversations on social inequalities in education necessary to truly uplift all learners aren't brought forth explicitly. The work towards equity and justice is extensive and requires everyone's attention through multiple pathways. There's no single solution to address the systems-level change needed. This guidebook can be used in service of the larger effort to teach for equity.

- **The learning content.** The guidebook emphasizes looking at learners' prior knowledge as places of strength, rather than deficit (Module 2). This mindset is critical in teaching for equity because all learners—not just those from privileged backgrounds—have useful resources that are the foundation for their learning. Helping learners to see and appreciate their own strengths and talents requires educators to see it, too. This guidebook stresses the significance of talking when learning, and fosters skills to invite learners to talk, and for educators to listen closely when learners speak (Module 3). This ability is crucial in teaching for equity because we can't truly care for all our learners' well-being and academic success if we don't know how to make space for them to express themselves or are unable to listen to them with whole-heart and open-mind. Without this understanding, we also won't be able to recognize this need within our community as we learn together. This guidebook pushes on using objects to support learning and urges critical examination of how we design learning experiences (Module 4). This habit of scrutiny is essential in teaching for equity because we need to know how to intentionally design our learning experiences to support all learners.

- **The culture and community.** This guidebook routinizes reflective practice, productive feedback, and collegial discourse as "the way we do things here." These routines as taken-for-granted patterns for work are fundamental to teaching for equity because the work towards justice requires self-interrogation, intergroup dialogue, and ongoing learning within a community. Encouraging educators to engage in the hard work needed to make equity and inclusion a part of their practice entails raising their social consciousness and normalizing critical dialogue among colleagues during their daily work where and when things happen, not relegating them to occasional, training meetings led by outside experts. This guidebook aims to build the culture and collegiality among peers that enables its members to grow and be responsive. A culture that's driven by growth and understanding may be more willing to confront the difficult questions needed to challenge systems of oppression and injustice. A community with strong social relationships among its members may be more resilient to the uncomfortable conversations that need to occur to actively challenge all forms of injustice through and in education.

Using this Guidebook, Implementing this Program

This guidebook is a comprehensive resource for busy professionals like yourself to enact with your colleagues. Additional examples and other planning documents for implementation are available as digital resources in the Companion Website.

Composition of your community. *Reflecting on Practice* is intended for educational professionals across the spectrum of experiences, qualifications, work-type, and audience-focus throughout your organization to learn together. Depending on the size, structure, goals, and needs of your organization, the community could comprise: five to 50+ people; staff from one or multiple departments; predominantly educators or include administrative support and animal caretakers. Be aware, if implementation is treated as "training" rather than professional learning, those with advanced degrees and/or years of experience will feel it's inappropriate for them, and then your community wouldn't benefit from their insights.

Facilitation team. Ideally, a team of at least two leaders and/or emerging leaders should facilitate the program. Given that the facilitators participate in the program while also leading and coaching their colleagues, it's best to have a learning buddy. The program has the potential to build leadership capacity from within your organization, so facilitators needn't be directors of departments exclusively. Consider an apprenticeship approach for the facilitation team, as well as using it to disrupt standard practices for leadership building. Be aware, facilitators need to be appropriately supported to be/become leaders by their colleagues, otherwise, undesirable social dynamics can overshadow the learning opportunities.

How the guidebook is organized. We've intentionally provided extensive details so that it's easy to use. Although not a recipe for how to teach, teaching is modeled throughout the experiences. Notes in the margins offer explanations and clarifications as you consider making adjustments; it'll be

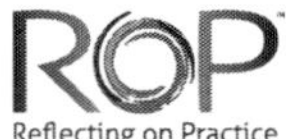

helpful for you to articulate why your changes make better sense for your community. Module 1 Session 1 has extra notes to elaborate further on components in all sessions. Details in each session include the following, in order of appearance.

- **Overview** and **Objectives**

- A suggested **Agenda**—annotated with descriptions and estimated time for each task

- **Materials**—what's needed, including all hands-on materials and copies of handouts

- **Getting Ready**—detailed instructions to prepare to facilitate, including:
  - readings and reminders for the facilitator to review,
  - reminders and suggestions for participants before the session,
  - preparing materials and web links for use during the session,
  - readings and assignments for participants to **Continue the Learning** (unless specified, all readings are free and downloadable from The National Academy Press, www.nap.edu), and
  - advance preparation for upcoming sessions.

- **Step by Step**—detailed directions for facilitating discussions and activities. The sidebar includes thumbnails of each handout for easy reference, and a projector icon denotes when to display corresponding PowerPoint slides (found on the Companion Website). Tasks and Routines are called out throughout the Step by Step, helping the facilitator to monitor and be mindful of the purpose and objectives of each part of the session and learning progression.

- **Handouts**—readings and examples specific for a session are included within the session; tools, processes, and worksheets used in multiple sessions are found in the appendices. All the handouts can be duplicated for educational use.
  - **Appendix A: Tools for Reflective Practice**—includes tools and processes for making, observing, and discussing videos and other work products during Video Reflections.
  - **Appendix B: Design and Teaching Tools**—includes tools (such as Thinking Routines) and worksheets for designing effective experiences based on the frameworks and pedagogy introduced, discussed, and practiced throughout the *Reflecting on Practice* program.

Digital resources helpful for facilitating this guidebook are stored on the publisher's Companion Website (www.routledge.com/CW/Tran). These items include digital copies of the handouts, basic PowerPoint slide deck, implementation examples, etc.

MODULE 1.
Learning, Reflections, and Science

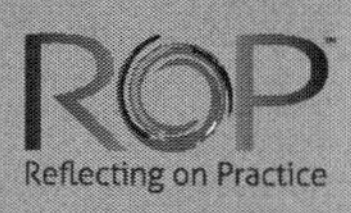

Overview

Module 1 introduces participants to the design and theoretical foundations of the *Reflecting on Practice*™ program. They learn that the program is an opportunity for practitioners with all levels of experience to learn from one another, develop shared language on educational practice, deepen their knowledge on the science of learning, and exchange ideas on how they teach.

Session 1 introduces the program and commitments, and presents **foundational ideas on learning** that will be explored throughout the program. It introduces tasks and routines integral to the program, including **research discussions** to relate their experiences and thinking to research; and **hands-on activities** designed to be engaging and challenging experiences for adults to think about learning and teaching.

Session 2 engages participants in **discussions about professional learning and reflective practice**. It introduces them to the **professional learning community** framework and the reflective tasks that are part of the program, with a special focus on observing and analyzing videos of one another's practice in **Video Reflections**. Participants practice using the **tools and processes** in their first Whole-Community Video Reflection.

In Session 3, participants gain insight into the **nature and practices of science** in the best way possible: by *doing* and *reflecting* on science. They explore, "What is science?" and "How does science work?" The conversation deepens with ideas on common misinterpretations of science, how science is one way of knowing, and young people's science aspirations and identities.

Module 1.
Session 1: Learning Beliefs, Behaviors, and Goals

Session Overview

This session serves as an introduction to the *Reflecting on Practice* program. It opens discussions about **how people learn** by introducing foundational ideas on learning. It also highlights the important contributions of **informal science learning environments** as places where people come to learn and develop interests in science. Participants learn about the major components of the program, and determine how they themselves will be involved and what their commitments will be.

Session Objectives

- Identify personal beliefs on learning, behaviors in learning, and goals for learning.
- Discuss the complexity of learning, and how this complexity pertains to learning and teaching science in informal environments.
- Determine personal goals and agendas for participating in the program.

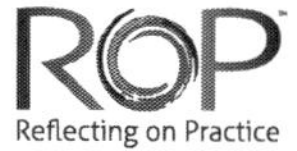

1.1

SESSION AGENDA		
Task Routine	**Description**	**Estimated Time (in minutes)**
Introduction (Part 1) *Goals and Objectives of Reflecting on Practice*	Goals and purpose of the program are introduced and Learning Journals distributed.	5
Retrieve & Connect Thought Swap: *Getting to Know Each Other*	First with a partner and then with the whole group, participants share ideas on a series of questions about (a) informal science education and (b) teaching and learning goals.	20
Introduction (Part 2) *The Program and Session 1*	Overview of the program and Module 1 is presented. Objectives of Module 1 Session 1 are introduced.	10
Retrieve & Connect 3-2-1-Bridge: *Ideas on Learning*	Participants jot down initial ideas on how they think people learn. They will revisit these thoughts in a later session.	5
Retrieve & Connect Turn & Talk: *Learning Beliefs and Behaviors*	Participants share their beliefs and assumptions about learning as they "turn & talk" to someone next to them. A whole-group share-out follows.	10
Break option		10
Hands-on *Cup & Card*	Small groups are presented with a hands-on challenge activity to promote discussion about learning and pedagogy.	20
From Learners' Perspective Table Talk: *What did you learn? How did you learn it?*	Participants examine the activity they just completed from their perspective as a learner.	10
Research Discussion Walkabout: *Five Foundational Ideas on Learning*	Participants briefly discuss why beliefs about the effectiveness of aligning teaching with learning styles are not actually supported by evidence.	55
Let's Talk Practice Minute Paper: *Personal Goals for the Program*	Participants gather their thoughts and write down personal goals in their Learning Journals.	3
Continue the Learning	Participants discuss the parameters and expectations for tasks outside of interactive sessions.	2
Total Estimated Time		150 mins. (2.5 hrs.)

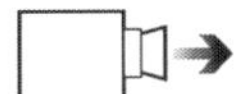

Note. Keep a lookout for this projector icon throughout the Step by Step in all sessions. Each appearance alerts you to "display" the corresponding slide from the slide deck.

"Fidget toys" are sensory objects that people can use to focus concentration, relieve stress, displace anxiety, quiet their minds, and for many other reasons.

Note. Handouts used in a session are located at the end of the session's Step by Step or the Appendices. All handouts can be duplicated for educational use (also see **Getting Ready** #1).

MATERIALS

Recurring for all Sessions

- Data projector
- White board or chart paper and markers
- PowerPoint slide deck for the session (see **Getting Ready** #1)
- Tape (masking or painter's)
- Sound-maker such as a bell, chime, and/or tambourine (for bringing groups back together)
- (Optional) Workshop materials kit for each table
  - Sticky notes, multiple sizes
  - Pens
  - Snacks
  - Fidget toys (e.g., modeling dough, Slinky®, etc.)

For this Session

- 5 chart paper posters for **Five Foundational Ideas on Learning** (see **Getting Ready** #11)
- 5 different-colored markers
- **Science Explanation for Cup & Card** (see Companion Website www.routledge.com/CW/Tran)

For each participant

- Handouts (1 copy each)
  - **Key Ideas from the Literature: Five Foundational Ideas on Learning**
  - **About the Program:** *Reflecting on Practice*
- 1 Learning Journal (see **Getting Ready** #5)
- (Optional) 1 folder to hold all handouts

For each small group

- 1 pan filled with water (about 13" x 9" x 6" deep or 33cm x 20cm x 15cm deep)
- 1 tray or box in which to place materials
- 5-7 pieces of thin plastic sheet, 2 different sizes (e.g., 3" x 5" and 5" x 8")
- 6-7 different types of clear containers, glass and plastic with different sized rim and volume (e.g., Erlenmeyer flask, wine glass, martini glass, water tumblers, small vase)
- Towel (for wiping up spills)
- ~ 50 coins of the same denomination (for the Coin Challenge)

Getting Ready

1. The publisher's Companion Website for this guidebook makes digital resources for facilitating these sessions easily retrievable. Find the website in the Support Material section for this guidebook at **routledge.com**. Digital resources include:

   - Basic **PowerPoint** slide decks for all sessions. Customize and insert additional slides as needed.

   - Digital copies of all the **handouts** (readings, tools, worksheets printed in this guidebook). Use for easier duplication when teaching.

   - Additional **Science Explanation** for hands-on activities. Reference as needed; determine whether to share with participants.

   - Additional **guidance documents** for implementing the program. Use for planning.

2. Send out a welcome invitation to all participants. The message should include the following types of information.

   - Gratitude. Thank participants for investing in themselves.

   - Support. Describe the organization's commitment to investing in staff.

   - Expectations. Share a brief description of time and anticipated commitment from participants. (This will be elaborated on further during the first session.)

   - Schedule. Provide a meeting schedule, if available (e.g., every other Wednesday from 2-5 pm from now until April).

   - Preparation. Distribute information to review prior to first meeting.

     ○ Read **About the Program:** *Reflecting on Practice* handout

     ○ Watch "**Learning styles & the importance of critical self-reflection**, Tesia Marshik" (TEDx Talks, 2015) (https://www.youtube.com/watch?v=855Now8h5Rs)

3. Review the handouts and slide deck for this session, and additional suggested readings.

   For the facilitator:

   - *How People Learn II* (Chapter 3: "Types of Learning and the Developing Brain").

Note. Basic slide decks are designed to complement the Step by Step details for each session. The slides include simplified instructions or text from the guidebook to use with participants during the sessions. These details are also just-in-time reminders for facilitators.

It might not be possible to "require everyone" to watch the video before the session. Make sure at least three participants (or 20% of the group, whichever is greater) watch at least the first 11 minutes of the video. This deliberate planning will make for a more productive collective recall to initiate the Research Discussion.

The learning content for each session is detailed in the text of the Research Discussions.

(continues)

Reflecting on Practice

(continued)

Unless otherwise specified, all readings are free and downloadable from the National Academies Press. Search for the following book titles at www.nap.edu:

- How People Learn (2000)
- How People Learn II (2018)
- Learning Science in Informal Environments (2009)
- Ready, SET, Science! (2008)
- Surrounded by Science (2009)
- Taking Science to School (2007)

The *Reflecting on Practice* professional learning program requires that both institutions and individuals invest time and resources for the professional growth of the participants. **Institutions adopt the program** and allocate time and space for staff to engage, learn, and test out new ideas. **Individuals participate in the program** and commit time and effort to reflect on, make public, and improve their practice.

- *Learning Science in Informal Environments* (Chapter 1: "Introduction").

Continue the Learning suggestion for participants:

- *Surrounded by Science* (Chapter 1: "Informal Environments for Learning Science").

4. Duplicate handouts.

5. If you've decided to supply the Learning Journals (spiral-bound notebooks, 3-ring binders, etc.), have them ready to distribute.

   The Learning Journal is a record book in which participants keep their written thoughts in one place. It's both a tool for <u>learning</u>, as participants make their thinking visible on paper, and for <u>reflection</u>, as participants review past entries to notice patterns and changes in their thinking.

6. Consider parameters and expectations for **Continue the Learning**—that is, assigned work for continued learning outside of the interactive sessions. This work generally consists of two kinds of assignments: reading and (starting in Module 2) doing Video Reflections.

   It's certainly helpful for participants to do reflective tasks outside of interactive sessions, so they actively engage with the ideas in their minds, share what they're trying, and seek help from one another. However, it's also understandable that these tasks may not be feasible for all participants. Some may be hourly paid staff; will they be paid for time spent doing these tasks outside of session hours? Participants likely already have too much work; where and how will time be made for this "extra work?"

   Before deciding on quantity of work beyond the sessions, the facilitation team should confer with participants' supervisors and senior management to determine whether these ongoing reflective tasks will be financially or otherwise supported.

7. Determine your requirements for the Video Reflections, and be ready to communicate them to the community in the next session (if not earlier). Participants will be anxious about the process; being upfront about expectations will help to alleviate their anxiety. Importantly, don't let their apprehension make you doubt the value of video reflections.

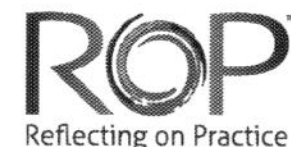

Video Reflections are driven by each individual's questions about their own practice, and as such may require participants to record a different video for each reflection. Every module includes at least one session in which participants engage in Video Reflection. There are two formats: first with the whole community, and then in small groups with their critical colleagues.

8. Make sure your room has sufficient space for participants to form two parallel lines for **Thought Swap**. It may be necessary to move into the hallway.

9. Prepare for the Hands-on task, **Cup & Card**. Prepare a tray of the items listed in materials "**For each small group**" of 4–6 participants. (<u>Keep the coins separate</u>; you'll distribute them to each group once it appears they are comfortable with the first part of the challenge.) Try out a few of the materials to make sure the plastic card is stiff enough to hold water when the container is inverted.

10. Review the **Science Explanation for Cup & Card** (see Companion Website) so you are comfortable with the content. Decide whether or not you will copy this optional handout for your participants. Keep in mind that the purpose of this task is not to reach a complete understanding of the physics concepts, but rather to initiate discussions about beliefs and attitudes about learning and what supports learning.

11. Prepare five chart-paper posters for the **Research Discussion**. Title each of the five sheets with a short description of one of the *Five Foundational Ideas on Learning*:

    - **Learning is an active process.**
    - **Learning builds on prior knowledge.**
    - **Learning that is authentic to the learner is more memorable.**
    - **Learning occurs in a complex social environment.**
    - **Learning requires motivation and cognitive engagement.**

    Small group sizes of 2–6 will work well for this activity. If you have fewer than 10 participants (which means that you can't start off a group of two at each poster), it's fine to have some of the posters unoccupied at any given time. Eventually, each group will be able to discuss and record at each of the posters. A smaller group just means that fewer participants will comment on any one poster.

Preparation for specific **Tasks** (e.g., Hands-on) and **Routines** (e.g., Thought Swap) are provided. Referencing corresponding details in the Step by Step will also be helpful as you plan.

Session 1 Step by Step

5 minutes

Introduction (Part 1)
Goals and Objectives of Reflecting on Practice

1. **Introduce goals and objectives of the program.**
   Display the program objectives and share the following:

   a. *Reflecting on Practice* is a modular program designed for adoption by informal science learning environments so that all educators in the organization can participate and learn together.

   b. The goal of the program is to advance the informal science education field by cultivating communities of learners among its professionals.

   c. The program has three primary objectives:

      - To <u>build shared language and understanding among professionals</u> by relating research to practice.

      - To <u>engage practitioners in habits of reflection</u>, including observing their own teaching, as a means to develop their practice and make it public.

      - To <u>nurture a tradition of continued professional learning</u> among participants, and thereby build a **professional learning community**.

   d. Emphasize that this program is **not** a blueprint for how to teach. Participants bring a wealth of pedagogical knowledge and experience to the program. *Reflecting on Practice* is an opportunity for practitioners with all levels of experience to learn from one another, develop a shared language on educational practice, deepen their knowledge on the science of learning, and exchange ideas on how they teach.

2. **Distribute and introduce Learning Journals.**
   Let participants know:

   a. **Learning Journals** are for their private use, to record in as they go through the program.

   b. The journals should be used to record their thoughts as they reflect on and learn about their own practice. Anything and everything can be entered in the journal. Use words, sketches, diagrams, etc. to express ideas. Encourage them to date their entries so they can see how their thinking changes over time.

A **professional learning community** is a group of practitioners who share and critically examine their practice in a way that is ongoing, reflective, collaborative, inclusive, learning-oriented, and growth-promoting.

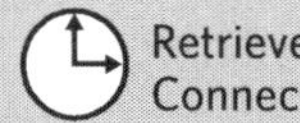

1·1

c. There will be opportunities to reflect and write during interactive sessions (e.g., "Quick Writes," "Minute Paper") as well as outside of the sessions.

The bar below denotes the Task (Retrieve & Connect), Thinking Routine (Thought Swap), and unique title and estimated time for this experience.

25 minutes — Retrieve & Connect

Thought Swap: *Getting to Know Each Other*

1. **A community builder first.**
   Tell participants before they proceed further, they will spend a few minutes bringing their minds into this session and getting to know one another.

2. **Form two lines.**
   Ask participants to form two lines facing one another, making sure each person in one line has a partner in the facing line. An easy way to ensure this is to have people across from each other make eye contact. Everyone needs a partner; if there is an odd number, someone from the facilitation team should join the line.

3. **Introduce the format.**
   Tell participants how the **Thought Swap** works:

   a. There is a set of questions for participants to discuss.

   b. You (the facilitator) will ask one question at a time.

   c. Participants will have 2–3 minutes to discuss their responses for each question with their partners.

   d. When time is up, you will initiate the "touch of silence" down the two lines until the whole group is quiet and ready for the next question.

4. **Display the first question:**
   What do you like about teaching science in your organization?

   As participants discuss the question with their partners, discreetly walk up and down the lines to get an overview of what partners are discussing.

   - **Listen to the noise.** Give participants 2–3 minutes to talk. If after 3 minutes the noise level is still fairly high, they may need an additional minute or two to talk. If the noise level drops (indicating that most discussion has stopped), initiate the "touch of silence" before the discussions spontaneously start up again.

   - **Ask for volunteers.** Depending on available time and size of community, encourage 2–4 volunteers to share what they discussed with their partner. Participants can share their own thoughts, what their partner shared, or what they talked about together.

Thought Swap is a whole-group discussion routine to accustom participants to sharing their ideas, first with a partner and then with the whole group. It's the first of many different **Thinking Routines** that will be used throughout the *Reflecting on Practice* program. The routines are designed to help make learners' thinking recognizable and visible to themselves and others. Thinking Routines are introduced in Module 4; details for how to use each one are located on the Companion Website.

Touch of silence: Facilitator touches the shoulder of the first person in each line. Those individuals stop talking and touch the shoulders of the persons next to them, who also stop talking and touch the shoulders of the persons next to them...and so on down the two lines.

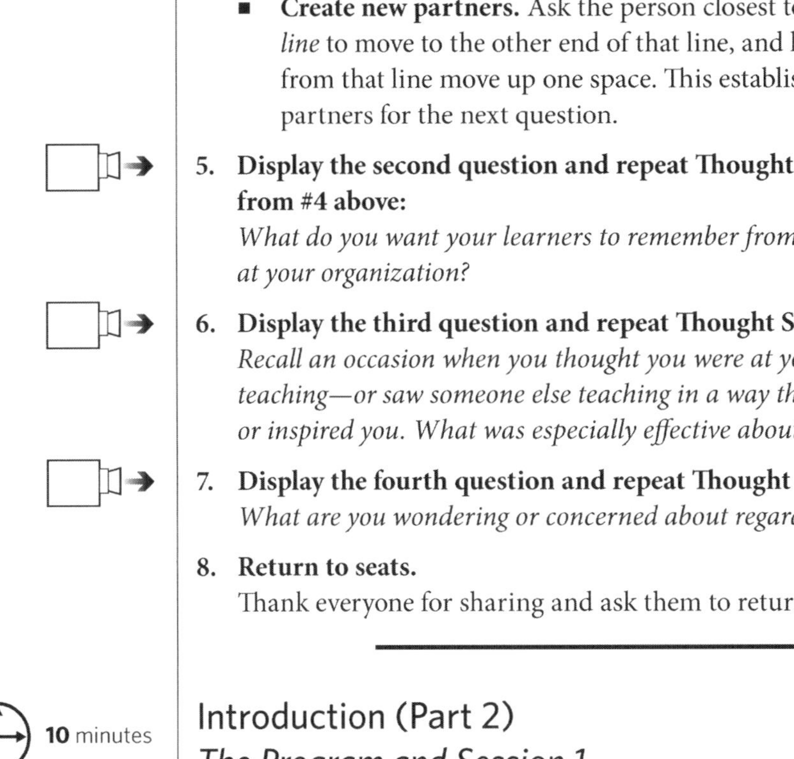

- **Create new partners.** Ask the person closest to you from *one line* to move to the other end of that line, and have everyone from that line move up one space. This establishes new sets of partners for the next question.

5. **Display the second question and repeat Thought Swap routine from #4 above:**
   What do you want your learners to remember from their experiences at your organization?

6. **Display the third question and repeat Thought Swap:**
   Recall an occasion when you thought you were at your best, teaching—or saw someone else teaching in a way that excited or inspired you. What was especially effective about it?

7. **Display the fourth question and repeat Thought Swap:**
   What are you wondering or concerned about regarding this program?

8. **Return to seats.**
   Thank everyone for sharing and ask them to return to their seats.

 **10** minutes

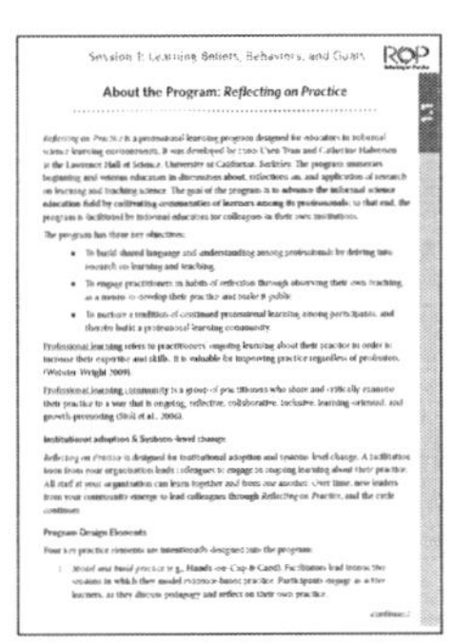

Introduction (Part 2)
The Program and Session 1

1. **Introduce modular design and the big ideas of *Reflecting on Practice*.**
   Distribute the **About the Program:** *Reflecting on Practice* handout and give participants a few minutes to read.

2. **Discuss handout.**
   Once most participants have had a chance to review the handout, invite questions, comments, and concerns for discussion. Some concerns expressed may include:

   - time it will take to participate and do the tasks on top of busy work schedules;
   - the logistics of making—and anxiety about sharing—videos of their teaching;
   - how this program is related to current/changing conditions at the institution;
   - how participation can/will affect job security and performance evaluations; and
   - value and purpose of the program for them, for the department, and for the institution.

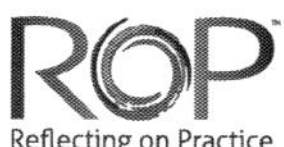

3. **Emphasize learning opportunity and professional recognition.**
   Let participants know the extent to which senior management supports and values their participation. Tell participants that this program is an opportunity for learning about their own practice for professional growth. For those who are interested, there is also opportunity for acquiring professional recognition after completion of this program. Let them know the following:

   Reflecting on Practice is a Learning Partner with the **Association of Zoos and Aquariums (AZA)**; an approved continuing education provider for the **National Association for Interpretation (NAI)**; and a special partner with the **Association of Science—Technology Centers (ASTC)**. Participants will receive a Certificate of Completion at the end of the program. They can submit a copy of the certificate to the respective associations to earn appropriate credit.

4. **Display and introduce objectives of Module 1 Session 1** and invite questions for clarification.

   - Identify your beliefs on learning, behaviors in learning, and goals for learning.

   - Discuss the complexity of learning, and how this complexity is relevant to learning and teaching science in informal environments.

   - Determine personal goals and agendas for participating in this professional learning program.

3-2-1 Bridge: *Ideas on Learning*

1. **Introduce 3-2-1 Bridge.**
   Have participants take out their **Learning Journals**. Describe that in this routine they will be given a prompt and asked to record their thoughts about it in three phases. Ask them not to overthink their responses! This is an opportunity to freely brainstorm their initial thoughts.

2. **Display prompt:**
   How do people learn?

3. **Ask participants to "record 3 words."**
   Have them quickly write three words that come to mind when they think about <u>how people learn</u>.

5 minutes · Retrieve & Connect

3-2-1 **Bridge** is a Thinking Routine that invites learners to uncover their initial thoughts, ideas, questions, and understandings about a topic, and then, after engaging in learning experiences and receiving some instruction, to connect these to new thinking and ideas. The other side of the bridge will occur in Module 4 Session 1 (or sooner if you prefer).

4. **Have them "ask 2 questions."**
   Now ask participants to write two questions they wonder about when thinking about <u>how people learn</u>.

5. **Ask them to "develop an analogy or metaphor."**
   Display the definitions and have participants write one analogy or metaphor that encapsulates their ideas about <u>how people learn</u>. (These definitions are adapted from literarydevices.net.)

This response may take a little more time for participants to complete than the first two.

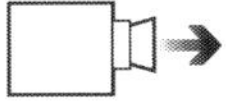

> An **analogy** is a <u>comparison</u> in which an idea or a thing is compared to another thing that is quite different from it. It "explains" that idea or thing by comparing it to something that is familiar.
>
> Example: How a doctor diagnoses diseases is like how a detective investigates crimes.
>
> **Metaphor** is a <u>figure of speech</u> that makes an implicit, implied, or hidden <u>comparison</u> between two things that are unrelated but share some common characteristics. In other words, a resemblance between two contradictory or different objects is made based on one or more common characteristics.
>
> Example: It's going to be clear skies from now on. (This implies that clear skies aren't a threat and life is going to be without hardships.)

6. **Conclude 3-2-1 Bridge routine.**
   Let participants know they will return to these ideas about learning later in the session and throughout the program. Before they put away their journals, remind them to date and title the entry "3-2-1 Bridge: Ideas on Learning," so they can easily find this entry later. Have them put their Learning Journals away for now.

Retrieve & Connect   🕙 **10 minutes**

Turn & Talk: *Learning Beliefs & Behaviors*

The **Turn & Talk** routine gives learners the chance to express their thinking immediately to someone because the topic, previous task, or novel situation sparked reactions and memories that learners want to share.

1. **Introduce Turn & Talk routine.**
   Tell participants they will now have the opportunity to share their beliefs and assumptions about learning as they "turn & talk" to someone next to them.

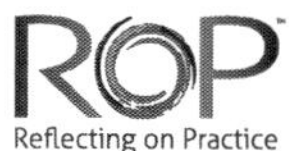

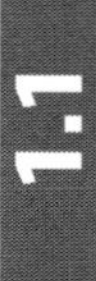

2. **Display questions for Turn & Talk discussion:**

   a. *What are the observable characteristics (signs) that learning is happening?*

   b. *What are the conditions that encourage and support learning?*

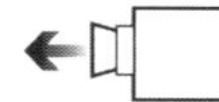

3. **Initiate "Turn & Talk."**
   Tell participants to turn to someone next to them and talk about the two questions. Remind participants to make sure both people have the opportunity to share their ideas over the next 5 minutes.

4. **Begin whole-group share.**
   Pay attention to the noise. When noise level drops, partner talk is mostly done. After 5–7 minutes of partner talk, call "time" and invite participants to continue talking in the whole group. Say that they can respond to one question at a time, or to a combination or synthesis of both questions. Facilitate the discussion using the following steps:

   - Invite and listen to their responses. Record ideas on the board as you facilitate the discussion.

   - Encourage participants to elaborate on their thinking by providing explanations, evidence, or clarifications.

     <u>Suggested probing questions</u>:
     - ☐ What makes you think that?
     - ☐ Please give an example from your experience.
     - ☐ What do you mean?

   - Wait sufficiently long for participants (especially the quieter ones) to share ideas, and stay neutral in your reaction to all comments.

   - Draw others into the conversation by asking them to think more deeply about what is being shared.

     <u>Suggested probing questions</u>:
     - ☐ Can anyone add something to that comment?
     - ☐ How can we consider that idea from another vantage point?

5. **Transition to Activity.**
   When no more new ideas are being shared, direct participants' attention to the board where their ideas were recorded. Tell them to keep these ideas in mind during the next task.

   The group will have generated many ideas on the board about how to tell if someone is learning and how to support learning. Refer to this list of ideas later, during the **From Learners' Perspective** task following the **Cup & Card** activity.

Facilitating whole-group share.
It takes practice and patience to facilitate whole-group discussions that are substantive and inclusive. The following steps and suggested probing questions provide guidance as you practice. It is important to be non-judgmental in your responses. Keep a "poker face," and refrain from saying "right!" to the response you were expecting and hoping to hear. Your neutrality encourages participants to contribute their viewpoints, which is critical for constructing meaning together. Take your time, be sure to pause after questions and comments, and encourage participants to listen attentively and consider the ideas being shared.

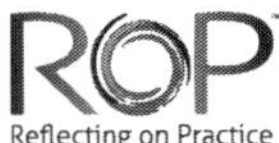

Module 1: Learning, Reflections, and Science

Hands-on 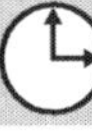 **20** minutes

| | ***Cup & Card*** |

The hands-on activities in this program are designed to be engaging and challenging science learning experiences for adults. As such, participants can engage 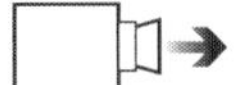 fully and authentically as learners. Tasks after the activity invite participants to consider the experience *from the learners' perspective* and use those insights to transfer the ideas on learning *into their practice.*

1. **Introduce the concept of being a learner.**
   Tell participants that the hands-on activities in the *Reflecting on Practice* program are designed to place them in the role of the learner. Display and share the following information:

   a. The hands-on activities provide a shared experience for participants to think about learning and teaching.

   b. These hands-on activities are designed for participants to be in the role of the learner. The activities reflect the types of hands-on experiences they lead for their learners, so encourage them to take advantage of the opportunity to engage fully as a learner.

   c. If they have done the activity before, or have a deep understanding of the concept, remind them to give space for their colleagues to mess with the ideas. They can also go deeper in their own understanding of the concept, and bring their colleagues along with them.

   d. Activities used in this program are not designed to be used "as is" with visitors to informal science learning environments. If interested, participants can design an appropriate variation of the activity to be used with visitors.

2. **Advise them to pay attention to the learning experience.**
   Tell participants they will be given a challenge to explore a physics concept with their colleagues in small groups. As they participate, they should also pay attention to what is being learned, and how.

3. **Display the instructions and challenge questions** and **demonstrate the steps** without actually doing the activity.

This activity is designed to spark conversation on learning beliefs, behaviors, and goals. As designed, it doesn't fully explore the physics concepts for complete understanding. Some participants are just fine with this, while others may feel dissatisfied. These various reactions foster an interesting discussion about whether learning was taking place, and if so, what was learned.

Cup & Card Instructions

Fill containers with water from the pan and place a plastic card over the top. Invert each container and observe what happens.

Challenge questions:

1. Can you get the card to stay on the inverted "cup?"
2. Does the amount of water matter?
3. How does the shape of the "cup" affect the results?

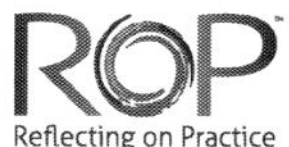

4. **Encourage exploration.**
   Encourage participants to explore using different amounts of water and differently shaped cups/containers. Ask them to learn all they can about the combination of cups, water, and card. Let them know they will have 15 minutes to investigate.

5. **Start the Activity.**
   Distribute the materials to each small group and have them work, explore, and discuss what they discover in their small group.

6. **Encourage discussion.**
   Circulating around the room, prompt group members to discuss their ideas, discoveries, and questions with each other. Suggested prompts:

   - Does it matter how much water is in the cup?

   - How does the shape of the container affect the results?

   - What generalizations can you make?

   - The other group made this observation. How does your experience compare?

7. **Introduce the Coin Challenge.**
   After participants have had an opportunity (~10 minutes) to explore various combinations of cups, water, and cards, distribute about 50 coins to each group. Do the following:

   a. Challenge each group to count how many coins can be added to a cup with water before the card falls off when the cup is inverted. What do you notice?

   b. Circulate around the room and ask some of the following guiding questions:

      - How many coins can you add? Who can add the most coins?
      - Is there a relationship between the containers' size/shape, amount of water, and number of coins each will hold?
      - Do you find anything puzzling about your discoveries?
      - What does and does not make sense to you?

8. **Wrap up the activity and clean up materials.**
   Have each group place all materials back on the tray and move the tray and pan to the side or back of the room.

<table>
<tr><td>

From Learner's Perspective   **10** minutes

</td><td>

Table Talk: *What Did You Learn?*

</td></tr>
</table>

Some participants will initially focus on explaining the scientific phenomenon, while others will dwell on the playfulness of the experience. Both elements are part of the activity, though neither are explored fully in the way the activity is designed for this session. Push on participants to think about and articulate their <u>thinking process in the learning experience</u> to consider what they learned and how they learned it.

Wait for participants to be ready to talk, especially the quieter ones, and stay neutral in your reaction to participants' comments.

1. **Display prompts and discuss in small groups.**
   Ask participants to think about the questions silently for a moment to collect their thoughts about the learning experience that just occurred.

   When they're ready, have them talk in their small groups and address the following questions about the **Cup & Card** activity:

   a. *What did you learn?*

   b. *How did you learn it?*

   c. *When did the learning occur?*

   d. *How did you know you were learning?*

2. **Begin whole-group share.**
   Encourage participants to share what they discussed in the small groups. Facilitate the discussion using the following steps:

   a. Invite and listen to their responses. Record ideas on the board as you facilitate the discussion.

   b. Encourage participants to elaborate on their thinking by providing explanations, evidence, or clarifications.

      <u>Suggested probing questions:</u>
      - What makes you think that?
      - Please give an example from your experience.
      - What do you mean?

   c. Draw others into the conversation by asking them to listen closely to what is shared.

      <u>Suggested probing questions:</u>
      - If that were the case, what assumptions are we making?
      - What's another way to phrase that comment/idea?
      - Can anyone add something to that comment?
      - What's another perspective on this idea?

   d. (Optional) Distribute the handout **Science Explanation for Cup & Card**.

3. **Connect back to ideas about learning.**
   Direct participants' attention to the previously discussed and recorded ideas about how to tell if someone is learning and how

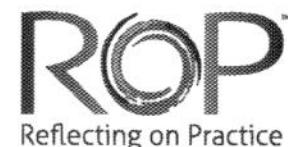

to support learning. Encourage participants to make connections between ideas shared in the **Retrieve & Connect** discussion and their experiences during the **Cup & Card** activity.

Walkabout: *Five Foundational Ideas on Learning*

55 minutes · Research Discussion

(A) Learning Styles (10 minutes)

1. **Hang the five chart papers for the *Five Foundational Ideas on Learning*** around the room. (If wall space is limited, the charts can be placed on five different tables.)

2. **Initiate a collective recall of video on learning styles.**
   Remind participants they were asked to watch a TEDx video on learning styles prior to this first session.

   a. Acknowledge that not everyone might have watched the video, and those who did watch might have questions and reactions.

   b. Display the prompt, "*What did the presenter talk about?*" to do a collective recall to remind everyone what the video was about, not their reactions to the content. They will get the chance to share their questions and responses afterwards.

   c. Allow participants to just call out what they remember, e.g., a woman gave a TED talk, she said learning styles are not real, she explained . . .

3. **Display the learning styles argument.**
   Following the collective recall, articulate the general argument for learning styles that Marshik explained had little to no evidence to support:

   - Learning styles refer to people having different preferences for

     □ processing certain types of information (e.g., visual, auditory, kinesthetic), and

     □ processing information in certain ways (e.g., intuitive or analytical thinker).

   - Learning styles theories argue that

     □ learners have different learning styles (or preferences), and

     □ their learning could be improved by matching our teaching with learners' preferred learning styles.

   - Studies in support of learning styles <u>would need</u> to provide evidence that there are greater learning gains if instruction aligned with the learner's learning style.

A collective recall is not a summary. If asking for a "summary," one or two people (likely the usual outspoken individuals) will give their take on what was presented. Participants with uncertainties or who didn't 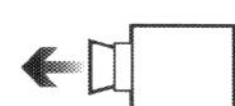 understand might be reluctant to speak. A collective recall asks everyone to contribute to piecing together the information and thus more individuals are retrieving their memories.

Pace yourself. Be sure to have at least 40 minutes for the Walkabout.

This conversation on learning styles is extremely brief, and your community will likely need/want more time. It's inserted here to address the idea upfront before proceeding further into conversations on how people learn. Otherwise, people who believe in learning styles will work to make it fit within the conversations to come. This acknowledgement at the beginning is intended to ease their embarrassment of supporting a popular idea publicly that has little or no evidence supporting it.

Participants who believe strongly in learning styles may have a really hard time with this conversation. They are experiencing cognitive dissonance. Give them space to process. Acknowledge to everyone that belief in learning styles is pervasive. There will be many opportunities throughout this program to revisit their beliefs, with readings, experiences, and discussion on learning. Reassure everyone that this is not the only conversation.

4. **Display and point out how those learning styles arguments are flawed, as presented by Marshik.**

   - There is little to no empirical evidence to support these learning styles theories.

   - Learning is the same regardless of how the content is presented to you.

   - Most of what we learn is stored in terms of meaning.

   - The best way to learn (or teach) something depends on the content itself (e.g., listening to bird songs to learn how to identify birds by their vocals).

   - Many things can be taught and learned using multiple senses.

5. **Brief discussion.**
   Invite participants to share their thoughts and questions from Marshik's talk.

   a. Invite and listen to their responses.

   b. Encourage participants to elaborate on their thinking by providing explanations, evidence, or clarifications. <u>Suggested probing questions</u>:

      ☐ What makes you think that?

      ☐ Please give an example from your experience.

      ☐ What do you mean?

   c. Draw others into the conversation by asking them to listen closely to what is shared. <u>Suggested probing questions</u>:

      ☐ Let's build on that idea.

      ☐ If that were the case, what assumptions are we making?

      ☐ Can we consider that idea from a different vantage point?

6. **Points to address if they don't emerge from the conversation.**

   - Recognizing and valuing individual differences in our learners is necessary, but this recognition doesn't translate into teaching according to learning styles.

   - Humans are multi-sensory organisms. Educators should provide learners with multiple experiences involving various modalities.

   - People may have preferences for learning in particular ways. The more often they favor those preferences, the more comfortable those ways become for them.

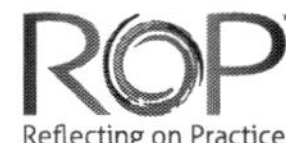

7. **Pose the rhetorical question to segue into the research discussion.**

    If learning styles don't exist, then how should we think about how people learn and how to teach in support of learning?

(B) Introduction and Directions (5 minutes)

1. **Introduce Research Discussion.**
    Explain to participants that their personal and professional experiences are valuable sources of knowledge for understanding learning and teaching. This knowledge, may be limited to what they've experienced personally. Share the following information.

    - Research Discussions give participants a chance to relate their experiences and thinking to ideas reported in "the literature."

    - "The literature" comprises published information, such as empirical studies, reviews, meta-analyses of studies, and theoretical papers. Ideas from these publications are synthesized into texts written for this program.

    - Participants are challenged to consider how ideas from the literature might be used to inform their practice. They are encouraged to agree or disagree with the ideas, extend the ideas into new and different directions, and want more details, perspectives, etc.

    - These reactions show they are thinking deeply about the ideas and are trying to reconcile them with their own understanding and experiences.

2. **Introduce Walkabout Routine.**
    Tell participants that throughout the *Reflecting on Practice* program they will have the opportunity to engage in research discussions in various ways. Describe the steps of this routine as follows:

    a. Five chart papers for the *Five Foundational Ideas on Learning* have been placed around the room. Each chart paper has one of the foundational ideas listed. More details about the idea are in the handout they will receive.

    b. They will work in small groups to consider and discuss each idea and record their conversation on the chart. They will have about 5 minutes for each idea.

    c. Each group will have a different-colored marker so that, as conversations are recorded from chart to chart, groups can compare their responses. Each group can decide whether everyone will take turns writing or if someone will be designated as a recorder.

Some participants may be familiar with the literature and concepts in these readings. Acknowledge their knowledge and invite them to continue to learn. Similar to the science content in the hands-on activities, if they have deep understanding of the concepts, remind them to give space for their colleagues to mess with the ideas. They can also go deeper with their own understanding, and bring their colleagues along with them.

The **Walkabout** is a Thinking Routine for learners to have progressively deeper conversations as they move from one topic to the next. While they move to a new idea every 4-7 minutes, the ideas are interconnected so, collectively, they will have discussed the main idea for a longer time.

> d. The facilitator will be time keeper. When time is called, rotate ("walkabout") clockwise to read, think, talk, and connect ideas between charts and between groups, until they have visited each chart in turn.

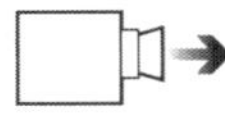

3. **Display the three question prompts.**
   Direct participants' attention to the questions they will respond to at each chart. <u>Leave these questions displayed throughout the task.</u>

   a. *What thoughts come to mind when you consider this idea?*

   b. *What connections can you make to others' responses?*

   c. *How can the idea be used in your practice?*

4. **Assign participants to groups for walkabout:**

   a. Divide participants into five groups (or fewer, if you have less than 10 participants), each stationed in front of one of the charts. Give each group a colored marker.

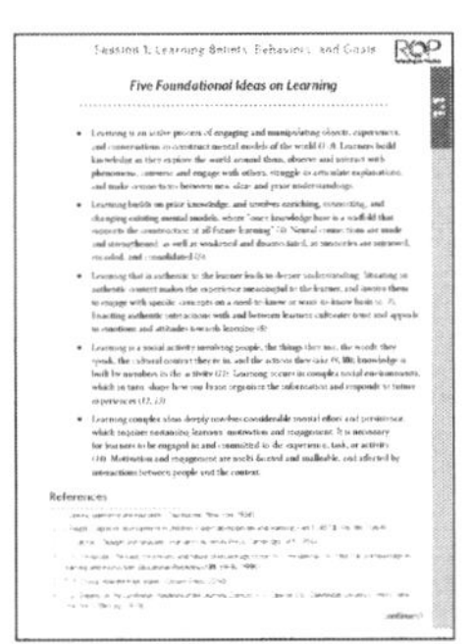

   b. Distribute the ***Five Foundational Ideas on Learning*** handout.

   c. Remind them to work with their small group to read, discuss, and then write on the chart paper with their assigned colored marker.

(C) Commence Walkabout (40 minutes)

1. **Keep note of the time.**
   Participants will need progressively more time as they walkabout because each group adds information to the charts for the next group to consider. Do the following:

   - Set a timer for yourself.

   - Provide less time for the earlier charts (~4 minutes) to have more time at the later charts (~6 minutes).

   - Give a 1-minute warning before calling time to rotate.

2. **Make observations and guide responses.**
   Circulate around the room as groups discuss each idea. Listen to what they say (and don't say), and notice what they write. Guide them on the types of responses they can make as they read the ideas on the chart.

   > <u>Suggested types of responses:</u>
   > - Connect ideas between charts and between groups
   > - Elaborate or comment on written ideas
   > - Ask for more information or clarification

Facilitators can be active participants to model how to listen thoughtfully, comment, question, and push.

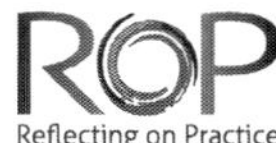

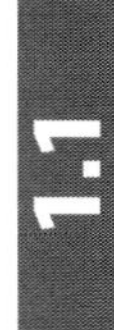

3. **Begin whole-group share.**
   Once all the groups have rotated through all five ideas, have groups
   return to the chart paper they originally started with to read what
   others have written. Allow time for participants to review the chart,
   and then invite each group to share its thoughts, using the following
   displayed prompts. These are designed to encourage synthesis of
   ideas as you lead the whole-group discussion:

   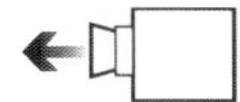

   - ❑ *What themes emerged?*

   - ❑ *Were there common issues and reactions across the groups?*

   - ❑ *What questions surprised them?*

4. **Facilitate group discussion.**
   As participants share their thinking, challenge them to elaborate
   on their ideas and encourage them to consider multiple viewpoints.
   Facilitate the discussion using the following steps:

   Remember to stay neutral and pause after asking a question. Wait for multiple participants before selecting someone.

   a. Invite and listen to their responses.

   b. Challenge participants to elaborate on their thinking by
   providing explanations, evidence, or clarifications.

   Suggested probing questions:
   - ❑ What makes you think that?

   - ❑ Please give an example from your experience.

   - ❑ What do you mean?

   c. Encourage participants to listen deeply to what their colleagues
   are saying.

   Suggested probing questions:
   - ❑ Can anyone add something to that comment?

   - ❑ How can we consider that idea from a different perspective?

   - ❑ If that were the case, what assumptions are being made?

   - ❑ What is another way to phrase that comment?

   d. Call out trends and challenges that emerge.

   e. Summarize ideas shared.

 Let's Talk Practice — 3 minutes

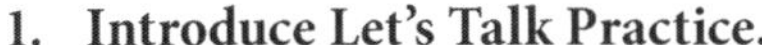
Minute Paper:
Personal Goals for the Program

1. **Introduce Let's Talk Practice.**
   Let participants know Let's Talk Practice is a designated time to make connections and to explore how to apply these ideas to their practice.

2. Tell participants that it's time for individual entries in their **Learning Journals.** Display the prompts and allow a few minutes for participants to gather their thoughts from this first session and then write plans for themselves.

   a. *What are your personal goals for participating in this professional learning program?*

   b. *Why are these particular goals personally or professionally important to you?*

   c. *How will you commit yourself to achieve these goals?*

2 minutes

Continue the Learning

1. **Discuss parameters and expectations for Continue the Learning tasks.**
   Based on your decision for assigning tasks outside of these interactive sessions, explain those expectations to the participants.

2. **Suggested reading.**
   If you've decided to assign the reading, share it now.

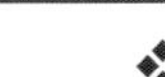

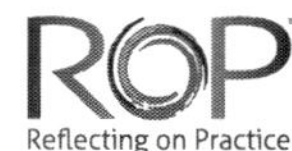

About the Program: *Reflecting on Practice*

Reflecting on Practice is a professional learning program designed for educators in informal science learning environments. It was developed by Lynn Uyen Tran and Catherine Halversen at the Lawrence Hall of Science, University of California, Berkeley. The program immerses beginning and veteran educators in discussions about, reflections on, and application of research on learning and teaching science. The goal of the program is to **advance the informal science education field by cultivating communities of learners among its professionals**; to that end, the program is facilitated by informal educators for colleagues in their own institutions.

The program has three key objectives:

- To build shared language and understanding among professionals by delving into research on learning and teaching.

- To engage practitioners in habits of reflection through observing their own teaching, as a means to develop their practice and make it public.

- To nurture a tradition of continued professional learning among participants, and thereby build a professional learning community.

<u>Professional learning</u> refers to practitioners' ongoing learning about their practice in order to increase their expertise and skills. It is valuable for improving practice regardless of profession (Webster-Wright, 2009).

<u>Professional learning community</u> is a group of practitioners who share and critically examine their practice in a way that is ongoing, reflective, collaborative, inclusive, learning-oriented, and growth-promoting (Stoll et al., 2006).

Institutional Adoption & Systems-Level Change

Reflecting on Practice is designed for institutional adoption and systems-level change. A facilitation team from your organization leads colleagues to engage in ongoing learning about their practice. All staff at your organization can learn together and from one another. Over time, new leaders from your community emerge to lead colleagues through *Reflecting on Practice*, and the cycle continues.

Program Design Elements

Four key practice elements are intentionally designed into the program:

1. *Model and build practice* (e.g., **Hands-on: Cup & Card**). Facilitators lead interactive sessions in which they model evidence-based practice. Participants engage as active learners, as they discuss pedagogy and reflect on their own practice.

(continues)

This page may be duplicated for educational use.

About the Program: *Reflecting on Practice* (continued)

2. *Relate research to practice* (e.g., **Research Discussions**). Participants read key ideas from the literature and discuss how to use the research in their practice. Research discussions provide participants with opportunity to talk with their colleagues about—and make sense of—ideas from research.

3. *Talk about and experiment with practice* (e.g., **Let's Talk Practice**). Participants think deeply about how they do their work now and different approaches they would like to try. They are encouraged to experiment with their practice, and to share their experiences with one another.

4. *Observe and reflect on practice* (e.g., **Learning Journals, Video Reflections**). Participants engage in activities designed for reflection both individually, in journals, and as a community, through review of videos and work samples.

Format, Content, and Shared Practice

Reflecting on Practice is a modular learning program. Each module focuses on topics relevant to educational practice in informal science learning environments. Each module consists of **interactive** sessions and **reflection** sessions.

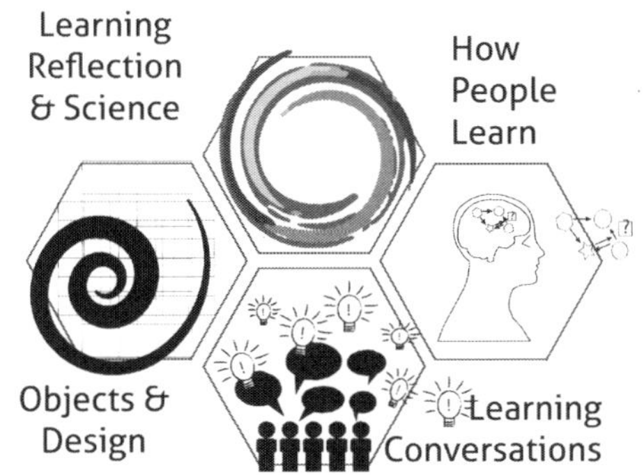

Interactive sessions are designed to engage participants in activities and discussions on learning and teaching—both drawing on their personal and professional experiences and integrating research from many disciplines, including psychology, education, and sociology. Participants express their beliefs and understanding, review their habits and assumptions, consider applicability of ideas from research, and determine ways to change their practice.

Reflection sessions are occasions when the community meets to present samples of their work and provide feedback on that work, in both small and whole group. Video is the primary type of work sample reviewed in Modules 1 to 4. By using video clips of their real-world interactions to reflect on their practice and draw on facilitated observations and feedback from colleagues, participants experience profound change in their teaching habits and mindsets.

     This page may be duplicated for educational use.

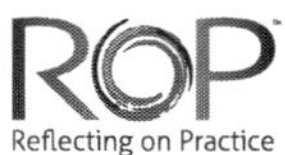

Five Foundational Ideas on Learning

- Learning is an active process of engaging and manipulating objects, experiences, and conversations to construct mental models of the world (*1-3*). Learners build knowledge as they explore the world around them, observe and interact with phenomena, converse and engage with others, struggle to articulate explanations, and make connections between new ideas and prior understandings.

- Learning builds on prior knowledge, and involves enriching, connecting, and changing existing mental models, where "one's knowledge base is a scaffold that supports the construction of all future learning" (*4*). Neural connections are made and strengthened, as well as weakened and disassociated, as memories are retrieved, encoded, and consolidated (*5*).

- Learning that is authentic to the learner leads to deeper understanding. Situating in authentic context makes the experience meaningful to the learner, and invites them to engage with specific concepts on a need-to-know or want-to-know basis (*6, 7*). Enacting authentic interactions with and between learners cultivates trust and appeals to emotions and attitudes towards learning (*8*).

- Learning is a social activity involving people, the things they use, the words they speak, the cultural context they're in, and the actions they take (*9, 10*); knowledge is built by members in the activity (*11*). Learning occurs in complex social environments, which in turn, shape how our brain organizes the information and responds to future experiences (*12, 13*).

- Learning complex ideas deeply involves considerable mental effort and persistence, which requires sustaining learners' motivation and engagement. It is necessary for learners to be engaged in and committed to the experience, task, or activity (*14*). Motivation and engagement are multifaceted and malleable, and affected by interactions between people and the context.

References

1. J. Dewey, *Experience and Education*. (Touchstone, New York, 1938).

2. J. Piaget, Cognitive development in children: Piaget development and learning, Part 1. *J. Res. Sci. Teach.* **2**, 176-186 (1964).

3. L. Vygotsky, *Thought and Language*. (Harvard University Press, Cambridge, MA, 1986).

4. P. A. Alexander, The past, the present and future of knowledge research: A reexamination of the role of knowledge in learning and instruction. *Educational Psychologist* **31**, 89-92 (1996).

5. D. A. Sousa, *How the Brain Learns*. (Corwin Press, Thousand Oaks, CA, 2016).

6. J. G. Greeno, in *The Cambridge Handbook of the Learning Sciences*, R. K. Sawyer, Ed. (Cambridge University Press, New York, NY, 2006), chap. 6, pp. 79-96.

(continues)

Five Foundational Ideas on Learning (continued)

7.   J. L. Kolodner, in *The Cambridge Handbook of the Learning Sciences*, R. K. Sawyer, Ed. (Cambridge University Press, New York, NY, 2006), chap. 14, pp. 225-242.

8.   L. Cozolino, *The Social Neuroscience of Education: Optimizing Attachment and Learning in the Classroom.* (Norton, New York, 2013).

9.   J. D. Bransford et al., in *Handbook of Educational Psychology*, P. A. Alexander, P. Winne, Eds. (Erlbaum, Mahwah, NJ, 2006), chap. 10, pp. 209-244.

10.   B. Rogoff, in *Cognition, Perception and Language: Handbook of Child Psychology*, D. Kuhn, R. S. Siegler, Eds. (John Wiley & Sons, New York, 1998), vol. 2, pp. 679-744.

11.   M. Scardamalia, C. Bereiter, in *The Cambridge Handbook of the Learning Sciences*, R. K. Sawyer, Ed. (Cambridge University Press, Cambridge, UK, 2006), chap. 7, pp. 97-118.

12.   D. C. Park, C.-M. Huang, Culture wires the brain: A cognitive neuroscience perspective. *Perspectives on Psychological Science* **5**, 391-400 (2010).

13.   M. D. Lieberman, *Social: Why our Brains are Wired to Connect.* (Broadway Books, New York, NY, 2013).

14.   J. A. Fredricks, P. C. Blumenfeld, A. H. Paris, School engagement: Potential of the concept, state of the evidence. *Review of Educational Research* **74**, 59-109 (2004).

Module 1.
Session 2: Effective Reflective Practice

Session Overview

This session introduces participants to the conceptual framework that informs this program as a whole: **reflective practice** and a **professional learning community**.

Participants discuss the value of "making practice public" by developing a shared language to describe their practice and goals, and to determine ways in which they can and will do both in this program. They learn more about the different reflective tasks that are part of *Reflecting on Practice*, with a special focus on observing and analyzing videos of one another's practice.

Video Reflections are a critical component in the program. This session introduces making, observing, and discussing videos; establishes the significance of this reflective task in the development of reflective practice; and introduces the process and tools for these video discussions.

Session Objectives

- Discuss the value of professional learning, reflecting on practice, and cultivating a professional learning community.
- Practice a major reflective task of the *Reflecting on Practice* program, Video Reflection, in support of those fundamental concepts.

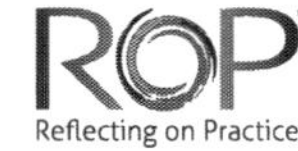

SESSION AGENDA		
Task **Routine**	**Description**	**Estimated Time** **(in minutes)**
Introduction (Part 1) *Session Objectives*	An overview of Session 2 and its objectives is presented.	2
Retrieve & Connect Think-Pair-Share: *Professional Learning & Professional Learning Community*	The ideas of professional learning and professional learning communities are introduced, and participants discuss the personal and institutional commitments needed.	18
Research Discussion Expert Jigsaw: *Reflective Practice*	Participants review and become "community experts" on some key ideas on professional learning and reflective practice. They consider how they already apply the ideas and what more they can do, as they deepen habits of reflection in their practice.	30
Break option		10
Introduction (Part 2) *Critical Friendships & Using Videos*	The concept and importance of critical friendships is introduced.	10
Hands-on *Practice Video Reflection*	Participants are introduced to the program's major reflective task, Video Reflection, by practicing it in whole community.	70
Let's Talk Practice Minute Paper: *Inquiring into My Practice*	In their Learning Journals, participants reflect on which aspect(s) of their practice they'd like to focus on during the program.	8
Continue the Learning	Participants are assigned a chapter on reflective practice.	2

Total Estimated Time | 150 mins. (2.5 hrs.)

MATERIALS

Recurring for all Sessions

- See Module 1 Session 1

For this Session

- External speakers
- Posters for Roles & Responsibilities, and Community Agreements (see **Getting Ready** #4)

For each participant

- Handouts (1 copy each)
  - **Key Ideas from the Literature: Reflective Practice and Professional Learning**
  - **Reflection Sessions in** *Reflecting on Practice* (Appendix A)
  - **Observation Instrument** (Appendix A)
  - **Protocol for Video Reflection** (Appendix A)
  - **Feedback Chart for Reflective Practice** (Appendix A)

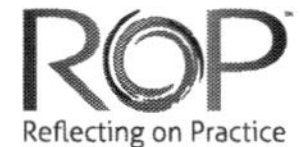

Getting Ready

1. Consider sending out two handouts **Key Ideas from the Literature: Reflective Practice and Professional Learning** and **Reflection Sessions in *Reflecting on Practice*** a few days before the session so that participants have the option to review the reading material ahead of time.

2. Review the handouts and slide deck for this session, and additional suggested readings.

   For the facilitator:

   - Tran, L. U. (2019). The how and why of reflective practice for science museum professionals.

   Continue the Learning suggestion for participants:

   - **Key Ideas from the Literature: Reflective Practice and Professional Learning**

   > Only sections A–D in the **Key Ideas** handout need to be reviewed in preparation for this session. The whole document is suggested for reading in **Continue the Learning** at the end of this session.

3. Duplicate handouts.

4. Make two posters from details in the **Reflection Sessions in *Reflecting on Practice*** handout in Appendix A.

   - <u>Community Agreements for Reflection Sessions</u>
   - <u>Roles & Responsibilities</u>

   > Save these posters to be reused in Whole-Community Video Reflections in Session 3 of Modules 2 to 4.

5. Determine which facilitator will be the educator-presenter (presenter) for this first Video Reflection practice; another facilitator will be the presenter for the second practice in Module 2 Session 3. As leaders of the community, facilitators are modeling how to be vulnerable in front of colleagues.

6. The presenter decides what aspect of their teaching practice to put forward for the community to provide feedback and help problem solve. The facilitation team can help to coordinate and focus the reflections on different aspects of practice they want the community to consider.

   - The presenter completes the **Reflection Exercise Worksheet to Prepare for Feedback** (in Appendix A). For this first practice, the presenter doesn't need to focus on the specific aspects of practice/topics of Modules 2-4 as mentioned in the Worksheet, but may decide to do so.

- The presenter can choose to seek help from another member of the facilitation team to prepare for this first practice. Remember that the purpose of video reflections is to "fine-tune" an aspect of our practice, meaning we tweak and refine our practice incrementally and over time.

7. Before the participants arrive:
   - Have the video clip queued up.
   - Know when to stop the video.
   - Test the audio in the room.

Preparing for Video Reflections

The community is introduced to Video Reflections in this session, but participants won't convene in small groups for their own Video Reflections until the end of Module 2. It is, however, necessary to plan the experience and procedures before then.

1. Small groups of four people are ideal for Video Reflections. Group these critical colleagues (or "reflection teams," or other term of your team's choosing) based on these factors: dynamics between and among your participants; the culture at your institution; and how the facilitation team wants the program to unfold. It's valuable for participants to be involved in deciding how the groups work, so prepare a plan and be open to modifications. Conditions to consider:

   a. *Group size.*
   - People notice different details for different reasons; having at least four people per group brings multiple perspectives to the conversation and keeps things interesting and informative.
   - Having too many people (more than four) in a group can make Video Reflections difficult to schedule. A large group may also feel more intimidating to those sharing their videos.

   b. *Group composition.*
   - You may decide to keep the same groups throughout the program or change group members as needed. Each strategy has its strengths and weaknesses:

     Option 1: Keep the same groups throughout the program. Advantage: teams really get to know one another, build a strong relationship, and strengthen their understanding of each other's practice. Weakness: the group can become insular and cliquey.

> Depending on the dynamics and size of the community, *Reflecting on Practice* facilitators may join small groups so that they're learning and reflecting together.

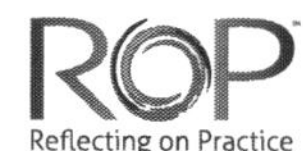

Option 2: Group members change as needed, usually based on availability. Advantage: the group is tremendously flexible and critical friendships are strengthened across the whole team. Weakness: there may not be consistency and commitment among team members.

c. *Assignment to groups.*

Option 1: You assign the groups. Advantage: you can distribute participants (knowing their strengths, weaknesses, personalities, knowledge, experience, etc.) across the groups. Weakness: participants don't get to choose.

Option 2: Participants choose groups. Advantage: participants have agency and may be more likely to engage. Weakness: participants may tend to choose people they already know and trust—which has benefits to start with, but may slow the strengthening of relationships across the whole community.

2. Decide how many Video Reflections you expect for each person per module going forward.

    As written, *Reflecting on Practice* designates a Video Reflection for each participant in Modules 2, 3, and 4. These video sessions are crucial to the program. Although making, observing, and discussing videos takes time, each step of this task lets participants see and talk about their practice in a concrete way. Video Reflections become easier and more familiar as participants do them regularly.

3. Consider participants who may not teach regularly. (These individuals may offer administrative support, develop curriculum, coordinate programs, design exhibits, handle animal care, or manage staff.) They bring different perspectives and experiences to your *Reflecting on Practice* community—remember, learners are not exclusively children; they include teens, adults, volunteers, and other professionals—so it will be important to support these participants' needs. They may feel that, since they don't engage directly with children, this reflection task doesn't address their needs. Consider the following:

    - The *Reflecting on Practice* reflection tools aren't exclusively for use with video; they are designed for reviewing and providing feedback for any type of work. For instance, a curriculum developer might present a working version of a curriculum that is being designed or redesigned. Administrative staff can bring email templates or phone scripts for consideration. They would each go through the same reflective process to select and present their work to the group.

- Create opportunities for everyone to engage directly with the public so that they can make a video and participate.

4. Decide on procedures for collecting, storing, and watching videos. The best way to ensure that participants follow through is to have an easy and streamlined process. Things to consider:

a. *Video-collecting process.*
Establish a process for how videos will be collected; be flexible. The facilitation team can propose a system and invite the community to modify it in consultation with their institutions. Possibilities:

- designate a small team of volunteers within the community to do all the video recording;
- have colleagues record one another;
- recruit your institution's AV team to do the job.

b. *Video-collecting equipment.*
Videos can be made with a variety of devices: camcorder, smartphone, mobile tablet, etc. Digital technology changes rapidly, lowering the costs for these video-recording devices and making them more readily available. These devices may be owned by the institution or belong to participants personally. The facilitation team needs to determine what equipment is available at (or can be purchased for) the institution, and decide on viable options for the community.

c. *Sound & tripod.*
Video Reflections are not productive if participants can't hear conversations, or if the video image is too shaky or taken from a bad angle. We highly recommend investing in external microphones (wired or wireless) and tripods.

d. *Familiarity with equipment.*
Whatever the final array of equipment, plan to give participants time to mess around with every device—institutional or personal—so they become familiar with the process from beginning (e.g., setting up the camcorder) to end (downloading, cueing up, and watching the video clips). We recommend that at least one person on the facilitation team be familiar with using each kind of device, to help participants troubleshoot.

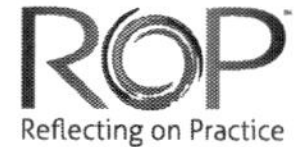

e. *Check-in/check-out procedure.*
If using shared equipment, it's important that it be both easily accessible to participants and carefully accounted for at check-in and check-out. Consult with the institution, if needed, to determine procedures.

f. *Personal storage.*
To build a video collection of their practice, participants need a means of storage. Several options are fairly low- to no-cost: flash drives, and external harddrives or upload onto a cloud-storage service (e.g., Box, Google Drive, Dropbox) or a private YouTube channel.

g. *Privacy to watch and discuss.*
There needs to be a way for groups to have privacy when they do their Video Reflections. Factors to keep in mind:

- At least one member of the group needs to have a laptop or tablet to display the videos so that the group can meet anywhere.

- Known list of private locations that are accessible to everyone.

- Headphone splitters are convenient for offering added privacy, enhancing audio quality, and increasing flexibility for locations (e.g., outdoors, large room with multiple groups). Individuals bring their own headphones to plug into the splitters to listen to the video.

5. Find out about your organization's policy for video recording the public, especially minors and school groups. These recordings are used for professional learning purposes, and will not be made public. For the most part, this use isn't an issue. However, defer to your organization's policy on how to handle recordings. It's good practice to ask learners' permission before recording (e.g., checking with the school teacher at the start of a lesson), posting a sign in public spaces to notify the public what you're doing, and placing the recording equipment in a visible location so those who don't want to be recorded can simply avoid the camera.

6. Steps 8, 10, and 11 in the **Protocol for Video Reflection** invite participants to document the community's evolving practice (**Educator Moves** and **Institutional Practices**). Decide on how these ideas will be gathered across the Video Reflections in whole- and small-groups, and when and how insights and concerns will be addressed.

Educator Moves and Institutional Practices push the community to specify how it's integrating the language and meanings that members negotiate in their various conversations and reflections. Experiences in the sessions offer opportunities to add to these documents. It's up to the facilitation team to ensure they're made available and updated.

Session 2 Step by Step

Introduction (Part 1)
Session Objectives

 **2** minutes

Throughout the program we commonly shorten "professional learning community" to "community." You may introduce this shorthand whenever you feel comfortable that all participants are well versed in the concept.

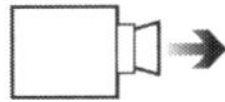

1. **Introduce the session.**
   Share that the purpose of this session is to review and come to an understanding of the three concepts fundamental to the design of the *Reflecting on Practice* program:

   * Engaging in **professional learning**
   * Understanding why and how to **reflect on practice**
   * Cultivating a **professional learning community**

2. **Display and review the session objectives:**

   * Discuss the value of <u>professional learning</u>, <u>reflecting on practice</u>, and cultivating a <u>professional learning community</u>.

   * Practice a major reflective task of the *Reflecting on Practice* program, Video Reflection, in support of those fundamental concepts.

Retrieve & Connect **18** minutes

Think-Pair-Share: *Professional Learning & Professional Learning Community*

1. **Introduce two components of investing in professional learning critical for success of the *Reflecting on Practice* program.**
   Share the following:

   Professional learning requires both *institutions* and *individuals* to invest time and resources in the professional growth of the community's members.

   a. **Institutions adopt the program**, and allocate time and space for staff to engage, learn, and test out new ideas.

   b. **Individuals participate in the program**, and commit time and effort to reflect on, make public, and make changes to their practice.

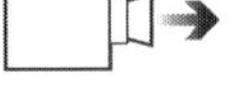

2. **Introduce the first step of Think-Pair-Share: think individually.**
   Display the following ideas and prompts, and ask participants to think about them silently for a few moments. Encourage participants to record their thoughts in their **Learning Journals**.

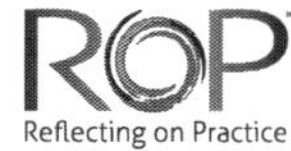

Professional learning refers to practitioners' ongoing learning about their practice in order to increase their expertise and skills, and is valuable for improving practice regardless of the profession (Webster-Wright, 2009).

A professional learning community is a group of practitioners who engage in professional learning together. They critically examine their practice in a way that is ongoing, reflective, collaborative, inclusive, learning-oriented, and growth-promoting. Members develop shared values and visions to support learning; take collective responsibility for their learners' learning; and promote individual and group learning (Stoll et al., 2006).

a. *What could ongoing learning look like in your institution?*

b. *What are some challenges and what suggestions do you have to overcome them?*

c. *What specific outcomes would you like to see come out of this professional learning program?*

d. *How would you like your **institution** to demonstrate commitment to the success of the program?*

Each question has a purpose:

a. Invites participants to imagine the possibilities.

b. Acknowledges that there will be challenges, and asks participants to problem solve.

c. Specifies desired goals for themselves and the facilitation team.

d. Gathers information for facilitation team and senior managers on how to support the community.

3. **Ask participants to pair up with a partner.**
Have participants discuss their thoughts with a partner. Emphasize *pair* discussion, so everyone gets a chance to talk out loud. (Form a group of three only if there is an odd person out.)

4. **Begin whole-group share.**
Make four sections on the board or chart paper; label each with a title: "Look Like," "Challenges/Suggestions," "Desired Outcomes," and "Institutional Commitment," to correspond to the displayed prompts. Invite participants to bring their partner conversation into the whole group for the community to consider. Record their ideas in the appropriate sections.

Expert Jigsaw: *Reflective Practice*

30 minutes · Research Discussion

1. **Introduce Research Discussion.**
Acknowledge that participants likely already give thought to what they do and how they do it, both in private and maybe with a few colleagues. Clarify that reflecting on practice (or reflective practice) is more than just "what to do differently next time," which might linger on the logistics without digging into the reasonings and mechanisms underlying the actions. Engaging in reflective practice is integral to professional learning, but what does it mean to engage

in "reflective practice?" Let participants know that this Research Discussion will give them information to consider. Explain that the reading is an excerpt from a chapter on reflective practice for informal science educators. The first two sections of the reading are for their reference (or assigned for Continue the Learning); they elaborate on the ideas of professional learning and reflective practice they discussed at the start of this session. This Research Discussion will focus on the last four sections of the reading (A-D) to consider what's involved in engaging in reflective practice.

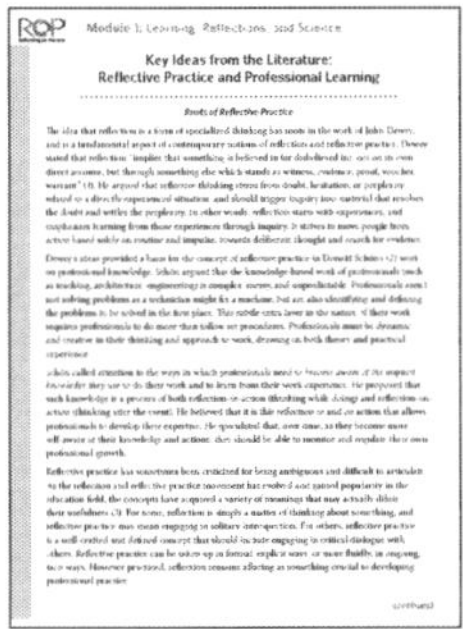

2. **Introduce Expert Jigsaw routine.**

   Tell participants that they will do an expert jigsaw for this Research Discussion. They will be divided into small groups, with each group assigned to become the "community expert" on one section of the reading.

If your numbers are too small to form a group of at least two people, then skip one of the topics.

   If small groups exceed six people, then make multiple groups for the same idea.

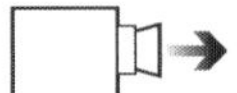

3. **Divide participants into four groups and distribute the "Key Ideas from the Literature: Reflective Practice and Professional Learning" handout.**

   Once the Expert Groups are in place, assign each group one of the key ideas (A-D) on which to become the community expert. Tell them they have about 15 minutes to read and discuss with their group.

4. **Display the discussion prompts.**

   Tell participants to read their group's assigned key idea, and then discuss the following prompts:

   a. *How does the idea relate to what you already know or believe?*

   b. *What are we doing, thinking, saying **now** that embodies this idea?*

   c. *What more could we be doing? What could we do differently?*

5. **Disperse experts to form teams.**

   When all community experts are ready, disperse them to form teams; each team comprises at least one expert for each idea. Tell participants that they'll work in their new teams as follows:

   - expert(s) reports out the key idea and the examples in practice, and

   - everyone pays attention to and makes connections between the ideas.

6. **Aid idea sharing across groups.**

   It's optional to lead a whole-group discussion after the second small-group conversation. If there is no whole-group discussion, then you will be the messenger to carry ideas between groups. As you circulate, be sure to listen for highlights, examples, and connections across the groups.

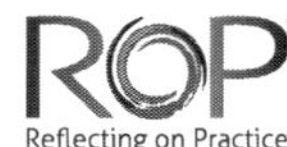

7. **Display and emphasize the following key points.**

    ■ **Problem setting is the iterative process of naming and framing.**
    Naming describes the situation and specifies how it's troublesome. **Framing** is how you "look" at the situation and involves articulating your position, attitude, or perspective. Too often, we generate solutions without truly understanding what's the problem to be solved in the first place. Attempting to solve problems before they are defined can leave us frustrated, exhausted, and no closer to any solutions. In *Reflecting on Practice*, the authors set the problem the program tries to solve. *The problem is that informal educators needed:*

    □ *a way to develop their educational practice;*

    □ *common language and knowledge to talk about their work with one another and beyond; and*

    □ *something specific to educational practice in informal STEM learning environments.*

    ■ **Reflective practice requires deliberately framing and re-framing one's practice in light of the consequences of one's actions, principles, beliefs, values, biases, and assumptions.**
    Your frame of reference affects how you perceive or understand the situation, and affects your approach towards finding solutions for a problem. In *Reflecting on Practice*, the authors frame the situation of staff development to be *educators as learners*, meaning those who teach are learners of their own teaching practice. From this reference frame,

    □ *participants are positioned in this program as learners engaging in ongoing learning together about how they do their work; and*

    □ *the program is designed for educators to lead themselves through the learning experiences.*

    ■ **Reflective practice means a willingness to *change* (behaviors, frame of reference, beliefs, etc.) based on those reflections.**
    Otherwise, we are rationalizing, while pretending to be reflective. When we rationalize, we risk making biases, false beliefs, and stereotypes justified and permanent.

8. **Articulate the reflective practice process used in *Reflecting on Practice* that emerges from these ideas.**
    Display the infographic and describe the process participants will take to set the problem and seek help from peers (critical friends)

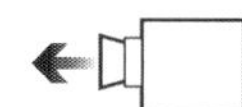

Frame of reference is a set of ideas, conditions, or assumptions that determine how something will be approached, perceived, or understood (Merriam-Webster Dictionary). How a situation is framed affects how people make decisions (Tversky & Kahneman, 1981).

Often, staff development is framed (by both management and staff) as "an expense to train people because they don't know better." From this reference frame, people approach these trainings as a burdensome chore to be checked off.

It's necessary for the facilitation team (along with senior management) to decide and articulate your frame of reference on staff development. This framing affects how *Reflecting on Practice* is perceived and carried out, which in turn affects how staff participate.

Participants began articulating their frame of reference towards engaging in professional learning in the Let's Talk Practice in Module 1 Session 1. As the program proceeds, they will specify frames of reference for different aspects of their practice as a part of each Video Reflection.

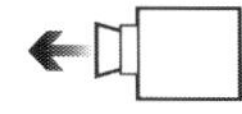

towards solving problems in their practice. Emphasize how the journey is both individual and collaborative. Assure participants they will be guided through this process. Point out two important components in the process: Peers and Videos.

 **10** minutes

Introduction (Part 2)
Critical Friendships & Using Videos

1. **Introduce the concept of "critical friendships."**
   Let participants know that, as a part of community-building to support professional learning, it's important that they think and talk with one another about what they teach, how they teach, and how their teaching supports their learning goals. As they learn and talk about their practice, they develop shared meaning and language, and work together to devise ways to make improvements. In this program, they build their learning community through *critical friendships*.

2. **Display descriptions of critical friendships.**
   Give everyone a moment to read the statement and consider what it means before asking them to share their responses to it. Remember to ask follow-up questions that encourage participants to elaborate on their comments and to listen closely to what's shared.

> "A critical friend is a trusted [colleague] who asks provocative questions, provides data to be examined through another lens, and offers critiques of a person's work as a friend. A critical friend takes the time to fully understand the context of the work presented and the outcomes that the person or group is working toward. The friend is an advocate for the success of that work" (Costa & Kallick, 1993, p. 50). As a learning community, critical friends "believe that together they are more capable of knowing what they need to know, and learning what they need to learn, than they are alone" (Baron, 2007, p. 57).

3. **Display and elaborate on how critical friendships show up in reflective practice.**

   - Critical friendships are <u>mutual relationships</u> that require colleagues to trust each other, share their sense of purpose, and reflect on and discuss their values, beliefs, ideas, and assumptions.

Critical (adjective) means expressing or involving an analysis of the merits and faults of a work of literature, music, or art: *professors often find it difficult to encourage critical thinking in their students* (New Oxford American Dictionary).

Analyzing the work is investigative, and will entail challenging habits and assumptions. The *Reflecting on Practice* Tools for Reflective Practice used in this process focuses the analysis towards **productive feedback** and away from judgment and disparagement.

Participants are cultivating critical friendships when they convene as a community for the interactive sessions, though this point is not called out explicitly. During the reflective sessions (e.g. Video Reflections), they are tasked to be more intentional about their critical friendships.

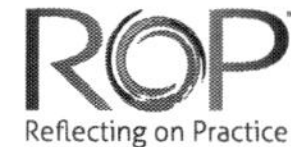

- In providing work samples (e.g., videos, instructional materials, prototypes) for examination, critical colleagues <u>make their practice public</u> to one another.

- In making detailed observations and asking provocative questions, critical colleagues <u>provide substantive feedback and push for deeper thought</u>. These discussions provoke and challenge assumptions and habits.

- In taking the time to reflect and advocating for learners' success, critical colleagues <u>hold each other accountable</u> for meeting the needs of their learners.

4. **Introduce the power of videos.**
    Remind participants that thinking and talking with one another about their practice, and how their work supports their learning goals, requires making their practice public. In this program, they do this using the major reflective task of **reviewing and discussing videos of their teaching**.

    Display and explain how **videos** can be a powerful tool for reflective practice. Videos serve as reference points for personal reflection and conversations with peers. They can be paused, reversed, and watched again. Only through videos can we observe ourselves.

5. **Explain the role of critical friendships in Video Reflections.**
    Tell participants that they will watch and discuss videos of their practice in small groups. Explain that the facilitation team will initially form these groups, and these groups can be modified with input from the community.

6. **Explain video commitment and use of tools.**
    Let the community know that each participant will record their own practice and watch clips from the video with assembled colleagues.

    a. Every module has opportunities for Video Reflection. The first video session in every module is conducted as a whole-group experience, to review and discuss video of one volunteer educator's practice.

    b. Most everyone in the community will do their Video Reflections within their small groups. (Either all groups meet at the same time during a video session, or the small-group members meet at different times arranged amongst themselves.)

    c. To guide this process, participants will use the *Reflecting on Practice* Tools for Reflective Practice. These tools make the process, roles, and responsibilities known to everyone. This transparency is intended to let everyone know what to expect and what is expected of them.

Be very clear that Video Reflections are disassociated from any performance reviews. The purpose of Video Reflection is to help professionals reflect on and improve their practice, not to evaluate the quality of their work.

Be clear and consistent that everyone needs to participate. All participants should commit to collecting, observing, and discussing their videos in order for the exercises to be meaningful and fair, and for the program to succeed. If any members will *not* be making videos, be upfront with the community about the reason and the alternative solution. **Ensure that all participants are perceived to be making their practice public in ways that are meaningful and productive for themselves and the community.**

    d. The tools will be introduced in the next routine as participants practice a Video Reflection as a community.

> The social relationships among participants are critical to the success of this program, especially Video Reflections. Members in a community won't trust one another if they don't know each other. If there is little to no trust, then the experience will feel judgmental and punitive. With trust, they may be more willing and able to receive the feedback and grow from it. Convening regularly to talk, think, read, and examine their practice together cultivates these social relationships.
>
> The tools are structured to give participants as much control as possible and to reduce uncertainty about what will take place. Practicing as a community before they meet in small groups has the same intentions. These efforts are important for diminishing participants' fear response to the experience. No doubt, making their practice public in this way may still feel scary and be out of their comfort zone. An acknowledgement of the participants' bravery and commitment to the process would be appropriate.

    e. Acknowledge that for some participants, video might not be the most appropriate work sample to review, e.g., those who develop exhibits. In those cases, the process will be the same, but the sample reviewed will be different.

7. Solicit reactions to use of videos.

Invite participants to respond to the idea of making videos of their practice and sharing them with their colleagues.

    a. Ask if anyone has video-recorded their own teaching before, or watched videos of others teaching. Encourage them to share their experiences with the community.

    b. Share your personal experience of showing your own recording and/or watching videos of others when you participated in *Reflecting on Practice* yourself to learn how to facilitate this professional learning program.

    c. Inform participants that, although everyone must make and provide videos (or other work products, for those who don't teach), this will happen in small, incremental steps to help all participants become comfortable with talking about and sharing their practice with others.

Participants may be apprehensive and skeptical about supervisors watching them teach and then talking about it. It is profoundly important to be sensitive to their potential concerns, and to reassure them that Video Reflections are separate from any human resources–related performance reviews.

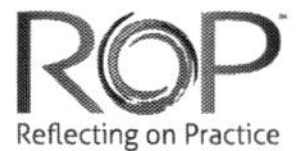

Practice Video Reflection

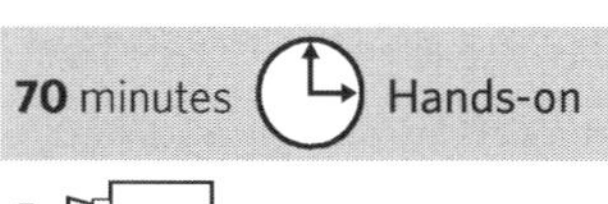

Reassure participants there will be an opportunity to ask questions at the end of this practice session. Hopefully, many of their questions will be answered by the end of this experience.

1.2

1. **Introduce the Video Reflection practice.**
   Display the list of Tools for Reflective Practice and tell participants it's now time to become familiar with doing Video Reflections through a practice session. They will practice using three tools in this session; additional tools will be introduced in Module 2.

2. **Display Roles & Responsibilities poster and introduce the three roles.**
   Point out that there are three roles in a Video Reflection. Over time, everyone will take on all the roles. Describe each role and their corresponding responsibilities.

   - The **Educator-Presenter** presents a selected sample of work for feedback in the Video Reflection. The presenter **decides** what aspect of their practice will be the focus of a Video Reflection discussion, **sets** the problem for the group to consider, and **selects** a 5- to 7-minute video clip for observation. More instructions will be provided in Module 2 when everyone will have their first chance to be an educator-presenter in small groups. For this session, Facilitator 1 will be the educator-presenter.

   - The **Guide** is responsible for leading the group through the Video Reflection, and can participate in the conversation. The guide keeps the conversation on track and focused on the work presented, and makes sure the presenter's focus questions are addressed. For this session, Facilitator 2 will be the guide.

   - **Peers** observe and contribute substantive feedback on the presenter's work.

3. **Display poster and introduce Community Agreements for Reflection Sessions.**
   Do the following.

   a. Direct participants' attention to the community agreements.

   b. Ask for volunteers to read each item out loud for the community to consider.

   c. Assure them that additions and revisions can be made to the agreements, but not until the community has had a chance to do some Video Reflections.

   d. Invite questions and clarifications about what the agreements mean or how they might look in practice.

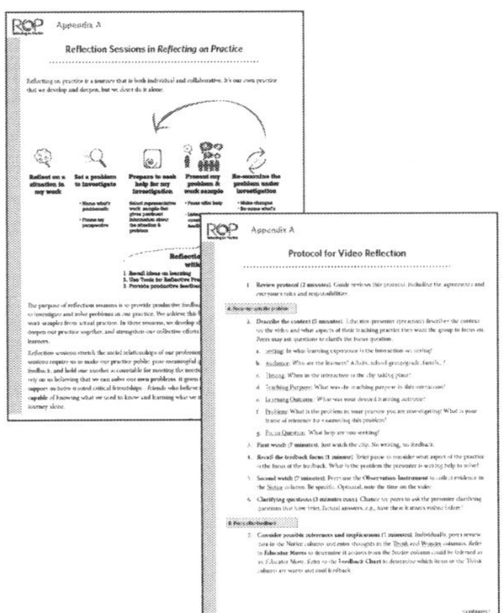

Point out connections between this section and the point about "problem setting" in the Research Discussion.

4. **Distribute "Reflection Sessions in *Reflecting on Practice*" handout.**
   Tell participants that this handout has details from the posters for them to keep. The **Responsibilities** and **Agreements** should be displayed and reviewed during every reflection session.

5. **Distribute the "*Reflecting on Practice* Protocol for Video Reflection" handout and provide a brief overview.**
   Tell participants this protocol lays out the steps of a Video Reflection, which the Guide uses to lead the group. Step 1 is to do a brief review of the protocol. Call out the following:

   - A full hour is dedicated for completing the entire protocol to provide substantive feedback to the presenter, though the whole hour doesn't need to be used.

   - Two additional tools will be distributed (in Steps 5 and 8 of the protocol) as they are about to be used.

   - There are 11 steps in the protocol, organized into 4 sections. Each step will be described separately and in more detail as we are about to do them.

   - Section A: **Presenter sets the problem**.

     □ The educator-presenter sets up the problem in Step 2, offering the frame of reference from which the presenter views the problem.

     □ The video clip is watched twice in Steps 3 and 5.

     □ There will be a chance to ask the presenter clarifying questions in Step 6.

   - Section B: **Peers provide feedback**.

     □ Peers get quiet time to prepare their feedback in Step 7.

     □ Peers discuss their feedback in Step 8, based on observation data. The presenter listens, but isn't part of the conversation.

   - Section C: **Presenter responds**.

     □ The presenter responds to the conversation in Step 9.

   - Section D: **Group reflection**.

     □ Step 10 is a quiet writing step for everyone to reflect on the conversation: something personal to take into their own practice and something for the community to reconsider regarding institutional norms.

     □ Step 11 gives the group the chance to put forth ideas and practices they want the whole community to reconsider.

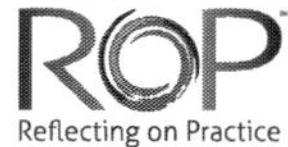

Section A: Presenter sets the problem

1. **Educator-presenter sets the problem for Step 2.**
   Thank Facilitator 1 for being the first presenter to share work with the group. Remind participants Section A is focused on the presenter **setting** a problem in their practice they're investigating and **presenting** a representative work sample to get help from colleagues to address this problem. Ask the presenter to describe the context and the focus question for the feedback they seek.

2. **Ask questions to clarify the focus question.**
   Allow participants to ask questions to clarify the presenter's requested feedback focus, and what the presenter views to be the problem.

3. **Steps 3 and 4 of the protocol.**
   Do the following:

   - In Step 3, tell participants to just sit back and watch the clip. Remind them there's no need to write down anything at this time.

   - Play the video clip.

   - In Step 4, ask for someone to remind the group of the problem that the presenter is investigating and their specific focus question for the help being requested. Encourage peers to write down the focus question.

4. **Distribute the "Observation Instrument" handout and discuss Side 1 for Step 5 of the protocol.**
   Point out that for Step 5, they will use the **Observation Instrument**. This instrument is open-ended, with questions to guide each section. Introduce the questions and explain when and how to fill in each section.

   a. **Notice.** Participants collect this information in Step 5 as they watch the video clip for the second time. *What do you see and hear happening in the video clip?* Record the observed actions from the video clip in this column. What is the educator saying and doing? What are learners saying and doing? Who else appears in the video, and what are they doing?

   b. **Think.** This information is generated in Step 7 *after* watching the video clip. *What do you think is going on?* In the middle column, record inferences, assumptions, speculations, and/or reasoning that comes to mind from the observation data. In this step participants <u>think</u> about what they noticed, and write down their thoughts. An additional tool will be introduced starting in Module 2 for use to support this step.

The purpose of Section A is to help the community understand the presenter's frame of reference and grapple with the problem to be solved—"This is the perspective from which I am viewing the situation. From this frame of reference, this is what I think is the problem. Please help." Peers consider the problem from the presenter's reference frame as they observe and think about the video clip. In Section B, peers *might* reframe the problem for the presenter to consider—"From our reference frame, this is what we notice. Consider the problem in this perspective. Here's why, and our suggestions."

1.2

Record the focus question on the board for participants to refer back to throughout the Video Reflection.

This observation instrument is a Thinking Routine. It intentionally teases apart observations (notice) from inferences (think), which is critical for doing science and engineering.

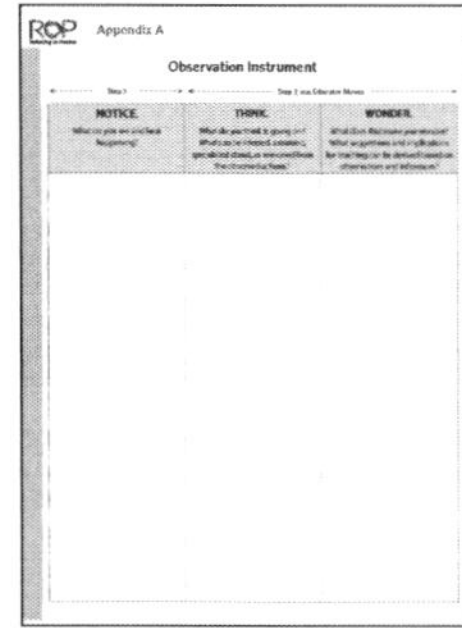

Participants aren't tasked to make interpretations of what's going on while the video is playing. During Step 5, they just write down what they notice—what they see and hear happening in the clip.

   c. **Wonder.** This information is also generated in Step 7, after thinking about their observations. *What does that thinking make you wonder?* Based on the observations and inferences, in the third column record questions that come to mind and suggestions for consideration. The wonderings, implications, and suggestions can be a synthesis of the observations or be directly related to specific observed actions.

5. **Flip over to Observation Instrument—Examples (Side 2) and describe.**

   Explain that this chart offers examples of details that might be entered in each column of the **Observation Instrument**. Give participants a moment to review the chart, and then tell them they will now use the instrument to gain concrete experience with it. Address any questions from participants on how to use the tool.

6. **Play video again for Step 5.**

   Remind everyone of the presenter's focus question before playing the video.

7. **Ask for clarifying questions for Step 6.**

   Invite participants to ask the presenter questions to help clarify the video clip. Do the following:

   ■ Emphasize that these questions are factual and logistical in nature to help them understand what's going on in the clip.

   ■ Call out when a question is appropriate, and explain how it's a suitable clarifying question, meaning it should help make the context coherent. For example,

      □ "What are those adults doing in the back?" is an appropriate clarifying question. It seeks information to understand how the additional adults in the clip are engaged.

      □ "Why aren't those adults tasked to work with the children?" is not an appropriate clarifying question. It assumes something about the adults and the educator's role in their engagement that involves a more elaborate conversation.

   ■ Be conscious of time—don't exceed three minutes.

Protect the presenter from having to answer questions that get into pedagogical moves and reasoning. Thank the peer for the question, and ask that they either rephrase the question or save it for Step 8.

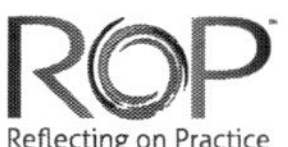

Section B: Peers provide feedback

1. **Offer quiet time for Step 7.**
   Instruct participants to complete the rest of the **Observation Instrument** for Step 7, and to keep in mind the presenter's feedback focus question. While peers work, direct the presenter to move to the back of the room.

2. **Prepare for the conversation in Step 8.**
   Tell participants in Step 8, they will spend approximately 20 minutes talking about the presenter's focus question, based on their observations from the video clip. Explain the following:

   - The presenter is now sitting in the back of the room—to listen closely, but not to participate in the conversation.

   - This conversation works towards providing the presenter with productive feedback to use in their practice. The presenter will be busy taking notes.

   - Distribute the **Feedback Chart for Reflective Practice** handout. The Feedback Chart offers guidance on how to offer and receive productive feedback.

3. **Review how to provide productive feedback (Side 1 of handout).**
   Acknowledge that giving and receiving feedback isn't easy for anyone—how do we say it without it being taken the wrong way, how do we hear it without being on the defensive? As with anything we learn to do, it gets easier with regular practice. In *Reflecting on Practice*, feedback is defined and used in the following ways.

   - Feedback is a gift of information given to colleagues because you care for them and their ongoing learning and self-improvement. Show them you care by your actions and thoughtful feedback during Video Reflection. **Be curious** about the work. **Be kind** to the presenter. **Be appreciative** of the experience.

   - Feedback focuses on the work presented, not the person. It's based on observation data from the video clip, and is informed by the ideas on learning that the community explores together.

   - Feedback is a learning opportunity for oneself. You learn to notice details and go deeper into your own practice.

   - Feedback is organized into two categories: Warm and Cool.

     - *Warm feedback* focuses on helping the presenter build on the strengths in their work. This doesn't mean general statements of approval, such as "That was good. You did a

The presenter should be in the room to listen to the conversation, but out of the line of sight so peers aren't tempted to direct comments towards the presenter.

1.2

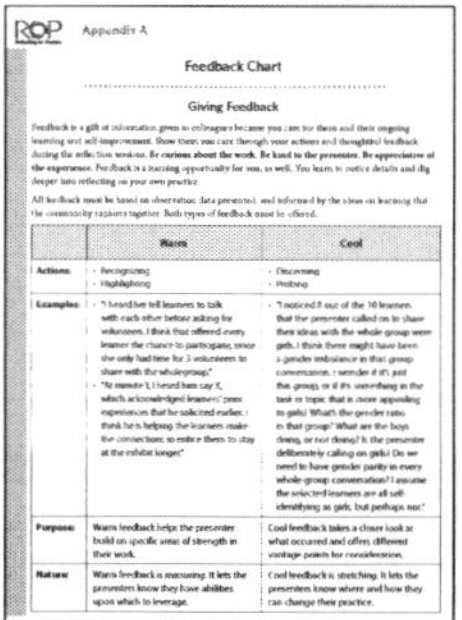

The **Feedback Chart** provides examples of warm and cool feedback statements that model how peers can effectively share their observations with the educator-presenter.

Value-laden language implies judgment and evaluation. Using this language when offering warm feedback might not feel threatening. However, it primes the presenter for being judged, which becomes problematic when it's time for cool feedback. Presenters might sense negative evaluation, even if the words "bad" and "not like" aren't spoken.

Moreover, feedback is not about what peers "like" or "not like" about the presenter's practice. Feedback is giving information about what happened from the peers' perspectives for the presenter to consider.

nice job." Warm feedback provides specific comments to the presenter that <u>highlight and recognize strengths</u>, drawing attention to what is effective. Refrain from using value-laden language, such as "I like …," "That was good use of …"

▢ *Cool feedback* analyzes the instances from a critical perspective. It isn't about calling out mistakes or offering judgment, as in "You should have spoken louder or walked to the back of the room so everyone could hear." Cool feedback takes a closer look at what occurred and offers different vantage points for consideration.

4. **Review how to receive feedback (Side 2 of handout).**
Share the following sentiments:

- Feedback can be a gift of information about ourselves from another person's vantage point, through their lens of experience, worldviews, identities, and knowledge. It's a perspective that we can't see for ourselves. It's difficult to regulate our own learning and change our practice without this information.

- Hearing the feedback can be uncomfortable. It might make us aware of the biases, assumptions, traditions, blind spots, and habits that shape our practice. We might not like what we hear. We might not be ready to hear the information.

- Be self-aware of how the information might make you react, despite the *good intentions*. Your emotional responses are valid. Acknowledge them and then strive to understand them.

  ▢ Status. Do I feel the information isn't true? Why? What does it mean if there are aspects of truth to it? Am I feeling concerned about what my colleagues think of me if this is true? How sensitive am I to how people think of me?

  ▢ Relatedness. Am I feeling this way because of the person giving the information? How would I feel if the same information is given by someone else? Am I conflating the information with my relationship with my colleagues? Am I feeling my colleagues think I don't belong?

  ▢ Fairness. Do I feel I'm being treated unfairly? Why? Do I feel I'm being singled out for something everyone does or is out of my control to change? Am I feeling that I'm being compared to other people?

Point out that listening with curiosity and considering the corresponding "questions for reflection" when receiving feedback connects to Dewey's notion of **"readiness"** in the Research Discussion Section A.

The reflection questions are intended to focus the presenter's attention to process emotions that potentially hinder their experience when they hear feedback.

The questions are organized into three major categories of human social experiences: status, relatedness, and fairness (Rock, 2008). These experiences can trigger threat or reward responses. When we feel threatened, it's natural for us to freeze, fight, flee, flail, etc.

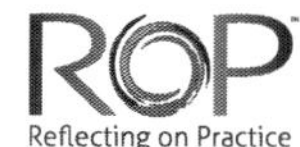

- Listen with curiosity to understand the perspectives being offered. *What's at the heart of all the details discussed?*

- Remember the feedback is based on only 5-7 minutes of your entire practice.

- Not all feedback needs to be acted upon. If you pass on taking action, be mindful of your reasons. Is it something beyond your ability to address? Is it something you're avoiding? Is it something you need more help to be able to achieve?

5. **Review Step 7 in light of the feedback chart, then begin Step 8.** Do the following:

   a. Look at what's written in the Think column that would be considered a *warm* feedback—highlighting strength.

   b. Start with warm and provide the observation that supports that interpretation.

      □ Be welcome as a facilitator to contribute to the conversation, but be careful not to dominate. It may be helpful to model using the sentence stems: I notice …, I think …, I wonder …..

      □ Insist that participants state what they saw or heard in the clip that underlies their feedback. This habit will be important when cool feedback is given. Someone else might interpret that same observation differently, or they might be making assumptions beyond what was observable.

   c. After a few turns of warm feedback, allow participants to begin offering *cool* feedback.

      □ Be sure the cool feedback addresses the presenter's focus question. For example, "The focus question is X, so how does this comment help to address that question?"

Section C: Presenter responds.

1. **Welcome the presenter back into the circle.**
   When all feedback is provided, or after about 20 minutes, tell participants that the presenter is now invited to rejoin the group. For Step 9, it's the peers' turn to stay silent while the presenter speaks. The presenter is welcomed to address any, some, or none of the comments, ideas, and suggestions shared in the conversation.

If the conversation gets off track, use the protocol to redirect everyone's attention back to the task. For example, "The protocol doesn't allow the presenter to be part of the conversation, so please redirect and rephrase your comment to the group."

Explain to the group how items for further consideration will be gathered across the Video Reflections in whole- and small-groups, and the process by which they will be addressed.

It takes time and practice to give and receive feedback, and to guide the community through the process. Be patient. The relationships among people in the community will make a huge difference in this journey. We strengthen those bonds during the interactive sessions, and then rely on and stretch those bonds during reflection sessions. The *Reflecting on Practice* Tools for Reflective Practice provide a process people can trust while they learn to trust one another.

Section D: Group reflection

1. **Steps 10 & 11.**
   Tell participants they now have three minutes of quiet time to write down two action items that they want to remember. Share the following prompts:

   - *What's one thing I want to take into my own practice?*

   - *What's one thing I want to discuss further regarding our institutional practices?*

   Step 11 invites participants to offer items for the community to discuss in future meetings regarding institutional practices.

2. **Begin whole-group question and answer.**
   Ask participants for questions about the Video Reflection process and experience. Some concerns that might arise:

   - All of the Tools for Reflective Practice can be customized but not until the community has used them a few times. Those conversations should involve the whole community.

   - There may be participants in the community who mostly engage with learners at exhibits, activity carts, or other such spaces where the interaction is ephemeral and there's no set start and end to the engagement. To collect enough footage for the video sessions, have participants video record 30–40 minutes of interaction at the exhibit or cart. This will provide a selection of interactions from which to choose for their clips.

   - "Learners" don't have to be children. If participants primarily work with adults (e.g., teachers, volunteers), then it's those interactions that should be reflected on.

   - Follow the organization's policies for video recording visitors, especially children.

   - There is at least one opportunity for Video Reflection for everyone in Modules 2–4. The community will first convene as a whole group, and then meet in small groups.

   - Explain the process for collecting, storing, and watching videos.

   - If someone's work doesn't lend itself to recording a video, they should talk with the facilitation team individually.

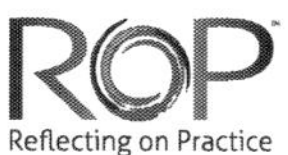

Minute Paper: *Inquiring into My Practice*

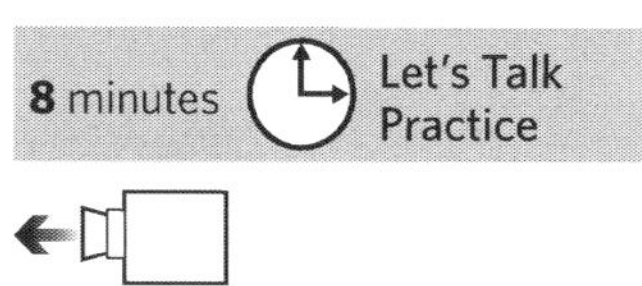

Display the following prompts. Allow participants quiet time to gather their thoughts from this session, and to make plans for themselves in their **Learning Journals**. Encourage them to discuss their reflections with a peer, if time permits.

a.   *What aspect(s) of your practice do you want to examine more closely throughout this program? Why is this important to you?*

b.   *What questions or wonderings do you have about your teaching practice?*

Continue the Learning

Suggested reading.
If you've decided to assign the reading, share it now.

2 minutes

Key Ideas from the Literature: Reflective Practice and Professional Learning

Roots of Reflective Practice

The idea that reflection is a form of specialized thinking has roots in the work of John Dewey, and is a fundamental aspect of contemporary notions of reflection and reflective practice. Dewey stated that reflection "implies that something is believed in (or disbelieved in), not on its own direct account, but through something else which stands as witness, evidence, proof, voucher, warrant" (*1*). He argued that reflective thinking stems from doubt, hesitation, or perplexity related to a directly experienced situation, and should trigger inquiry into material that resolves the doubt and settles the perplexity. In other words, reflection starts with experiences, and emphasizes learning from those experiences through inquiry. It strives to move people from action based solely on routine and impulse, towards deliberate thought and search for evidence.

Dewey's ideas provided a basis for the concept of reflective practice in Donald Schön's (*2*) work on professional knowledge. Schön argued that the knowledge-based work of professionals (such as teaching, architecture, engineering) is complex, messy, and unpredictable. Professionals aren't just solving problems as a technician might fix a machine, but are also identifying and defining the problems to be solved in the first place. This subtle extra layer in the nature of their work requires professionals to do more than follow set procedures. Professionals must be dynamic and creative in their thinking and approach to work, drawing on both theory and practical experience.

Schön called attention to the ways in which professionals need to *become aware of the implicit knowledge they use* to do their work and to learn from their work experience. He proposed that such knowledge is a process of both reflection-*in*-action (thinking while doing) and reflection-*on*-action (thinking after the event). He believed that it is this reflection *in* and *on* action that allows professionals to develop their expertise. He speculated that, over time, as they become more self-aware of their knowledge and actions, they should be able to monitor and regulate their own professional growth.

Reflective practice has sometimes been criticized for being ambiguous and difficult to articulate. As the reflection and reflective practice movement has evolved and gained popularity in the education field, the concepts have acquired a variety of meanings that may actually dilute their usefulness (*3*). For some, reflection is simply a matter of thinking about something, and reflective practice may mean engaging in solitary introspection. For others, reflective practice is a well-crafted and -defined concept that should include engaging in critical dialogue with others. Reflective practice can be taken up in formal, explicit ways, or more fluidly, in ongoing, tacit ways. However practiced, reflection remains alluring as something crucial to developing professional practice.

(continues)

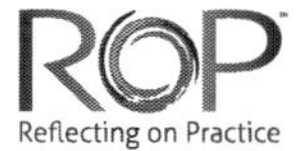

Key Ideas from the Literature:
Reflective Practice and Professional Learning *(continued)*

* * *

Reflective Practice is Integral to Professional Learning

Professional learning refers to practitioners' ongoing learning about their practice in order to increase their expertise and skills, and is valuable for improving practice regardless of the profession (*4*). Given that reflective practice is the process of learning from one's work experience (*2*), it is an integral part of professional learning. For professional learning to be meaningful and effective, it's best that the learning is continuing, social, active, related to practice, and positioned within a community that supports learning (*5*). Organizing professional learning activities within real-world work environments gives practitioners the opportunity to make connections between knowledge from the field and their practice with colleagues (*6*). It can engage them in actively working with one another on genuine issues within their professional practice.

The need for reflection and collaboration as part of professional learning is not surprising given that learning is social, and reflection is a part of making sense of one's experiences. As a social activity, learning involves people, the artifacts they use, the words they speak, the cultural context they are in, and the actions they take, and participants build knowledge as they interact (*7*). Embedding learning experiences in their daily work give professionals a *reason to understand*, which makes the learning meaningful and authentic to them. They generate memories with a frame of reference, which facilitates the retrieval and application of these memories in new situations (*8*). Reflection is being metacognitive; it's knowing one's thinking and how to regulate it (*9*). Doing so enables a person to detect inconsistencies of thought and identify connections between areas of conceptual understanding (*10*).

In an already overextended workday, however, constraints such as time, resources, logistics, and additional job responsibilities often prevent educators from having opportunities to think about and try out new teaching approaches—or even to reflect with peers. Similar to teaching in schools, teaching in informal science learning environments can be an isolated and solitary experience; the "egg-crate model" of instruction results in educators spending most of the day in a single room, separated from other adults. This situation is especially the case as financial constraints reduce teaching staff even as tightening budgets push institutions to offer more programs for the money.

Is the reality of curtailed budgets keeping us from investing in professional growth through professional learning (and reflection)? It needn't be that way. Although cultural norms are not easily changed, research in schools shows that when schools are strategic in creating time and productive working relationships among teachers, and in fostering professional learning communities, both educators and students benefit (*5*).

(continues)

This page may be duplicated for educational use.

Key Ideas from the Literature:
Reflective Practice and Professional Learning *(continued)*

(A) Reflective Practice is Influenced by Motivation

It takes considerable mental effort and persistence to learn complex ideas deeply (*7*). Motivation exerts crucial influence on learning, whether the learning is about understanding climate change or one's own practice. Motivation is a dynamic, multifaceted phenomenon that influences a person's participation, persistence, and performance in learning (*11*). People can be motivated in many ways, and their motivation varies depending on the situation, context, and topic. Although research in this area has predominantly focused on K–16 students in school settings, there is increasing application of the ideas in the workplace (*12*).

Motivation to participate may be driven by learners' **goal orientation** (*12, 13*). Learners may be stimulated to advance their own understanding and abilities or to outperform others; they may be reaching for mastery or performance goals—or avoiding them, for fear of looking less capable. Motivation is also affected by **mindset**, or feelings of competence in being able to succeed. Mindsets are people's lay beliefs about the nature of human attributes, such as abilities, intelligence, and personality (*14*). People with a *fixed* mindset believe these attributes (such as ability to teach, math intelligence, playing tennis) are innate and can't be changed. In contrast, people with a *growth* mindset believe these attributes are learned and can be developed. While we need experience and information to think and learn, our mindset influences the extent to which we learn from the experience in a *thoughtful* way, as opposed to the inclination to judge based on custom and prejudice.

Dewey captures the idea of motivation and mindset in reflective thinking as "**readiness.**" The essential characteristics of readiness include: open-mindedness, whole-heartedness, and responsibility. *Open-mindedness* is "freedom from prejudice, partisanship, and such other habits as close the mind and make it unwilling to consider new problems and entertain new ideas" (*1*). To have open-mindedness, the reflective practitioner listens to more than one side of an issue, gives attention to alternative viewpoints, and recognizes that even one's firmest held beliefs can be questioned. *Whole-heartedness* is being "thoroughly interested in some object and cause, [and throwing oneself] into it … with a whole heart" (*1*). This characteristic applies to the content to be taught, learners' learning of the content, and the educator's teaching and how it is affecting learners' learning. *Responsibility* is "to consider the consequences of a projected step. [Practitioners must ask themselves] the meaning of what they learn, in the sense of what difference it makes to the rest of their beliefs and to their actions" (*1*). For instance, a reflective educator who realizes that her English language learners are not lazy but instead are not at the reading level she assumed or wanted them to be, must then be responsible, and change her teaching approach and attitude towards her learners (*15*).

(continues)

     This page may be duplicated for educational use.

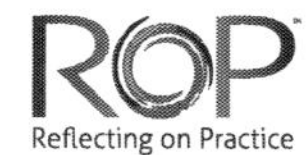

Key Ideas from the Literature:
Reflective Practice and Professional Learning *(continued)*

(B) Reflective Practice Focuses on Understanding Puzzling, Troubling, or Interesting Situations

Reflective practice inquires into the "how," "why," and "for whom" questions underlying the ways practitioners do their work. It is *driven by questions* emerging from situations that are puzzling, troubling, or interesting. Professional, knowledge-based work such as teaching is complex, messy, and unpredictable. Examining practice in a reflective way pushes the work of professionals beyond problem solving. One is not just solving problems, but also identifying and "setting" the problems to be solved in the first place. Thus, professionals need to do more than strictly follow set procedures.

Precisely because professionals notice and set up the problems to be solved, they need to be aware of their beliefs, values, assumptions, and biases that influence what they notice as problematic situations in the first place (*2*). Problem "setting" is putting boundaries around a situation and determining what needs to be examined more closely. It involves recognizing what draws one's attention to the situation because, then, the practitioner can begin to frame and reframe their practice.

A person might not "see" as much as one might assume, even when it's right in front of them because humans have capacity limitations in focusing attention (*16, 17*). Such limitations are fine (and necessary) as we go about our daily lives; we can modify our behavior without being aware of the reason (or even the need) for adjustment, which helps us from getting overstimulated or easily distracted (*17*). But this facility comes at a price—we have no idea how much we're missing.

Groups of individuals with similar goals and experiences, like those within a profession, develop distinct patterns and sensitivities to what they notice. The specialized knowledge and skills of a discipline shape the way its practitioners pay attention to details and view complex situations (*18*). For example, archeologists develop abilities to discern and talk about variations in color, texture, and consistencies of soil, which are essential aspects of being able to reason about a landscape (*19*). Farmers would notice different details in that same soil, while someone with little to no experience with soil may not even see *any* differentiating details. In short, individuals learn to notice in particular ways through practice and as they develop expertise in their profession (*18*).

(C) Reflective Practice is About Making Change

Reflective practice requires a deliberate process of framing and reframing one's practice because of one's actions, principles, beliefs, values, expectations, and experiences. Reflective exercises stimulate practitioners to observe, review, and talk with one another about what they do. It

(continues)

© 2021 Regents of the University of California ▪ **63**

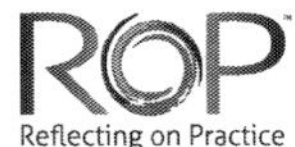

Key Ideas from the Literature:
Reflective Practice and Professional Learning *(continued)*

encourages them to learn from their work experiences by questioning why and how they do their work. Very importantly, reflective practice ultimately involves taking action; effecting change as a result of these reflections (*1, 2*).

It's vital to be attentive to the change-making aspect of reflective practice. When practitioners go through all the steps and exercises of reflecting on their practice but do so without scrutinizing their assumptions and actions and then doing things differently *based* on that scrutiny, they aren't engaging in reflective practice. This unexamined reflection becomes rationalization, which masquerades as reflection, and existing norms and biases become reasonable and reified (*3*).

How do we support change in practice when change is not easy or quick?

First, space is needed for staff to discuss alternative viewpoints, to try doing things differently, and to fail and try again. Setting norms for experimenting and allowing setbacks is a strong start in creating this space. Structures for participation in discussion offer transparency and make compliance a group responsibility. These types of actions acknowledge failure is part of learning, even for professionals, which can ease concerns about being judged.

Second, multiple settings and occasions for sharing their practice should be provided. The variety is helpful for staff, offering myriad ways to reflect and engage with one another. Some people are more expressive in writing, while others are talkative; some are comfortable challenging viewpoints in large groups, while others prefer conversations with a few. For example, meeting in small groups to review and discuss videos of one another's practice provides a friendlier way to make practice public. Digital platforms (e.g., Piazza) enable conversations to move between in-person and online spaces.

Third, documentation and follow-up are important for keeping track of changes in practice and for accountability. Urging educators to record their reflections in journals helps them chronicle their evolving thinking and actions, which serves as a tool for future reflection.

Finally, change will vary. It may be large-scale (e.g., an institution changes its mission statement to focus on learner-centered experiences) or small-scale (e.g., individual educators change the way they facilitate conversations with visitors). It may be immediate or take several years of incremental revisions. Change can involve shifts in behaviors, attitudes, perceptions, values, etc. Recognizing the *need* for change, however, is especially important to the social and political dimensions of learning and professional practice. Critical reflection asks us not only to move beyond the immediate situation, but also to reflect on actions we may want to take in situations in the future.

(continues)

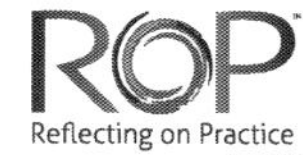

Key Ideas from the Literature:
Reflective Practice and Professional Learning *(continued)*

(D) Reflective Practice is Deliberately Learned Over Time within a Community

Reflection is activated by awareness of a disruption, perplexity, or curiosity in one's practice, but it takes more than provocation to make the reflection on practice productive and effective. Reflective practice calls for analysis and integration of experience. It calls for ongoing conversations within a community. Integrating new understandings into one's existing knowledge base involves adding to, distinguishing between, and making connections among ideas (*20*). It requires practitioners to articulate and examine their ideas with others—a habit that must be learned, and which develops over time.

It may be tempting to think that as educators gain experience, they automatically become better at analyzing and reflecting on their practice. In truth, expertise in "noticing and reflecting" is neither an inherent competence nor something that emerges solely from years of teaching. Sherin and her colleagues (*21*) have shown that the ability to notice behaviors in their classrooms, and then interpret and build from what they notice is a skill learned over time and through supported experiences. Reflections using videos take time to do and learn, but they are consistently more targeted on instruction and focused on learners than reflections based on recall. In comparing teachers' written reflections based on memory recall, versus video clips of those same teaching sessions, Rosaen and her colleagues (*22*) found that video-supported reflection enabled teachers to write more detailed comments about their teaching. Teachers shifted the content of their reflections from attention to classroom management to a focus on student thinking.

A common thread across the various models and activities for reflective practice is discourse within a community. Group dialogue gives practitioners the opportunity to make sense of ideas together because talking is a way for learners to make neural connections more memorable (*23*). Talking is integral to learning (*23-25*). We learn and make connections *during* the process of expressing with others how we think and reason about what we're learning (*26*).

This discourse within a community shapes the community's practice. As a group of individuals work together regularly, their endeavor builds *mutual engagement* in activities; *joint enterprise* towards goals and purposes; and development of a *shared repertoire* of habits, language, rules, and traditions (*27*). A consensus for talking, doing, and thinking emerges over time. This community of practice becomes a **professional learning community**, as the work it does together focuses on critically examining practice (*28*). The discourse gives practitioners the chance to develop common understanding about their work and shared language to talk about it; these assets are important for developing the profession (*29*).

(continues)

This page may be duplicated for educational use.

Key Ideas from the Literature:
Reflective Practice and Professional Learning *(continued)*

References

1. J. Dewey, *How We Think: A Restatement of the Relations of Reflective Thinking to the Educative Process.* (D.C. Heath & Company, New York, NY, 1933).

2. D. A. Schön, *The Reflective Practitioner: How Professionals Think in Action.* (Basic Books, New York, NY, 1983).

3. J. J. Loughran, Effective reflective practice: In search of meaning in learning about teaching. *Journal of Teacher Education* **53**, 33-43 (2002).

4. A. Webster-Wright, Reframing professional development through understanding authentic professional learning. *Review of Educational Research* **79**, 702-739 (2009).

5. L. Darling-Hammond, R. C. Wei, A. Andree, N. Richardson, S. Orphanos, "Professional learning in the learning profession: A status report on teacher professional development in the United Sates and abroad," (National Staff Development Council, Dallas, TX, 2009).

6. L. U. Tran, M. Werner-Avidon, L. R. Newton, Successful professional learning for informal educators: What is it and how do we get there? *Journal of Museum Education* **38**, 333-348 (2013).

7. J. D. Bransford et al., in *Handbook of Educational Psychology*, P. A. Alexander, P. Winne, Eds. (Erlbaum, Mahwah, NJ, 2006), chap. 10, pp. 209-244.

8. D. A. Sousa, *How the Brain Learns.* (Corwin Press, Thousand Oaks, CA, 2016).

9. G. Schraw, K. J. Crippen, K. Hartley, Promoting self-regulation in science education: Metacognition as part of a broader perspective on learning. *Research in Science Education* **36**, 111-139 (2006).

10. National Research Council, *Taking Science to School: Learning and Teaching in Science in Grades K-8.* R. A. Duschl, H. A. Schweingruber, A. W. Shouse, Eds., (National Academies Press, Washington, DC, 2007).

11. C. Senko, C. S. Hulleman, J. M. Harackiewicz, Achievement goal theory at the crossroads: Old controversies, current challenges, and new directions. *Educational Psychologist* **46**, 26-47 (2011).

12. A. Tanaka, T. Okuno, H. Yamauchi, Longitudinal tests on the influence of achievement goals on effort and intrinsic interest in the workplace. *Motivation and Emotion* **37**, 457-464 (2012).

13. A. J. Elliot, Approach and avoidance motivation and achievement goals. *Educational Psychologist* **34**, 169-189 (1999).

14. C. S. Dweck, Mindset: The New Psychology of Success. (Random House Publishing Group, New York, NY, 2007).

15. C. Rodgers, Defining reflection: Another look at John Dewey and reflective thinking. *The Teachers College Record* **104**, 842-866 (2002).

16. W. Schneider, R. M. Shiffrin, Controlled and automatic human information processing: I. Detection, search, and attention. *Psychological Review* **84**, 1-66 (1977).

17. D. J. Simons, Attentional capture and inattentional blindness. *Trends in Cognitive Sciences* **4**, 147-155 (2000).

18. V. R. Jacobs, L. L. Lamb, R. A. Philipp, Professional noticing of children's mathematical thinking. *J. Res. Math. Educ.* **41**, 169-202 (2010).

19. C. Goodwin, Professional vision. *American Anthropologist* **96**, 606-633 (1994).

20. E. A. Davis, Characterizing productive reflection among preservice elementary teachers: Seeing what matters. *Teaching and Teacher Education* **22**, 281-301 (2006).

21. M. G. Sherin, S. Y. Han, Teacher learning in the context of a video club. *Teaching and Teacher Education* **20**, 163-183 (2004).

(continues)

     This page may be duplicated for educational use.

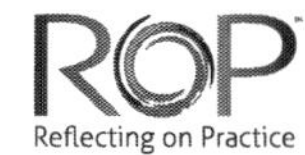

Key Ideas from the Literature:
Reflective Practice and Professional Learning *(continued)*

22. C. L. Rosaen, M. Lundeberg, M. Cooper, A. Fritzen, M. Terpstra, Noticing noticing: How does investigation of video records change how teachers reflect on their experiences? *Journal of Teacher Education* **59**, 347-360 (2008).

23. N. D. Forrin, C. M. MacLeod, This time it's personal: The memory benefit of hearing oneself. *Memory* **26**, 574-579 (2018).

24. F. M. van Blankenstein, D. H. J. M. Dolmans, C. P. M. van der Vleuten, H. G. Schmidt, Which cognitive process support learning during small-group discussion? The role of providing explanations and listening to others. *Instructional Science* **39**, 189-204 (2011).

25. M. K. Smith *et al.*, Why peer discussion improves student performance on in-class concept questions. *Science* **323**, 122-124 (2009).

26. L. Vygotsky, *Mind in Society: The Development of Higher Psychological Processes*. (Harvard University Press, Cambridge, MA, 1978).

27. E. Wenger, *Communities of Practice: Learning, Meaning, and Identity*. (Cambridge University Press, Cambridge, UK, 1998).

28. L. Stoll, R. Bolam, A. McMahon, M. Wallace, S. Thomas, Professional learning communities: A review of the literature. *Journal of Educational Change* **7**, 221-258 (2006).

29. National Research Council, Ed., *Learning Science in Informal Environments: People, Places, and Pursuits*, (National Academies Press, Washington, DC, 2009).

Module 1.
Session 3: Nature and Practices of Science

Session Overview

In this session, participants gain insight into the nature and practices of science in the best way possible—by **doing** and **reflecting** on science.

Participants take a critical look at what is and is not an accurate view of science by attempting to define it. They experiment with a "mystery tube" in an investigative activity designed to provide an explicit sense of the practices of science: what scientists do and how science works. They come to appreciate the public's perceptions of science and why clarifying their own understanding of the nature and practices of science affects—and is in fact critical to—their practice as informal science educators.

Session Objectives

- Deepen understanding of "what is science" and the nature and practices of science.
- Consider what knowledge about science and how science works the general public brings to informal science environments.
- Discuss how understanding the nature and practices of science influences how educators teach and communicate science to their learners.

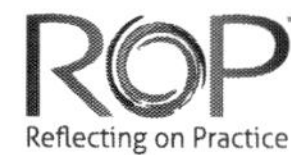

SESSION AGENDA		
Task Routine	**Description**	**Estimated Time (in minutes)**
Introduction *Session Objectives*	An overview of Session 3 and its objectives is presented.	3
Retrieve & Connect Quick Write: *What Is and Is Not Science?*	Participants respond to two prompts about the definition of science in a Quick Write, followed by a whole-group sharing to record ideas.	15
Hands-on *Mystery Tube*	Participants collaborate in small groups to work out the interior structure of a mystery tube—without looking inside.	40
From Learners' Perspective Table Talk: *How Science Works*	Participants review *How Science Works* flowchart and relate this to their experience investigating the Mystery Tube.	20
Break option		10
Research Discussion Walkabout: S*cience & Knowing*	Participants read and discuss four key aspects of science from the literature.	45
Let's Talk Practice Connect, Extend, Challenge, & Apply: *Tell a Friend*	In their Learning Journals, participants connect, extend, challenge, and apply the session ideas about the nature of science in regards to their practice.	15
Continue the Learning	Participants may be assigned two chapters on the nature of science.	2
Total Estimated Time		150 mins. (2.5 hrs.)

1.3

MATERIALS

Recurring for all Sessions

- See Module 1 Session 1

For this Session

- 4+ sheets chart paper
- 4 different-colored markers

For each small group

- 1 completed Mystery Tube (See **Getting Ready** #3)
- 1 enlarged copy (11" x 17" or A3) of **How Science Works** flowchart (See **Getting Ready** #4)
- 1 sticky-note pad or dry-erase pen
- 1 blank sheet of paper

For each participant

- Handouts (1 copy each)
  - **How Science Works** flowchart
  - **Key Ideas from the Literature: Science & Knowing**

Time-saver: In the interest of time during the activity, consider providing larger sheets of paper on which groups can draw their diagrams, then have them post or hold these up for the community share, rather than drawing them again in real-time.

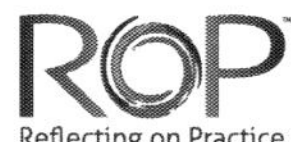

Getting Ready

1. Consider sending out the handout **Key Ideas from the Literature: Science & Knowing** a day or two before the session so that participants have the option to review the reading material ahead of time.

2. Review the handouts and slide deck for this session, and additional suggested readings.

    For the facilitator:

    - *Taking Science to School* (Chapter 6: "Understanding How Scientific Knowledge is Constructed")

    - Iaccarino, M. (2003). Science and culture. *Embo Reports,* 4:220–223.

    - Mazzocchi, F. (2006). Western science and traditional knowledge—Despite their variations, different forms of knowledge can learn from each other. *Embo Reports, 7:*463–466.

    Continue the Learning suggestion for participants:

    - *Surrounded by Science* (Chapter 2: "Science and Science Learning")

    - *Next Generation Science Standards* (Appendix F: "Science and Engineering Practices")

3. Duplicate handouts.

4. Prepare for **Hands-on: Mystery Tube**. Construct a Mystery Tube for each small group of 2–3 participants. Instructions for making the Mystery Tube can be found at http://undsci.berkeley .edu/lessons/mystery_tubes.html or search the Internet for "Mystery Tube."

    The Mystery Tube has become a popular activity to teach the nature of science. The design is the same, with many different authors making minor revisions to the teaching instructions. Participants in your community may already be familiar with the construction and activity. Alternative options include:

    - Change the name to "Perplexing Pipes" or "Conundrum Columns" and don't allow participants to search the Internet during the activity.

The articles by Iaccarino and Mazzocchi are retrievable from the Internet. They're particularly critical to read for the "Ways and Knowing" topic in the **Key Ideas from the Literature: Science & Knowing**. Consider sharing these two articles with participants after this session.

1.3

Go to **nextgenscience.org** to find Appendix for Next Generation Science Standards.

Retrieve a printable, digital copy of the flowchart from the Understanding Science website (https://undsci.berkeley.edu/article/scienceflowchart).

- Use "Mystery Box" variation. Instructions can be found at https://undsci.berkeley.edu/lessons/mystery_boxes.html or search the Internet.
- Use a complex design for the Mystery Tube that involves three ropes and two rings. (See Briggs, 2019)

5. Enlarge and duplicate one copy of *How Science Works* flowchart for each small group of two or three participants. Copies should be at least 11" x 17" (A3) and preferably **in color**. Consider laminating the enlarged copies so that they can be reused. (Laminating permits fine tipped dry-erase pens to be used instead of sticky notes in **Table Talk: How Science Works**.)

6. Prepare four chart paper posters for the **Research Discussion**. Label each sheet with the title of one of the sections from the **Key Ideas from the Literature: Science & Knowing**:

   - Nature & Practices of Science and Engineering
   - Ways of Knowing
   - Misinterpretations of Science
   - Science Identity, Aspirations, & Careers

Session 3 Step by Step

Introduction
Session Objectives

1. **Introduce the session.**
   Share that this session focuses on *what science is*—what defines it
   and how it's done—and the significant impact that understanding
   science can have on the practice and programs of informal science
   educators. Participants explore perceptions of science (their
   own and the general public's), and discuss the role that informal
   educators and institutions can play in shaping those perceptions.

2. **Display and review the session objectives:**

   - Deepen understanding about "what is science" and the nature
     and practices of science.

   - Consider what knowledge about science and how science works
     the general public brings to informal science environments.

   - Discuss how understanding the nature and practices of science
     influences how educators teach and communicate science to
     their learners.

3 minutes

1.3

Quick Write: *What Is and Is Not Science?*

1. **Introduce the question, *What is and is NOT science?***
   Explain that by closely examining what science is and how it works,
   we gain an understanding of its strengths and limitations. Share the
   following.

   a. Taking a critical look at what are and are not accurate views of
      science can help hone our own understanding of science.

   b. In teaching about science, it's vital to spend time thinking about
      the nature of science and how it can be communicated.

   c. If learners don't have a good understanding of what science is,
      they'll have a difficult time determining which questions science
      can and can't answer.

   d. Understanding *what science is* and *what it is not* helps us become
      critical consumers of information, better able to engage in
      decision-making as knowledgeable citizens.

15 minutes — Retrieve & Connect

The **Quick Write** routine prompts
learners to retrieve memories
relevant to the topic and make a
tangible record of those memories
for closer consideration. This act
pushes learners to externalize
their understanding, regardless
of whether that understanding is
accurate or ill-formed.

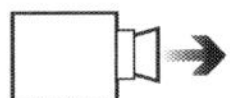

2. Display the Quick Write prompts.
Give participants about two minutes to consider and record their ideas in their Learning Journals.

a. *What is science?*

b. *What is it not?*

3. Have everyone pair up.
Give them a few minutes to talk about their responses with their partner.

4. Broaden the discussion to a whole-group share.
Create two columns on the board or chart paper: one titled *Science IS*, the other, *Science IS NOT*. Invite participants to share their thinking. Record their ideas on the board as you facilitate the discussion. Remember to listen to participants' responses; ask for *evidence*, *explanation*, or *clarification*; and encourage them to consider what their peers share. Participants may share some or all of the ideas listed below. (If needed, revisit the definition of empirical evidence: *evidence obtained by direct observation or experimentation.*)

If possible, one facilitator should lead the whole-group discussion while another records participants' ideas.

If it comes up, you might specify why science is defined as the study of the natural, not the supernatural world. Science doesn't weigh in on the subject of the supernatural one way or the other, because the supernatural isn't based on **empirical evidence**. It can neither be proven nor disproven using science.

Science IS	Science Is NOT
Study of the natural world	Study of the supernatural world
Observations and measures of phenomena	Opinions
	Stagnant
Based on empirical evidence	Oblivious to evidence that counters other evidence
Replicable	
Falsifiable	A platform for moral judgments
Peer reviewed	
Durable (robust, resilient), but open to change	

5. Organize and summarize.
Sum up the ideas that were shared. Let participants know that they will think and talk about these ideas throughout this session.

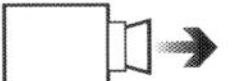

6. Display prompts for reflection and another quick write.
Ask participants to contemplate the following questions and record what images come to mind (either sketch or describe them). They

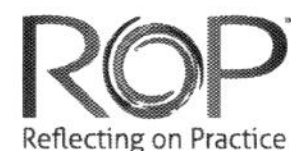

will return to these entries later in the session; no sharing out at this time.

- *How does a person "do science?"*

- *Who does science?*

- *What is a scientist?*

- Recall your experiences with your learners. *What is your impression of how they would depict science and scientists?*

Mystery Tube

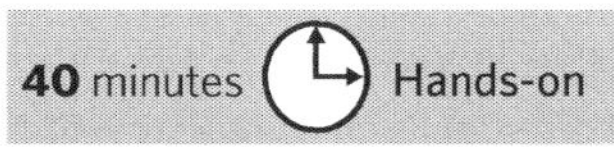

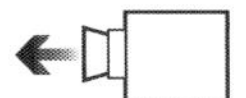

1.   Display instructions and introduce activity.
Tell participants that they will work in small groups to do this investigation. The purpose of the task is to have a common experience to explore what is science and how it's done. Review the instructions:

- *Explore the tube and share ideas.*

- *Determine what the interior structure looks like.*

- *Draw a diagram to model your hypothesis.*

2.   Distribute materials to each group.
Give each group a mystery tube and sheet of paper and markers. Remind them that these diagrams are *models* that can be used to explain their understanding about the internal structure of the tube. Encourage them to label what they think is going on inside.

3.   Circulate and encourage deeper exploration.
Listen to what participants talk about. Use the following suggested prompts:

- How might you test your ideas and the accuracy of your model without looking inside or destroying the tube?

- What evidence have you gathered to support your model?

- How confident are you in your model?

- What observations make you wonder whether or not your model is an accurate representation of the internal workings?

4.   Pair up groups for sharing.
After about 10-12 minutes, or when most groups have a model drawn, have each group share with another group: (a) their model (diagram), and (b) what they would say if they were going to publish their findings about the mystery tubes right now. Encourage them to comment and ask questions about one another's models.

If you decided to use a different name for this activity, substitute that name throughout.

1.3

Although this activity is about "solving" a mystery, don't reveal the answer during the activity itself. But be prepared: Being denied the "right answer" may distract and frustrate some participants. If that happens, it can be helpful to ask follow-up questions, such as, *How certain can scientists be of getting the "right answer" in their investigations? What do scientists do in their investigations to be more certain? What can you do to be more certain of your model?*

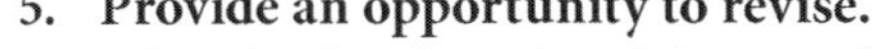

5. **Provide an opportunity to revise.**
   After they've shared and discussed their models, give each group a few minutes to revise their diagrams if they wish. Tell them that revision based on new evidence is part of the practices of science.

6. **Begin whole-group share.**
   Ask a representative from each group to come up (all at the same time) and draw their group's model on the board for everyone to see. Use some or all of the prompts below to encourage groups to comment on each others' models. Remember to listen to participants' responses; ask for *evidence, explanation,* or *clarification*; and encourage them to listen closely to what is shared. Let them know that vigorous discussion among scientists is also part of how science is done.

   - Did something you heard from sharing with another small group lead you to change your model?

   - How confident are you about your hypothesis?

   - What fits your observations and what doesn't? Which models are plausible? [Likely most, or all] Is there a way to decide which model most closely fits all the available evidence we have right now? [Possible answers may include: arguing, voting, revisiting our evidence/observations, choosing the most complicated/least complicated]

   - What additional evidence would we need to decide between the models? How could we be more confident? [Obtain more evidence; cut the tube open; develop a three-dimensional model to test each idea and compare to our observations of the mystery tube...]

7. **Collect the tubes.**

8. **Display and discuss reflection questions.**
   Challenge participants to provide explanations, evidence, or clarifications to elaborate on their thinking. Nudge them to listen closely to what their peers share.

   a. *Were you "doing science"?*

   b. *What were you doing that is similar to what scientists do?*

Time-saver: If you decided to provide larger pieces of paper to each group for sharing their models, have them post or hold these up now for the whole-group share.

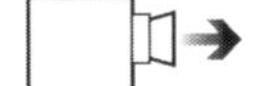

If you decide to divulge the "answer" to the inner workings of the mystery tube, it's best to provide participants the Internet search phrase ("mystery tube"), so that those who are interested and motivated can find the answer themselves.

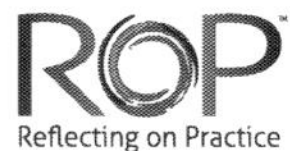

Table Talk: *How Science Works*

20 minutes — From Learners' Perspective

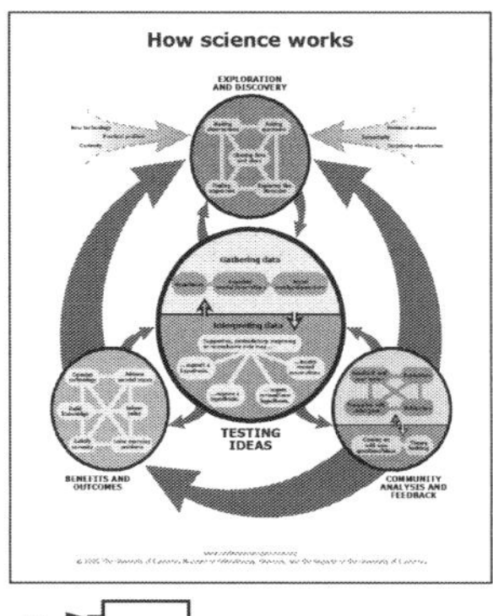

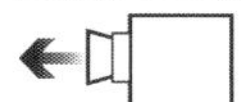

1. **Introduce the *How Science Works* flowchart.**

   a. To each small group for the Mystery Tubes task, distribute one flowchart and sticky-note pad (or dry-erase pen). Tell them this diagram shows another way to illustrate how science works.

   b. Give them a minute to review silently, before asking them to talk briefly with others in their group about their reactions and what they find interesting and relevant in the flowchart.

2. **Display and review instructions.**

   a. Ask participants to think about the steps their group went through when they explored the Mystery Tube.

   b. Have them mark on the flowchart the first step they took with a #1 (using pencil, sticky note, or dry-erase pen) to show where they began.

   c. Have participants proceed, in their small groups, to number all the steps they took and place those numbers on the flowchart.

3. **Pair up groups to compare flowcharts.**
   Have each small group share its completed flowchart with another group. Ask them to spend just a few minutes looking for similarities and differences, then choose a few groups with quite different flows represented on their charts and have them place their flowcharts on the wall where everyone can see them. (Optional: have the volunteers briefly share their pathway.)

4. **Expand the discussion to a whole-group share.**

   a. Ask participants to share their conversations regarding the *How Science Works* flowchart.

   b. Use some or all of the following prompts to lead a discussion about how what they did in this session relates to how science works.

      ◻ What do you notice about your pathway shown on the flowchart?

      ◻ How do your images of how science gets done fit within your pathway?

      ◻ How does what you did in this activity differ from what you were taught about the science process in school?

> □ Which part(s) of the science process and practices shown on the flowchart did we address today?
>
> □ Which part(s) did we not address?
>
> □ Which part(s) do you think are most often addressed and not addressed in informal institutions? Why do you think that is?

5. **Distribute a *How Science Works* flowchart handout to each participant.**
   Mention that they may want to use the flowchart as a resource when designing experiences.

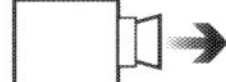

6. **Shift perspective to their learners.**
   Tell participants to revisit their impressions of how their learners might depict science and scientists. Display questions for participants to discuss with a partner:

   - *Compare your impressions of your learners with the **How Science Works** flowchart. To what extent do you imagine your learners think of science in this way?*

   - *What do you think is your role (and/or your organization's role) in helping learners come to a more accurate understanding of what science is and how science works?*

7. **Lead a whole-group share.**
   As participants share their thinking, challenge them to elaborate on their ideas and encourage them to consider multiple viewpoints. Facilitate the discussion using the following steps:

   a. Invite and listen to their responses.

   b. Challenge participants to elaborate on their thinking by providing explanations, evidence, or clarifications.

   <u>Suggested probing questions:</u>

   > □ What makes you think that?
   >
   > □ Please give an example from your experience.
   >
   > □ What do you mean?

   c. Encourage participants to listen closely to what their colleagues are saying.

   <u>Suggested probing questions:</u>

   > □ Can anyone add something to that comment?
   >
   > □ How can we consider that idea from a different perspective?
   >
   > □ If that were the case, what assumptions are being made?

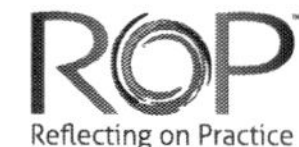

    d. Call out trends and challenges that emerge.

    e. Summarize ideas shared.

Walkabout: *Science & Knowing*

45 minutes   Research Discussion

1. **Revisit Walkabout Research Discussion.**
   Remind participants that they engaged in this research discussion routine in the first session. Review the procedure as follows:

   a. Four chart papers have been placed around the room. Each is titled with one big idea from the research discussion.

   b. They will work in small groups to consider and discuss each idea and record their conversations on the chart.

   c. Each group will be assigned a different-colored marker so that, as conversations are recorded from chart to chart, the responses of all groups can be compared. (Groups can decide for themselves how to rotate the recorder's role.)

   d. Once they're done discussing and recording ideas on one chart, they'll rotate ("walkabout") clockwise to read, think, talk, and connect ideas on every chart until they have visited each of the charts in turn.

2. **Distribute a "Key Ideas from the Literature: Science & Knowing" handout to each participant.**
   Tell participants that they should read the section of the handout corresponding to the title on the chart as they arrive there on their walkabout. After their group has had a chance to read over that section of the handout, they will respond to three question prompts directly on the paper charts, with their markers.

3. **Display question prompts.**
   Direct participants' attention to the questions they will discuss at each chart. Leave these questions displayed throughout the task. (Acknowledge that participants can't discuss **c.** until they begin circulating.)

   a. *What thoughts come to mind when you consider this idea?*

   b. *How can you use this idea in your work?*

   c. *What connections can you make to others' responses?*

4. **Assign participants to groups.**
   Place participants into one of four groups (unless you have fewer than 8 participants) and assign each group a pen color and a chart to start with. Remind them to work with their small group to read and

Small group sizes of 2–6 will work well in this routine. If your community has fewer than 8 members, you won't be able to start a group of two off at each poster, but it's fine if some of the posters are unoccupied at any given time.

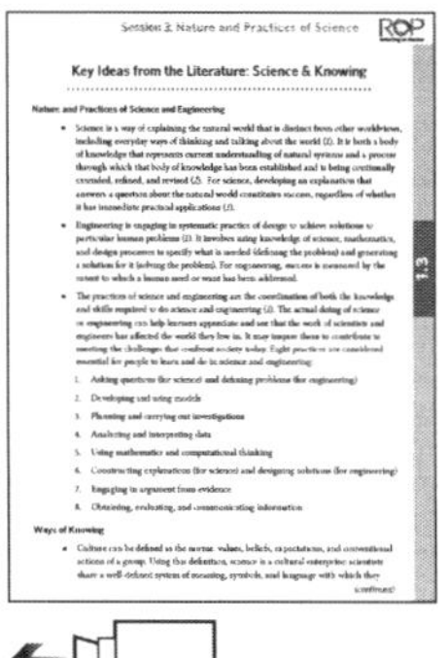

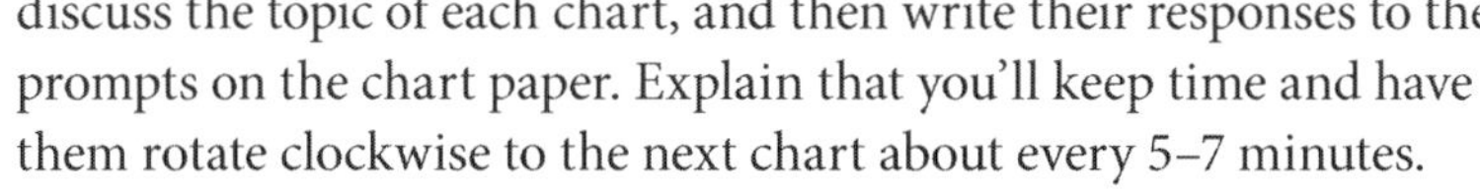

discuss the topic of each chart, and then write their responses to the prompts on the chart paper. Explain that you'll keep time and have them rotate clockwise to the next chart about every 5–7 minutes.

Facilitators may choose to be active participants, to model how to respond, comment, question, and push.

5. **Circulate, observe, and facilitate.**
Circulate around the room as groups discuss the four topics and write on the chart papers. Listen to what they say (and don't say) and notice what they write. Guide them on the types of responses they can make as they consider the topic on each chart. Some suggestions include:

- connect ideas between charts and between groups;

- elaborate or comment on written ideas; and/or

- ask for more information or clarification.

6. **Begin whole-group share by revisiting the charts.**
Once all the groups have rotated through all four ideas, have them return to the chart paper at which they started, to read what others wrote. Allow time for participants to review the charts, and then invite each group to share its thoughts with the community.

Display the following prompts to encourage synthesis of ideas as you lead the whole-group discussion:

a. *What themes emerged from everyone's responses?*

b. *Were there common issues and reactions across the groups?*

c. *What questions surprised you?*

7. **Facilitate group discussion.**
Remember to listen to participants' responses; ask for evidence, explanation, or clarification; and encourage them to listen closely to what is shared.

As participants share their thinking, challenge them to elaborate and encourage them to consider multiple viewpoints. Further thoughts to bring into the conversation if they don't emerge:

- *Ways of knowing.* Science educators commonly struggle with the idea that science is a way of looking at the world that may be distinct from other worldviews and ways of knowing. This tension could be due to lack of awareness about other knowledge systems, or an inability to imagine another way of knowing if one hasn't experienced it. Recognizing that there are other knowledge systems helps educators understand why some learners struggle with understanding and engaging in science.

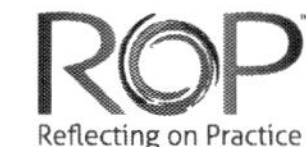

- *Culture of Science.* Educators often question whether science is a cultural enterprise. This struggle may be connected to recognizing Western science as a worldview, rather than as a static and universal concept. It may also reflect an erroneous assumption that "culture" is something possessed by "ethnic groups." If educators view science strictly in terms of facts, laws, and theories to explain the natural world, then how can it be a culture? And if they view it as a culture, is that contrary to the concept that science is objective?

Connect, Extend, Challenge & Apply: *Tell a Friend*

15 minutes — Let's Talk Practice

1. **Display the Let's Talk Practice chart.**
   Ask participants to use a blank page in their Learning Journals to make a **Let's Talk Practice** chart as shown.

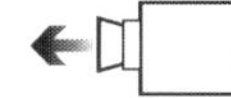

CONNECT	EXTEND	CHALLENGE
APPLY		

2. **Display the Let's Talk Practice questions.**
   Ask participants to **think about a friend** not in this community who would have enjoyed the conversation today (the friend need not agree with all the ideas discussed). Have participants consider the following prompts and record their thoughts on what they would want to tell their friend about how today's session pushed their thinking and practice.

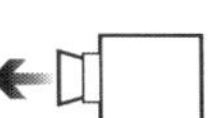

   a. **Connect.** *How are the ideas and information discussed in this session connected to what you already know?*

   b. **Extend.** *What new ideas extended or broadened your thinking in new directions?*

   c. **Challenge.** *What still seems challenging or confusing? What questions or wonderings do you have?*

   d. **Apply.** *What one or two changes will you make in your practice to apply your new ideas or understandings?*

If your community members work at a diverse range of tasks and programs, consider sorting them into small groups based on the work they do or the audience with whom they most interact (i.e., by types of programs or activities, subject matter, age group, etc.).

 **2** minutes

3. **Have participants share in small groups.**
   Ask participants to work in small groups of 3–4 to share their thoughts and exchange notions of how to integrate these new ideas into their practice.

4. **(Optional) Conduct a whole-group share.**
   If time permits, invite groups to share key points from their discussions with the rest of the community.

Continue the Learning

Suggested reading.
If you've decided to assign the readings, share those now.

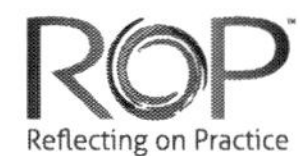

Key Ideas from the Literature: Science & Knowing

Nature and Practices of Science and Engineering

- Science is a way of explaining the natural world that is distinct from other worldviews, including everyday ways of thinking and talking about the world (*1*). It is both a body of knowledge that represents current understanding of natural systems and a process through which that body of knowledge has been established and is being continually extended, refined, and revised (*2*). For science, developing an explanation that answers a question about the natural world constitutes success, regardless of whether it has immediate practical applications (*1*).

- Engineering is engaging in systematic practice of design to achieve solutions to particular human problems (*1*). It involves using knowledge of science, mathematics, and design processes to specify what is needed (defining the problem) and generating a solution for it (solving the problem). For engineering, success is measured by the extent to which a human need or want has been addressed.

- The practices of science and engineering are the coordination of both the knowledge and skills required to **do** science and engineering (*1*). The actual doing of science or engineering can help learners appreciate and see that the work of scientists and engineers has affected the world they live in. It may inspire them to contribute to meeting the challenges that confront society today. Eight practices are considered essential for people to learn and do in science and engineering:

  1. Asking questions (for science) and defining problems (for engineering)

  2. Developing and using models

  3. Planning and carrying out investigations

  4. Analyzing and interpreting data

  5. Using mathematics and computational thinking

  6. Constructing explanations (for science) and designing solutions (for engineering)

  7. Engaging in argument from evidence

  8. Obtaining, evaluating, and communicating information

Ways of Knowing

- Culture can be defined as the norms, values, beliefs, expectations, and conventional actions of a group. Using this definition, science is a cultural enterprise; scientists share a well-defined system of meaning, symbols, and language with which they

(continues)

Key Ideas from the Literature: Science & Knowing *(continued)*

interact socially (*3-5*). The values of science include common commitments to questions, research perspectives, and ideas about what a viable scientific stance involves (*6*). Science is done by humans. Thus, evidence is collected, interpreted, and influenced by current scientific perspectives, the society in which the work takes place, and scientist's personal values and opinions (*7*).

- Groups of people across Earth and throughout human history have sophisticated knowledge systems to explain and thrive in their natural world (*8*). "Indigenous knowledge systems are the complex arrays of knowledge, know-how, practices, and representations that guide human societies in their innumerable interactions with the natural milieu: agriculture and animal husbandry; hunting, fishing and gathering; struggles against disease and injury; naming and explaining natural phenomena; and strategies for coping with changing environments" (p. *2*) (*9*). These ways of knowing do not: interpret reality based on linear, reductive conceptions of cause and effect (*10*); separate empirical and objective from spiritual and intuitive (*9*); or prefer written words as the mode of transmission. These features are foundational to Western science. However, there is no sound basis for deciding that one knowledge system has a better reference point for reality than another (*9*).

- Science is the socially constructed knowledge system *attributed* to Western societies that has become prevalent in the modern world. However, whether by choice or force, humans throughout history have traveled, mixed, and adapted. Human societies in one place thrived before new groups arrived. Islamic science flourished *in advance* of the European Renaissance; the scientific revolution occurred *after* colonization of the Americas and India. Considering human movement over time, "are we certain that what we call 'Western' science…is indeed Western in origin, ingredients, and rationality" (*11*)? Whether through appropriation, assimilation, imitation, or adaptation, "could the Indigenous peoples of Africa[, Asia, Australia, and Americas] be co-authors in the knowledge stores that have been monopolized through imperialistic power" (p. *2*) (*11*)?

Misinterpretations of the Scientific Process

- The following are all *inaccurate* statements that people commonly think about science, scientists, and the scientific process (*12*).

  1. Scientists are completely objective in their evaluation of scientific ideas and evidence.

  2. Scientists work without considering the applications of their ideas.

(continues)

   This page may be duplicated for educational use.

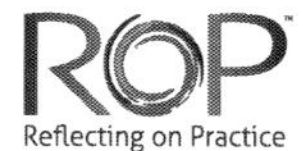

Key Ideas from the Literature: Science & Knowing *(continued)*

3. If evidence supports a hypothesis, it is upgraded to a theory. If the theory then garners even more support, it may be upgraded to a law.

4. The process of science is purely analytic and does not involve creativity.

5. Experiments are a necessary part of the scientific process. Without an experiment, a study is not rigorous or scientific.

6. The "hard" sciences (e.g., chemistry and physics) are more rigorous and scientific than "soft" sciences (e.g., psychology and sociology).

7. Scientific ideas are absolute and unchanging.

8. Because scientific ideas are tentative and subject to change, they cannot be trusted.

9. Scientific ideas are judged democratically.

10. The job of scientists is to find support for their hypotheses.

- Implicit instruction alone, about the nature of science (such as doing an investigation with no explicit reflection on the practices of science involved), is insufficient for learners to develop an understanding of what science is. The nature and practice of science must be intentionally and explicitly integrated into instruction (*13-15*). Even teachers whose views on the nature of science are consistent with those advocated by current reforms in science education struggle to convey their understanding through instruction. Teachers need help to develop skills and strategies to transform their knowledge into practice (*13, 16, 17*).

Science Identities, Aspirations, and Careers

- Young people's perceptions of and interests in science are influenced by the social relationships they value; activities that support their sense of agency; and their sense of belonging and connectedness with their community and its members (*18*). Additionally, family attitudes towards science and encouraging and fostering science in everyday family life (e.g., conversations, social networks, leisure activities) are important to children's science learning and aspirations (*19*), and perhaps more influential than the family's social structure (e.g., ethnicity, class) (*20*). These experiences and connections contribute to young people's science capital (*21*). However, families might not have all the knowledge and relationships they need for success in science within their household. Hence, they rely on their social network (e.g., family, friends, acquaintances) and formal institutions (e.g., schools, labor unions, libraries, museums) for support, services, access to information, and

(continues)

Key Ideas from the Literature: Science & Knowing *(continued)*

resources. These relationships contribute to the families' social capital. Young people's science and social capital influence their pursuits and access to science; these capitals are in turn influenced by the structures and sociopolitical context of society. In short, it isn't always due to a lack of interest in, or aspiration for, science that keeps young people from underrepresented groups out of science (*22*).

- Children and their parents have complex conceptions of who are science professionals, but there's a tendency to view those who do science as "others" and science careers as irrelevant to them. They're aware of the stereotypical image of those who do/go into science as nerdy/geeky, male and/or white (*20*). Sometimes, they simultaneously acknowledge and challenge these stereotypes (*23*). They ascribe special characteristics to scientists or science-interested peers, such as naturally clever, highly intelligent, innately science-minded, and hard-working. If they don't identify themselves with these characteristics, then these descriptions can contribute to young people positioning those who pursue science as "others," "not me/us." By age 10, children may report enjoying science (they find it fun, exciting, important, and interesting), and yet they may not choose to pursue it in higher levels of study, or take up a science identity (*24*).

- Language can influence learners' conceptions of science. Children, age 4-11, maintained more robust interest and efficacy about "doing science" when science was presented as something that people **do** and told **actions they can take** (action-oriented language) (*25, 26*). In contrast, they lost interest and thought they would be worse at science over time when they were invited to "be scientists" and told how scientists do science (identity-oriented language) (*25*). In short, children may rule out science altogether due to how science is talked about, despite learners having interests, aspirations, and successes in science.

References

1. National Research Council, *A Framework for K-12 Science Education: Practices, Crosscutting Concepts, and Core Ideas.* (National Academies Press, Washington, DC, 2012).

2. NGSS Lead States, *Next Generation Science Standards: For States, By States.* (The National Academies Press, Washington, DC, 2013).

3. D. Baker, P. C. S. Taylor, The effect of culture on the learning of science in nonwestern countries: The results of an integrated reseasrch review. *Intl. J. Sci. Ed.* **17**, 695-704 (1995).

4. M. Ogawa, Toward a new rationale of science education in a non-western society. *Eur. J. Sci. Educ.* **8**, 113-119 (1986).

5. D. Pomeroy, Science education and cultural diversity: Mapping the field. *Studies in Science Education* **24**, 49-73 (1994).

6. M. Fenichel, H. A. Schweingruber, *Surrounded by Science: Learning Science in Informal Environments.* (The National Academies Press, Washington, DC, 2009).

    This page may be duplicated for educational use.

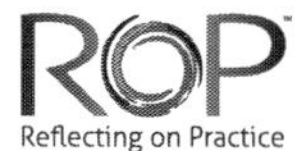

Key Ideas from the Literature: Science & Knowing *(continued)*

7. R. Schwartz, N. G. Lederman, What scientists say: Scientists' views of nature of science and relation to science context. *Intl. J. Sci. Educ.* **30**, 727-771 (2008).

8. M. Iaccarino, Science and culture. *EMBO Reports* **4**, 220-223 (2003).

9. D. Nakashima, M. Roue, in *Encyclopedia of Global Environmental Change*, P. Timmerman, Ed. (John Wiley & Sons, Ltd., Chichester, 2002), vol. 5, pp. 314-324.

10. M. M. Freeman, The nature and utility of traditional ecological knowledge. *Northern Perspectives* **20**, 9-12 (1992).

11. C. Mavhunga, *What Do Science, Technology, and Innovation Mean from Africa?*. (The MIT Press, Cambridge, MA, 2017).

12. "Misconceptions about science." Understanding Science. University of California Museum of Paleontology. 18 May 2020 https://undsci.berkeley.edu/teaching/misconceptions.php.

13. N. G. Lederman, Teachers' understanding of the nature of science and classroom practice: Factors that facilitate or impede the relationship. *J. Res. Sci. Teach.* **36**, 916-929 (1999).

14. S. K. Abell, M. Martini, M. George, "That's what scientists have to do": Preservice teachers' conceptions of the nature of science during a moon investigation. *Intl. J. Sci. Educ.* **23**, 1095-1109 (2001).

15. D. M. Moss, Examining student conceptions of the nature of science. *Intl. J. Sci. Educ.* **23**, 771-790 (2001).

16. J. H. Van Driel, N. Verloop, W. de Vos, Developing science teachers' pedagogical content knowledge. *J. Res. Sci. Teach.* **35**, 673-695 (1998).

17. N. G. Lederman, Students' and teachers' conceptions of the nature of science: A review of the research. *J. Res. Sci. Teach.* **29**, 331-359 (1992).

18. S. J. Basu, A. C. Barton, Developing a sustained interest in science among urban minority youth. *J. Res. Sci. Teach.* **44**, 466-489 (2007).

19. G. Nugent *et al.*, A Model of Factors Contributing to STEM Learning and Career Orientation. *Intl. J. Sci. Educ.* **37**, 1067-1088 (2015).

20. J. DeWitt *et al.*, Young children's aspirations in science: The unequivocal, the uncertain and the unthinkable. *Intl. J. Sci. Educ.* **35**, 1037-1063 (2013).

21. L. Archer, E. Dawson, J. DeWitt, A. Seakins, B. Wong, "Science capital": A conceptual, methodological, and empirical argument for extending bourdieusian notions of capital beyond the arts. *J. Res. Sci. Teach.* **52**, 922-948 (2015).

22. J. DeWitt *et al.*, High aspirations but low progression: The science aspirations-careers paradox amongst minority ethnic students. *International Journal of Science and Mathematics Education* **9**, 243-271 (2010).

23. J. DeWitt, L. Archer, J. Osborne, Nerdy, brainy and normal: Children's and parents' constructions of those who are highly engaged with science. *Research in Science Education* **43**, 1455-1476 (2013).

24. L. Archer *et al.*, "Doing" science versus "being" a scientist: Examining 10/11-year-old schoolchildren's constructions of science through the lens of identity. *Science Education* **94**, 617-639 (2010).

25. R. F. Lei, E. R. Green, S. J. Leslie, M. Rhodes, Children lose confidence in their potential to "be scientists," but not in their capacity to "do science". *Developmental Science* **22**, e12837 (2019).

26. M. Rhodes, A. Cardarelli, S. J. Leslie, Asking young children to "do science" instead of "be scientists" increases science engagement in a randomized field experiment. *Proceedings of the National Academy of Sciences* **117**, 9808-9814 (2020).

MODULE 2.
How People Learn

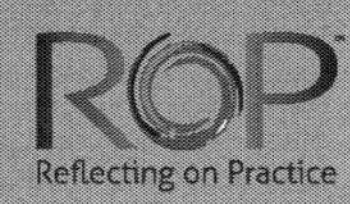

Overview

Module 2 delves into the Five Foundational Ideas on Learning introduced in Module 1. Participants consider the implications of these ideas on how they teach and design learning experiences.

Session 1 opens with a hands-on activity that places participants in the role of the learner, and then invites reflection on the experience to think more deeply about **how people experience learning**. Expanding on their reflections, they're given more information to consider, focused on three topics: cognitive system and prior knowledge, social cognitive system and conversations, and motivation and engagement in learning. Participants are then introduced to the research-based learning design framework and pedagogical practices used in the hands-on activity they just experienced.

Session 2 continues the discussion on learning with an emphasis on conceptual change and the **critical role of prior knowledge**. Another hands-on activity is used as a shared experience to explore the varying conditions of prior knowledge and how that affects learning. Participants are given a framework for thinking about learners' prior knowledge, and then self-interrogate their own experience in the hands-on activity to generate ways to transfer the framework into their practice.

Session 3 is the second whole-community **Video Reflection,** this time focusing on retrieving and connecting to prior knowledge. Participants revisit a previous Research Discussion on reflective practice and professional learning and are introduced to additional Tools for Reflective Practice to use in the video reflection. Participants are provided tools to prepare for their own Video Reflections in small groups.

Module 2.
Session 1: How People Learn

Session Overview

This session expands on the foundational ideas on learning introduced in Module 1 Session 1. Participants engage in a shared activity (about what causes the phases of the Moon) to experience how learners can work together to achieve understanding, and discuss the pedagogical practices that support sense making.

Knowledge from the science of learning is organized into three topic areas for participants to read and discuss: the brain system and prior knowledge; conversations and social activities; and engagement in learning. Participants make a concept map from this reading to visualize and articulate their evolving understanding.

Session Objectives

- Discuss three topics on learning to dig deeper into foundational ideas on learning.

- Make a visual representation of how educators think people learn.

- Discuss how educators' practice supports learning.

- Consider an instructional design framework to use in the design of experiences that support learning and sustain engagement.

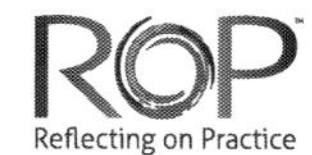

SESSION AGENDA		
Task Routine	**Description**	**Estimated Time (in minutes)**
Introduction *Session Objectives*	The goals and objectives of the session are introduced.	3

Hands-on
Phases of the Moon
	Participants have a shared science learning experience that will be used as a common reference point to consider how learning happens.
	  A. *Invitation*
	  B. *Exploration*
	  C. *Concept Invention, Shadows*
	  D. *Concept Invention, Phases & Light*
	  E. *Concept Invention, Explore New Connections*
	  F. *Application, Eclipses*
	  G. *Reflection, What I want to remember*
		50
		
		(3)
		(7)
		(10)
		(15)
		(5)
		
		(5)
		(5)

From Learners' Perspective Table Talk: *Reflecting on Phases of the Moon*	Participants reflect on what they did to build knowledge and make sense of what causes the phases of the Moon.	10
Break option		5
Research Discussion Reading Partner Jigsaw & Concept Map: *How People Learn*	Participants discuss research about how people learn, focusing on three topics: the cognitive system and prior knowledge; social cognitive system and conversations; and engagement in learning. They work in pairs to generate a concept map to organize and visualize their reading topic.	55
Let's Talk Practice Connect, Extend, Challenge, & Apply: *Learning Cycle Design Framework*	The community is introduced to the Learning Cycle Design Framework that is used throughout the program for designing experiences to support learning. Participants record and discuss how they can apply ideas about how people learn to their practice.	25
Continue the Learning	Participants are reminded about upcoming Video Reflections in this module.	2
Total Estimated Time		**150 mins. (2.5 hrs.)**

MATERIALS

Recurring for all Sessions

- See Module 1 Session 1

For this Session

- 1 corded light-bulb socket with plug—no shade (clamp recommended)
- 1 light bulb, 450 or 1100 lumens
- 1 extension cord, 25 feet or 8 m
- Room that can be darkened to make shadows
- **Science Content for the Facilitator: Moon Phases & Eclipses** (see Companion Website)

For each participant

- 1 polystyrene ball, 2 inches or 5 cm (See **Getting Ready** #5)
- 1 unsharpened pencil or thin dowel to insert into the polystyrene ball as a handle
- Handouts (1 copy each)
  - **Key Ideas from the Literature: How People Learn**
  - **Learning Cycle Design Framework** (Appendix B)
  - **Example of a Learning Cycle Designed Activity**

For pairs

- 1 sheet of chart paper
- 1 packet of small sticky notes, or index cards and tape
- 2 thin-write markers

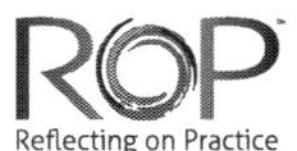

Getting Ready

1. Consider sending out the handout **Key Ideas from the Literature: How People Learn** a day or two before the session so that participants have the option to review the reading material ahead of time.

2. Review the handouts and slide deck for this session, and additional suggested readings.

    For the facilitator:

    - *How People Learn II*
        - Chapter 4 ("Processes that Support Learning")
        - Chapter 6 ("Motivation to Learn")

    Continue the Learning suggestion for participants:

    - *Surrounded by Science* (Chapter 4: "Learning with and from Others")

3. Duplicate handouts.

4. Decide whether you will use "Phases of the Moon" or a different activity to provide a shared learning experience on which participants can reflect during subsequent discussions in this session. **It's critical that participants who already understand the science concept of the activity engage and have a meaningful experience.** Consider having some additional challenge questions to pose to those who already understand what causes the phases of the moon.

5. Gather the "moonballs" and handles you will use for the Phases of the Moon activity. Just about any balls will work, as long as they're opaque and can have a handle inserted. Polystyrene balls are dense and will need to have a hole prepared before the activity. Extruded foam (Styrofoam™) balls will also work, if painted with white latex or other water-based paint. Without being painted, a shadow doesn't form well on Styrofoam. Check to make sure the balls you choose will show a contrast between dark and light areas (i.e., areas in and out of shadow) when testing your bulb and room in #6 below. Your goal is to be able to see the "phases of the moon" on your moonball.

If an alternative activity is needed, be sure to design it using the **Learning Cycle** and refer to the **From Learners' Perspective** task to ensure facilitation of your activity enables you to lead that discussion. Additionally, the Research Discussion in the next session on prior knowledge references shadows explored in the Phases of the Moon activity as an example of a critical false belief. If possible, be able to identify the critical false belief for the concept in the alternative activity.

6. Prepare for the moonball modeling activities.

   a. **Prepare the room.** Find a room that you can darken completely. You may need to tape black paper over windows, or convene in a different room to achieve the necessary conditions to do the activity. Place the lamp in the center of the room at about shoulder height. Use an extension cord, as needed, and tape the cord down to the floor for safety. Have a box of balls and handles ready to distribute.

   b. **Test to see which light bulb to use in your space.** Before the session, determine which light bulb will provide the best contrast for the room in which the activity will take place. Stand about 6 feet (2 m) from the bulb, the same distance from the lamp as the participants will stand. Hold a moonball by the handle at arm's length and move it to observe the contrast between dark and light sides of the ball. Brighter light bulbs usually provide more contrast if you have a large room, or in any room into which some light enters from outside. Dimmer bulbs provide greater contrast in smaller rooms with white walls.

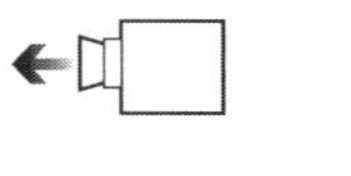

Session 1 Step by Step

Introduction
Session Objectives

🕐 **2** minutes

Let participants know that this session will drill deeper into learning and thinking, and to understanding how to support both. Display the session objectives:

- Discuss three topics on learning to dig deeper into foundational ideas on learning.

- Make a visual representation of how you believe people learn.

- Discuss how your practice supports learning.

- Consider an instructional design framework to use in the design of experiences that support learning and sustain engagement.

Phases of the Moon

50 minutes 🕐 Hands-on

1. **Introduce the Hands-on task.**
   Explain that this activity is about the phases of the Moon. Cover the following:

   a. Some participants may be familiar with all or parts of this activity (or something like it), or may already have a good understanding of what causes the phases of the Moon. Remind them that the purpose of the activity is to serve as a shared experience to learn a science concept *as their learners might,* while providing an opportunity to deepen discussions about their observations and ideas on learning and teaching.

   b. For those participants who may already have an accurate and deep understanding of what causes the phases of the Moon share the following:

      ▫ Challenge them to refrain from telling their colleagues all they know; rather, have them try to assist others to come to their own understanding.

      ▫ Encourage them to pay close attention to how others are making sense of the phenomenon.

      ▫ Invite them to wonder and extend their own thinking.

2. **Ask participants to be attentive.**
   Remind participants that while they engage in the activity, they should be attentive to what they are doing to make sense of the concepts, as well as what the facilitator is doing.

The most useful "science background" on moon phases is to periodically look at the Moon and explore the Moon/ball, Sun/light bulb, Earth/head model yourself. If participants ask content questions about moon phases or eclipses, the facilitator's best response may be to tell the participants to "ask the objects," and attempt to use the model to figure out the answers themselves.

During the activity, eavesdrop on their conversations and pose additional challenge questions to those who already have a deep understanding of what causes the phases of the Moon.

The **Invitation Phase** focuses on sparking curiosity and encouraging learners to begin determining how the experience can be meaningful to them.

The **Exploration Phase** is driven by learners' interests, questions, and prior knowledge. Learners gather new information from free exploration and consider how the information is related to what they know.

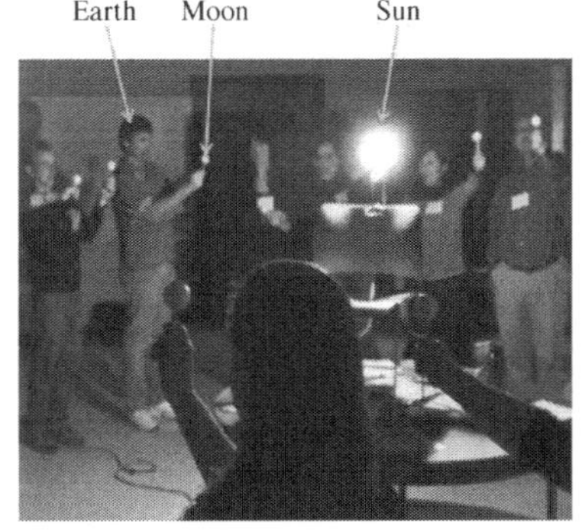

(A) Invitation (3 minutes)

1. **Ask everyone to recall memories of the moon.**
   Ask participants to take a few moments to first think about, then write in their Learning Journals, their responses to the following two prompts. Remind them that their journal entries won't be collected or read by others.

   Display the prompts:

   - *Think about the times you've looked at the Moon. What did it look like? Did you see it last night? What shape was it?*

   - *The different "shapes" of the Moon are referred to as the **phases** of the Moon. What do you think causes the phases of the Moon?*

2. **Have them talk with a partner.**
   Ask participants to turn to a partner and discuss their responses. Regain their attention after about 5 minutes. Let them know they're about to get the chance to mess around with their ideas in further detail.

(B) Exploration (7 minutes)

1. **Describe the set-up.**
   Ask participants to get up and form a large circle in the room. They should stand at least 6 ft (or 2 m) from the center of the circle. If needed, move furniture around so everyone can be in the circle.

2. **Position the light bulb.**
   While participants move around to form the circle, set up the light bulb in the center of the room at about shoulder height and turn it on. Darken the room; the bulb should be the only source of light.

3. **Introduce the Sun, Earth, and Moon model.**

   a. Pass out one "moonball" and one pencil or dowel "handle" to each participant.

   b. Demonstrate how to stick the handles into the balls, and explain that each of their moonballs represents the Moon.

   c. Explain the three parts of the model:
      - Each moonball represents the Moon.
      - The light bulb represents the Sun.
      - Each participant's head represents Earth.

   d. Ask participants to share a few inaccuracies they perceive in this model. Acknowledge these observations but let them know that the model is nonetheless a classic and useful one for demonstrating what they're about to do.

4. **Free exploration with the model.**
   Tell participants to use this Sun, Earth, Moon model to explore their ideas on what causes the phases of the Moon. Encourage them to work with others next to them and discuss what they notice.

5. **Wander and eavesdrop on the conversations.**
   Pay attention to what participants say as indicators of how they may be struggling with the concept. Inaccurate ideas that typically arise include:

   - Phases of the Moon are caused by a shadow from Earth, clouds, or the rotations of Earth or the Moon.

   - Different countries see different phases of the Moon on the same day.

   - The Moon makes its own light (the same way the Sun does).

   - The Moon doesn't rotate.

   - The Moon is only visible at night.

   - The same half of the Moon is in darkness all the time.

6. **Brief whole-group reporting of ideas.**
   Gather the group's attention and invite volunteers to report out the ideas they're discussing in two to three words. The point of this reporting out is to gather the broad range of ideas participants are considering, so dissuade anyone from offering a lengthy explanation. If there are ideas you overheard that aren't called out, like shadows, then share them with the group.

(C) Concept Invention, Shadows (10 minutes)

1. **Initiate an exploration of shadows.**
   Acknowledge that there seems to be the need for the community to clarify shadows before proceeding further. Ask participants to explore shadows in the room at various distances from the light-bulb sun for a brief moment; they can put aside their moonball or use it to make shadows. Ask them to share their discoveries with people next to them. After about a minute or so, regain their attention and ask a few participants to share their discoveries with the large group.

2. **Introduce three regions of a shadow.**

   a. Explain that shadows can be described as having three regions. Hold up your hand and point out the shadow of your hand on the wall. Tell them this is the region that most people notice: the shadow cast <u>by one object on another object</u>.

Don't worry just yet about inaccuracies in their language or use of the model. What's important here is that they try out some of their ideas and engage in discussions with their colleagues.

2.1

For more misconceptions about the Moon and the more accurate explanation, visit https://moon.nasa.gov/about/misconceptions/

In this activity, we explore the permutations, or regions, of a shadow that cause the Moon to appear to go through phases—not the attenuating parts of a shadow called the umbra, penumbra, and antumbra.

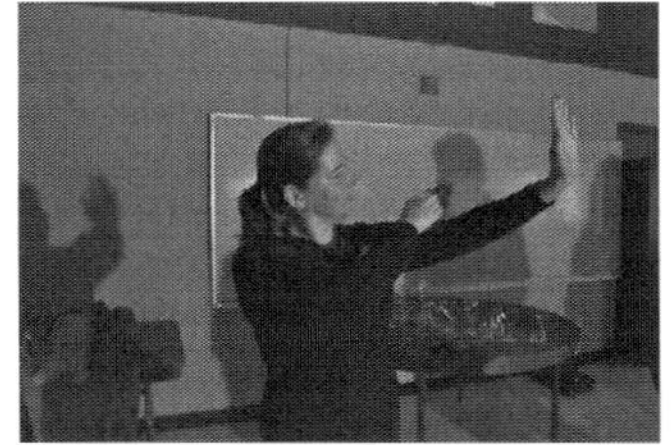

Extensive time is spent in the **Concept Invention Phase** (Sections C and D). The emphasis is providing opportunities for learners to integrate what they know with more scientifically accurate views from the discipline that you give them. To make these neural connections, learners need time and space to talk with peers and make the experience meaningful to them.

Facilitation within this phase oscillates between the following:

- facilitator guides exploration
- learners explore on their own, using their own way of speaking and testing out the language of the discipline offered by the facilitator and peers
- facilitator gives information, including terminology and explanations
- learners talk with their peers & facilitator about what they're thinking and doing

After giving participants relevant information to consider (e.g., shadows) they need opportunities to express their thinking as they work to connect and integrate the new information.

The most common errors learners make here are (1) not moving the moonball far enough to the left as they "orbit," and/or (2) looking at the light-bulb sun instead of at their moon. Help individuals as needed.

b. Ask if anyone can identify a second region of your hand's shadow. Someone will likely identify the darkness <u>at the backside of your hand</u>, facing away from the light bulb.

c. If they haven't identified it already, point out the third: the region <u>in the air on the side of your hand away from the light bulb</u>. Draw attention to the area by putting a finger from your other hand into the darkness; your finger is in the shadow. Point out that this region of the shadow can only be seen when you move an object into it.

3. **Have them explore the three regions of a shadow in pairs.**
Ask participants to play around with these three regions of shadows with a partner. Encourage them to explore parts of shadows cast by several different objects, including shadows on their moonballs. Make sure they include the regions of a shadow cast by their partner's head (this will be important later).

4. **Wander and eavesdrop on conversations.**
Give participants space to talk with one another. Listen closely to what participants say about shadows. Remind them to find the three regions of a shadow on their moonballs.

(D) Concept Invention, Phases & Light (15 minutes)

1. **Start first guided exploration to make the lunar phases.**
After a few minutes of playing with the moonballs and shadows, regain everyone's attention. Acknowledge the variety of ideas that were overheard. Tell participants you will guide them through two explorations to give them information relevant to their conversations. Ask them to face the light-bulb sun and hold their moons in their left hand, out in front of them at about shoulder height.

Model the movements with your own moonball as you facilitate.

2. **Orbit to a waxing crescent moon.**
Remind participants that the Moon orbits Earth.

a. Instruct them to move their moons to the left until a bright crescent of light shows on the ball, then hold it there. We call this sickle of light a **crescent moon**.

b. Explain that when the lighted part of the moon (as we see it from Earth) increases each night, the Moon is said to be **waxing**.

c. Have them demonstrate the crescent on their moons to someone beside them. Check to make sure everyone can see the crescent-shaped light on the ball.

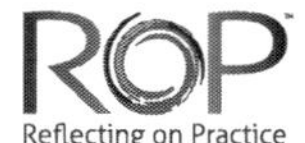

3. **Continue the orbit to the first quarter moon.**
   Tell participants to continue orbiting their moons at shoulder height in the same direction, until exactly half of their moon is lit. (They will, of course, need to turn their bodies to the left too.) This is a **quarter moon**. Ask them to discuss with the person next to them the following questions:

   a. *What have you noticed about the moonball and moon phases so far?*

   b. *As your moon gets fuller, is it moving towards the sun or away from it?* [Away from the sun, just like the real Moon]

4. **Orbit to a waxing gibbous moon.**
   Have participants continue orbiting their moon in the same direction until it's halfway between a quarter moon and a full moon. This is a **gibbous moon**, the stage at which less than all but more than half the Moon is lit.

5. **Orbit to a full moon.**
   Have them continue moving the moonball in its orbit until the part they can see is fully lit. (Their backs should now be to the light-bulb sun.) This is a **full moon.**

   a. Explain that they must hold the moonball just above the shadow of their heads in order to properly experience the full moon.

   b. If someone asks why they need to raise the ball up high during this phase, have participants discuss their ideas with the person(s) next to them.

   c. Ask for volunteers to share briefly their explanations, and invite others to agree/disagree or add on to the explanations. If it doesn't come up, tell them that most of the time the Moon is not on the same plane as Earth as our planet orbits the Sun. This means the Moon doesn't line up directly on the opposite side of Earth from the Sun each month, but just above or below it. Tell them we'll come back to this in the next part of the activity.

   d. Have them discuss briefly with a partner the following question:

      When the Moon is full, is Earth between the Moon and the Sun, or on the opposite side of Earth from the Sun? [On the opposite side of Earth from the Sun]

6. **Have them pause and discuss with a partner.**
   Ask them to discuss with the person next to them what they've noticed so far about the phases of the Moon and any questions they have. After a few minutes, ask for volunteers to share their ideas, questions, and observations about the phases of the Moon.

Why do we call what appears to be a half-lit moon a "quarter moon"? Because the Moon is one-quarter the way through its orbit around Earth. (The Moon is actually always half-lit, but we don't see that from our perspective on Earth.)

2.1

7. **Continue orbiting to a waning gibbous moon.**

   Tell participants to continue moving their moon in its orbit until it's gibbous (more than half lit, but less than full) once again. Tell them that when the lighted part of the Moon (as we see it from Earth) decreases each night, the Moon is said to be **waning**.

8. **Continue to a waning quarter moon.**

   Have participants continue orbiting their moon in the same direction until it is just half-full again, back to a quarter moon. Ask them to discuss the following with the person next to them:

   As your moon moves towards the sun, does it appear to get fuller or thinner? [Thinner]

9. **Model waning crescent moon and then new moon.**

   Ask participants to continue to move their moons until they see a very thin crescent again. Explain that the Moon usually doesn't pass directly in front of the Sun, but just above or below it. Ask them to discuss with the person next to them the following question:

   What do we call the lunar phase when we can see the Moon only faintly or not at all?

   After a moment or so, have someone tell—or share, yourself—that this is called the "new" moon, because it marks the beginning of its cycle.

10. **Guide a second orbit emphasizing light and shadows.**

    Tell participants you will lead another orbit of the moonball from waxing crescent to new moon. At various points in the orbit, instruct participants to pause and discuss the following questions about light and shadow, first with a partner and then to summarize for the whole group.

    a. *What is making the bright side of your moon bright?* [Light from the sun]

    b. *What is making the dark side/part of your moon dark?* [The beginning of the moon's own shadow]

    c. *Using a finger from your other hand, can you find places around your moon that are also in shadow?*

11. **Relate the model to the real world.**

    Describe that the movement of the Moon from waxing crescent to full takes about two weeks. It takes about a month (29.53 days) for the Moon to fully orbit around Earth. Tell participants that they have modeled one full cycle of the Moon's orbit around Earth.

Why do we call the Moon we cannot see, "new"? Some traditional peoples believed that a brand-new moon was being born at this time.

Lead this second orbit more quickly than the first one.

Question b., about the dark side/part of the moon, is particularly important to understanding the lunar phases; the most common misconception about the phases of the Moon is that the dark part is caused by Earth's shadow.

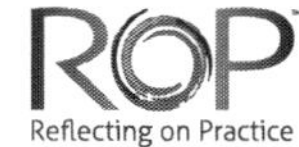

2.1

(E) Concept Invention, Explore New Connections (5 minutes)

1. **Have participants explore the model with peers.**
   Circulate as you encourage participants to move their moons in
   another orbit around Earth several times, exploring light, shadows,
   and moon phases. Ask them to share their discoveries, questions,
   and understandings with a partner or small group standing next to
   them, and to challenge each other with questions and statements,
   just as they would with their own learners. These might include:

   - Show me what you mean.

   - How can you be sure of that?

   - Can you help me figure this out?

   - I get this part of the explanation, but I'm still confused about
     _______.

   > *To engage participants who already understand what causes
   > the phases of the Moon, ask these additional questions as you
   > circulate, challenging them to figure each one out using evidence
   > from the model. Encourage these participants to work
   > and talk together. Then have individuals share their ideas and
   > evidence in the large group during the whole-group share.*
   >
   >   - Can we see a full moon during the day?
   >
   >   - Why do we only see one side of the Moon; doesn't it rotate?
   >
   >   - How can you use the model to show why the crescent is
   >   sometimes near the bottom of the Moon and sometimes
   >   near the top?
   >
   >   - Use the model to describe how you would know if the Moon
   >   were waxing or waning when you see it in the real world.

2. **Begin whole-group share.**
   Invite participants to share their ideas and understandings about
   what causes the phases of the Moon with the whole group. Encourage
   them to ask questions of each other and to use the objects to help
   explain their thinking. Ask guiding questions and answer questions
   from the participants where it seems appropriate, and to help them
   in their sense making.

Encourage participants to
make every effort to figure out
answers to their questions using
evidence from the objects in
the model and discussions with
each other.

As participants explore freely,
sometimes their observations
and questions don't help them
explain the concept. Allow them
the freedom to explore, but listen
closely to what they're saying. If it
seems like their inquiry is leading
to more confusion than clarity,
ask them to consider how the
information they're requesting would
help them explain what causes the
phases of the Moon. You might also
encourage them to ask their peer if
the explanation or question they're
posing makes sense. When they
realize it doesn't, then redirect them
towards a more fruitful path.

The **Application Phase** offers learners the chance to use their new understanding in a related, but different context. Similar to the previous phase, the facilitator can give information and guide exploration, but time must also be allotted for learners to talk and explore on their own.

This may be an audible "aha" moment, as participants truly register that Earth's shadow falls on the Moon only during a lunar eclipse (and not during phases of the Moon!).

(F) Application, Eclipses (5 minutes)

1. **Create a solar eclipse.**
   Now ask participants to move their moonballs directly in front of the light-bulb sun to try to create a solar eclipse (also called an eclipse of the Sun, as the Moon blocks the light from the Sun from reaching Earth). Tell them it works better if they close one eye. They'll know they've succeeded when the sun is completely blotted out by their moon.

2. **Observe the Moon's shadow on Earth.**
   As participants observe this solar eclipse, have them hold their moons exactly where they are and glance at the people standing on the opposite side of the circle. Ask them to consider the following questions:

   a. *Do you see the shadows over everyone's eyes?* (Remind them that each head represents Earth.)

   b. *The people who live where your shadowed eye is see an eclipse of the Sun, but how about the people who live on your chin? Or on your ear?* [Only the people who live on the shadowed eye will observe a solar eclipse—the people on their chin or ear can still see the Sun!]

3. **Create a lunar eclipse.**
   Ask participants to orbit their moons in a circle, as before, until they reach the phase of the full moon. This time, tell them to move their moons into Earth's shadow (the shadow of their heads). Explain that this alignment models an eclipse of the Moon, as Earth blocks the light from the Sun from reaching the Moon.

4. **Explain the lunar eclipse.**
   While the moons are still in the shadow of participants' heads, point out that everyone who lives on the side of Earth facing the Moon can see the Moon in eclipse. Earth is much larger than the Moon. The shadow that Earth casts on the Moon blocks light from the Sun from reaching the Moon. Remind them that, during an eclipse of the Sun, only the people *inside* the shadow of the Moon can see the Sun being eclipsed.

5. **Identify phases of the Moon during eclipse events.**

   a. Ask participants to continue moving their moons around their heads until they again see an eclipse of the sun. Ask them to consider and discuss with their partner the following question:

   At what phase does the Moon appear just before or just after an eclipse of the Sun? [Crescent or new phase]

b. Have them continue their orbit until they see another eclipse of the moon. Ask them to consider and discuss with their partner the following question:

At what phase does the Moon appear just before or just after a lunar eclipse? [Full]

c. Have participants return to their tables.

(G) Reflection: What I want to remember (5 minutes)

1. Take time for Learning Journal entries.
Ask participants to look back at their initial ideas about what causes the phases of the Moon. Ask them to draw a line below their initial entries. Display the questions below and ask participants to write for 3 minutes, answering the following prompts below the line.

a. *What did you learn and want to be sure to remember about what causes the phases of the Moon?* (Or for those who already had an accurate understanding, what did you learn about your peers' ideas?)

b. *What specifically changed in your understanding?* (Or for those who already knew, what did you learn about supporting others to come to a more accurate understanding?)

c. *What remaining questions do you have?*

2. Hold a brief whole-group share.
Allow a couple of minutes for a few volunteers to share their "Aha"s and remaining questions.

3. Identify the four key ideas about the phases of the Moon.
Display the summary slide of the four big ideas about the phases of the Moon. Explain that these are the concepts the community explored in this activity:

a. The phases of the Moon are caused by the relative positions to one another of the Sun, Earth, and the Moon.

b. As the Moon moves in its orbit, different regions appear (to us) to be lit up as we look at it from Earth.

c. The Moon itself doesn't change, nor does the amount of Moon that is lit by the Sun; half of the Moon is ALWAYS lit by the Sun, even if we can't see that from Earth.

d. Earth casts a shadow on the Moon ONLY during a lunar eclipse.

The **Reflection Phase** is time set aside for learners to look back on the experience to consider what they learned and how their understanding changed. It's an important moment to monitor their thoughts before moving on.

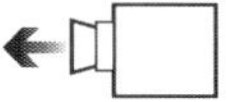

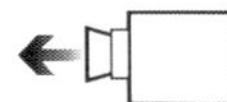

Summaries like these are helpful for learners to get a concise description of the key ideas explored. With all the exploring and talking they did, it might not be clear to some individuals what is the "answer" to the inquiry question. They can compare what they thought and wrote with a more expert explanation.

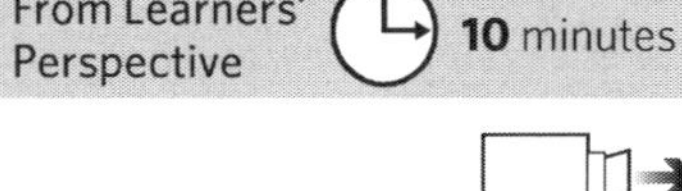

<table>
<tr><td>From Learners' Perspective   🕐 **10** minutes</td><td>**Table Talk:** *Reflecting on Phases of the Moon*</td></tr>
</table>

1. **Ask participants to think about their thinking process.**
   Explain that the activity they just participated in modeled how a community of learners can work together to generate explanations for a phenomenon and build understanding of a concept. They were thinking together as they learned together. Now it's time to be metacognitive—to think about thinking and make their thinking visible. Display the following questions:

   *What did you and your colleagues do to make sense of the science? What **action verbs** could you use to describe how you engaged in the activity?*

2. **Have them discuss with table group.**
   Have participants turn to someone next to them or at their table and discuss the question.

3. **Begin whole-group share.**
   After 2–3 minutes of talk, call time and invite participants to share with the whole group what they discussed. As they share, push on participants to specify what they were doing to learn, and **how** doing that action helped them make sense of the science. Remind them to use action verbs (e.g., talking, moving the moon ball around, retrieving prior knowledge, etc.) and acknowledge the actions that participants offer. After the whole-group share, call attention to items on the list below that aren't mentioned.

   - Manipulate a model

   - Listen to and talk with peers

   - Think on your own

   - Listen and talk with facilitator in whole-group discussion

   - Overhear other peers

   - Discuss and test out ideas that agree or disagree with your own understanding

   - Ask new questions

   - Explain your ideas to peers and the facilitator

   - Access and make connections to prior knowledge and experiences

4. **Provide quiet moment to contemplate.**
   Ask participants to take a moment to contemplate the extent to which they offer their learners opportunities to do those actions

in the learning experiences they facilitate or develop. Tell them the Research Discussion that we'll do next provides more information on learning. This will be followed by the introduction of an instructional design tool which can be used to incorporate those ideas into learning experiences.

Reading Partner Jigsaw & Concept Map: *How People Learn*

55 minutes — Research Discussion

1. **Connect back to "Five Foundational Ideas on Learning."**

   a. Point out that the **Phases of the Moon** activity offered an experience to learn together. In this Research Discussion, they'll be given more information to consider and discuss how the learning happened. This discussion will dig deeper into the ideas introduced in Module 1 Session 1.

   b. Remind participants they were introduced to the **Five Foundational Ideas on Learning**. Prompt participants to retrieve their handout from that session, then recap highlights from those conversations—particularly connections and questions that emerged.

2. **Introduce "Key Ideas from the Literature: How People Learn."** Distribute the handout to each participant. Explain that in this research discussion they will read and discuss three big topics on learning, first with a reading partner and then in small groups. Those topics are:

   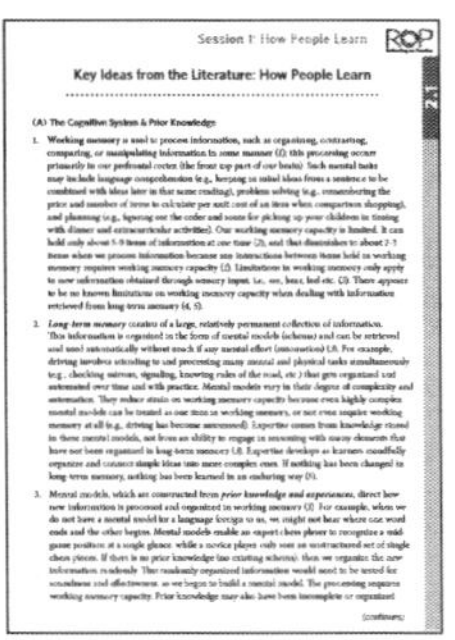

   a. The cognitive system and prior knowledge

   b. Social cognitive system and conversations

   c. Engagement in learning

3. **Form small communities of six, then pairs.**
   Form groups of six or more each, or let participants form the groups on their own. Ask everyone in each group to pair up with a reading partner. Emphasize that all three topics will be read and discussed within the groups.

4. **Display and review the instructions on the slide:**

   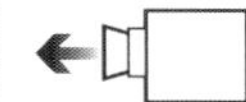

   a. Choose a partner with whom to read and (when ready) discuss one topic. Divide up the information so that each pair of reading partners in the group reads and discusses a different topic.

   b. With your partner, discuss what the ideas mean and how they relate to what you already know.

> c. Make a concept map for a visual representation of your topic. You'll have about 25 minutes for this portion; some instructions will be provided.
>
> d. Come back together in your small group to share big ideas from your topic discussions. Use your concept map as a visual aid.
>
> e. Discuss the following prompts in your small-group discussion:
>
>   □ Where do the Five Foundational Ideas show up in the three topics?
>
>   □ How are the ideas in the three topics connected?

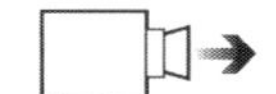

5. **Display and introduce the concept map.**
   Tell participants they'll make a **concept map** to help them articulate and visualize the ideas in their selected reading section on how people learn. Share that a **concept map is a visual way to show one's evolving thinking about a topic**, identifying the relevant ideas and their connections. Their concept maps aren't expected to capture all the details of their readings. Display and describe the example.

6. **Introduce focus question.**

   a. Tell participants to use this task as an opportunity to map out their sense making of their assigned reading. Their concept maps should address the focus question: **What is your reading topic about?**

   b. Distribute a sheet of chart paper, two markers, and a sticky-note pad (or index cards and tape) to each pair.

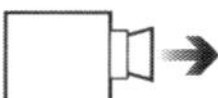

7. **Explain the six steps for creating a concept map.**
   Point out that there are different ways to make concept maps, and acknowledge that participants may be familiar with some of those. Display the directions and point out that the following structured routine will be used here. Explain the six steps and keep them displayed throughout the task:

   a. <u>Generate</u> a list of ideas from the reading: words or short phrases that seem to capture what the topic is about. Move on to the next step once partners have about five or six ideas; they can add more later.

   b. <u>Record</u> your pair's ideas, one per sticky note, and pile them up on the chart paper.

The rationale for introducing the concept map at this juncture is to apply the ideas on learning about which participants are reading. Participants aren't left to make sense of the complex reading alone and in silence. Instead, the combined map making and conversation allows participants to mess around with the ideas and their connectedness visually with a partner.

You may want to have participants record their lists of ideas (7a) in their Learning Journals. This list is not meant to be exhaustive; it's just a starting point, and they will add more ideas when they get to the last step.

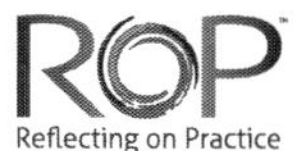

c. <u>Sort</u> and cluster your notes according to how central or tangential they are to *explain your topic*. Pile central ideas near the center of the paper and more tangential ideas towards the outside.

d. <u>Consider</u> why you clustered the ideas as you did. What do they have in common? How are they related? As you consider the related ideas, separate them on the chart paper.

f. <u>Connect</u> ideas that have something in common by drawing lines between related sticky notes. Refer back to the reading, as needed.

g. <u>Write a short sentence, phrase, or word</u> on each connecting line to explain how the ideas are connected.

h. <u>Elaborate</u> on any of the ideas/thoughts you've written so far. Select a few central ideas and break them down into sub-categories and connections. Add ideas that expand, extend, or add to your initial thinking. Consider the following questions:

- How can the ideas and connections be fleshed out further?
- What additional details come to mind for each idea?
- What other ideas can be added?

8. Start the task.
Provide about 25 minutes for participants to read, talk with their partner, and work on their concept maps. For groups that really struggle with making the concept map for their whole topic, encourage them to just focus on one segment.

9. Bring the activity to a close.

a. Ask pairs to write both names on their chart-paper concept map.

b. Photograph each pair's map and send to individuals for their records (or have participants photograph their own maps).

c. Invite them to reflect on their own thinking after the session. They can either transfer their maps into their Learning Journals by hand, or print the photo and affix it to a page in their journal.

10. Circulate and monitor the pace of discussions.
Check that all the topics are addressed within each group. Ensure there is silence at the beginning for participants to read their sections. As people finish reading, let them know partners can start discussing their topic whenever they're ready.

- Monitor each group to ensure that partners are working at the same pace as others in their group.

- Across multiple small groups, monitor to ensure that all groups are working at about the same pace, so that all will be ready for the whole-group share.

11. **Nudge connections within and across small groups.**
There will be no whole-group discussion, so the facilitation team should split up across the small groups to ensure groups are making connections across the topics. Do the following:

a. Listen closely to what they are saying.

b. Stay neutral.

c. Be attentive to quieter individuals, and encourage them to express their thinking.

d. Encourage and challenge participants to provide explanations, evidence, or clarifications to elaborate on their thinking. Suggested probing questions:

   □ What makes you think that?

   □ Tell me more about what you mean.

   □ Was there anything in the reading that you can cite for supporting evidence?

e. Encourage participants to listen more closely to the ideas shared. Suggested probing questions:

   □ If that were the case, what assumptions are we making?

   □ What's another perspective on this idea?

   □ What's another way to phrase that sentiment?

   □ What do others think about that idea?

   □ How is that similar to/different from what was said earlier?

f. If the following ideas or connections don't emerge in the discussion, bring them up for participants to consider.

   □ **Learning is making connections between ideas and organizing these connections** into mental models for storage in long-term memory. Experiences and examples strengthen the connections because they offer details and context that can be useful in other situations. [Foundational Idea #1 and #2]

   □ **Learning is social.** Learning is affected by our social environment and enhanced by our social interactions. [Foundational Idea #4] Learners need opportunities to express their thoughts and feelings as a part of making sense of experiences.

□ **Learning requires time, effort and persistence**, which requires learners to be engaged. There are different ways for learners to be engaged (behavioral, emotional, cognitive), and these modes of engagement can be intertwined. For instance, an English Language Learner may not like having to learn English, but she'll do so because of its value for future goals. The value lightens the negative feelings, prompting the learner to be behaviorally and cognitively engaged to learn English. [Foundational Idea #5]

12. **Summarize their conversations.**

(Optional) If time permits, bring the whole community back together for a brief summary of the ideas they discussed.

Connect, Extend, Challenge, & Apply: *Learning Cycle Design Framework*

25 minutes   Let's Talk Practice

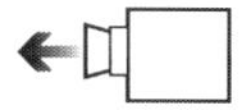

(A) Learning Cycle Design Framework (15 minutes)

1. **Transition with an entry in Learning Journals.**
Point out that the Phases of the Moon activity modeled ways to incorporate research on learning into a hands-on experience. Have participants respond in their Learning Journals to the following question:

Based on what we know now about how people learn, how do we design experiences to support learning?

2. **Introduce the Design & Teaching Tool.**
Share that, as participants consider how to design or redesign learning experiences for their learners, you'll share a tool that is fundamental to the program.

3. **Reflecting on the design of Phases of the Moon.**
Tell participants to recall the Phases of the Moon activity, again. This time they will look at the educator's perspective. That hands-on experience was designed with numerous opportunities for engaging with the science content in a particular sequence for the learning to unfold. The sequence followed the **Learning Cycle Design Framework**.

4. **Provide background about the framework.**

a. Ask for a show of hands from those who've heard of a learning cycle, such as the 5E instructional model. Explain that the learning cycle you'll introduce today is similar to the 5E model.

Both models evolved from the original learning cycle developed in the 1960s at the University of California, Berkeley's Lawrence Hall of Science (Bybee et al., 2006; Fuller, 2002).

Prior knowledge will be explored more extensively in Module 2 Session 2. A reference point from this activity to mention in that later session is how the facilitator handles learners' prior knowledge. The facilitator invites learners to retrieve their prior knowledge, and connect and modify it throughout the experience. The facilitator didn't need to hear what everyone knows throughout the entire activity for talking about their prior knowledge with peers to be valuable to the participants.

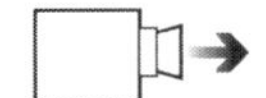

The information in this section comes from two handouts, **Learning Cycle Design Framework** and **Example of a Learning Cycle Designed Activity**. The examples offer a useful reference for designing learning experiences.

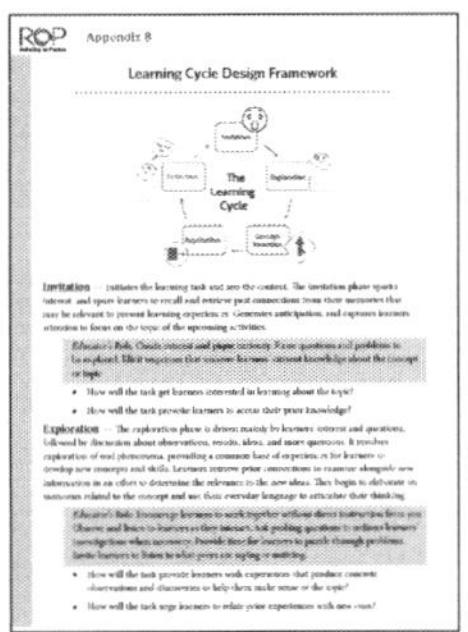

b. The learning cycle is modeled after how scientists inquire, and offers a structure for educators in guiding learners' discovery. The essential feature of the learning cycle is **providing the opportunity for learners to explore *before* introducing conceptual explanations and technical terms.** This ordering is intentional. It gives learners the opportunity to retrieve relevant prior knowledge, which invites them to make the experience more authentic to them.

5. **Distribute a "Learning Cycle Design Framework" handout to each participant**.

   a. Point out that the handout describes each phase and provides directions to educators regarding the important, necessary, and strategic roles they can play in each of the phases to support learning.

   b. Emphasize that the bulleted questions are checks to assess whether the design of the experience is modeling the learning cycle framework.

6. **Display and describe the Learning Cycle phases and examples.** Briefly describe each phase of the learning cycle with an example from the Phases of the Moon activity, using some or all of the following information.

   a. **Invitation**: initiates the learning task, sets the context, captures learners' attention, and sparks interest. Learners:

   - recall and retrieve memories that may be relevant to the learning experience.

     - **Example:** Learners are asked to recall memories of the Moon and answer the question, "What do you think causes the phases of the Moon?" They record ideas in their Learning Journals and then discuss them with a partner

   b. **Exploration**: is driven mainly by learners' interest and questions as they explore phenomena, which sustains their attention to pursue their inquiries further. It provides a common base of experiences for learners to develop new concepts and skills. Learners:

   - continue to retrieve prior knowledge and determine the relevance to the new ideas as they explore; and

   - begin to elaborate on existing memories related to the concept and use their everyday language to articulate their thinking.

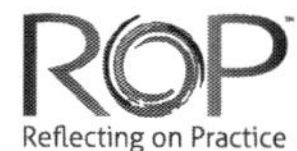

□   **Example:** Learners use a Sun, Earth, and Moon model to explore the phases of the Moon with peers. They express their ideas and uncertainties to peers, and answer one another's questions.

c.   **Concept Invention**: provides opportunity for learners to actively process the experience. Learners review evidence and data *they themselves have gathered* through exploration and try to make sense of it through discussions with peers and those with more knowledge. Learners:

□   describe their observations and generate explanations;

□   make new and/or different connections to existing memories; and

□   articulate their evolving thinking using newly introduced technical language and scientific views.

□   **Example:** Learners are given information about shadows, and are guided through two complete orbits of the Moon model, stopping often to ask questions. Learners discuss with peers and use their models to answer questions. At the end of the activity, the educator summarizes the key takeaways.

d.   **Application**: offers opportunities to strengthen newly formed connections through transfer of new knowledge to unfamiliar contexts in the world. Learners:

□   apply new knowledge and skills to solving a problem or meeting a challenge; and

□   retrieve newly formed connections and elaborate on them further as they apply them to novel contexts.

□   **Example:** Learners are asked to determine what causes solar and lunar eclipses, and how the cause of eclipses is distinct from what causes the phases of the Moon.

e.   **Reflection**: thinking about how original notions have been or need to be modified. These opportunities serve as a resource for future experiences. Learners:

□   consciously consider the new connections they've made to existing memories, and how their memories have changed; and

□   cultivate metacognitive skills as they think about what they learned and how they learned it.

The tension between scientific and everyday views is eased as the educator introduces technical language for learners to use as they articulate their evolving thinking.

> ❑ **Example:** Learners revise their writing and compare their new entry with their previous ideas. They reflect on what they learned and want to remember, what specifically changed in their thinking, and what helped them come to a more accurate understanding.

7. **Distribute "Examples of a Learning Cycle Designed Activity" handout.**

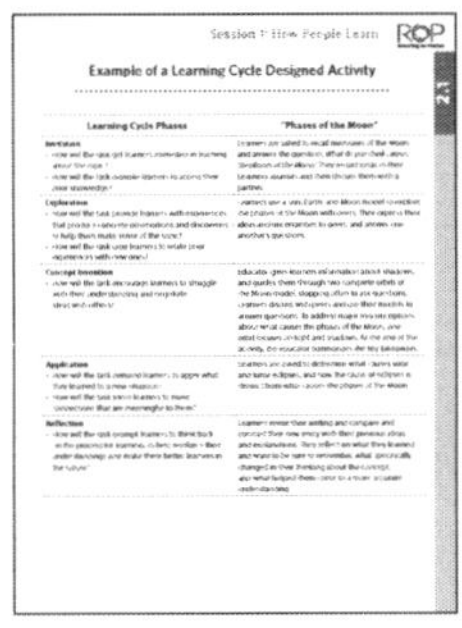

Point out that this example illustrates how the Learning Cycle was incorporated into the Phases of the Moon activity as was just discussed.

Give participants a few minutes to look over the handouts, talk with peers, and ask any questions they might have.

8. **Optional (depending on time): Extend the discussion.**
Share some additional ideas on the Learning Cycle for the community to consider.

   a. **Each phase and their sequence serves a purpose for learning.**

   > ❑ *Engage in discovery through an opening <u>invitation</u> and subsequent <u>exploration</u> phases.*
   > For example, an educator might want to introduce concepts and new vocabulary at the start of an activity, then ask learners to do an activity to explore those concepts. However, front loading the information in this way provides learners little opportunity to activate relevant prior knowledge and become interested in the topic.

   > ❑ *Deepen understanding by negotiating new connections and existing memories through <u>concept invention</u>.*
   > For example, an educator might provide multiple opportunities for discovering and exploring, but little or no opportunity for the learners to articulate connections between ideas for themselves with other learners. Without time for learners to explain and connect ideas, the experiences are only "hands-on" without being also "minds-on."

   > ❑ *<u>Apply</u> and <u>reflect</u> on the experiences.*
   > For example, an educator may view this part of the learning experience as something to be provided if time remains after the hands-on or explanations are completed. For learners, the new connections are likely still tenuous; application to a different context offers the chance to recall and strengthen those new connections. Moreover, the metacognition encourages learners to form the habit of monitoring and regulating their own learning.

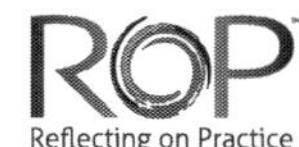

2.1

b. **Cycle within cycles**. In this introduction of the Learning Cycle Design Framework, a familiar hands-on activity was used as a concrete example. However, the learning cycle is not limited to discrete activities, but also used in the design of interactive sessions throughout the *Reflecting on Practice* program. For instance, the introduction of this framework itself served as the Concept Invention phase of the Learning Cycle. The Invitation and Exploration phases occurred during the Phases of the Moon activity. The Application and Reflection phases will come next and also surface in subsequent efforts to use the Learning Cycle.

(B) Application and Reflection task (10 minutes)

1. **Display the Let's Talk Practice chart.**
   As before, ask participants to use a blank page in their Learning Journals to make a **Let's Talk Practice** chart as shown.

CONNECT	EXTEND	CHALLENGE
APPLY		

2. **Display the Let's Talk Practice questions.**
   Ask participants to think about the following prompts and record their thoughts about the ideas discussed in today's session.

   a. **Connect.** *How are the ideas and information discussed in this session connected to what you already know? What aspects of this tool are already in your work?*

   b. **Extend.** *What new ideas extended or broadened your thinking in new directions?*

   c. **Challenge.** *What still seems challenging or confusing? What questions or wonderings do you have?*

   d. **Apply.** *What one or two changes will you make in your practice in order to apply your new ideas or understandings? Where and how can you use this tool in your practice?*

3. **Have participants share in small groups.**
   Ask participants to work in small groups of 3–4 to share their thoughts and determine how to apply these new ideas into their practice.

4. **(Optional) Conduct a whole-group share.**
   If time permits, invite groups to share key points from their discussion with the rest of the community.

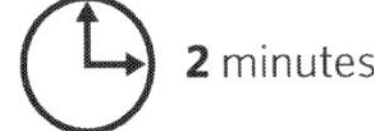 **2** minutes

Continue the Learning

1. **Reminder about Video Reflections in this module.**

   a. Remind participants they will be doing Video Reflections in this module—first as a whole community, and then in small groups with their critical colleagues. The whole-community Video Reflection will occur in the third session in this module. The small-group video reflections will take place afterwards at a time convenient to each critical colleague group.

   b. Solicit a volunteer (if you don't already have one lined up) to be the next presenter for the whole-community Video Reflection. The focus this time will be on prior knowledge, since that's the topic of this module.

2. **Suggested reading.**
   If you've decided to assign the reading, share it now.

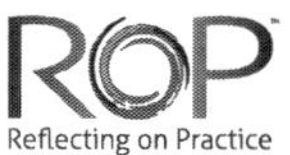

Example of a Learning Cycle Designed Activity

2.1

Learning Cycle Phases	"Phases of the Moon"
Invitation • How will the task get learners interested in learning about the topic? • How will the task provoke learners to access their prior knowledge?	Learners are asked to recall memories of the Moon and answer the question, *What do you think causes the phases of the Moon?* They record ideas in their Learning Journals and then discuss them with a partner.
Exploration • How will the task provide learners with experiences that produce concrete observations and discoveries to help them make sense of the topic? • How will the task urge learners to relate prior experiences with new ones?	Learners use a Sun, Earth, and Moon model to explore the phases of the Moon with peers. They express their ideas and uncertainties to peers, and answer one another's questions.
Concept Invention • How will the task encourage learners to struggle with their understanding and negotiate ideas with others?	Educator gives learners information about shadows, and guides them through two complete orbits of the Moon model, stopping often to ask questions. Learners discuss with peers and use their models to answer questions. To address major misconceptions about what causes the phases of the Moon, one orbit focuses on light and shadows. At the end of the activity, the educator summarizes the key takeaways.
Application • How will the task demand learners to apply what they learned to a new situation? • How will the task invite learners to make connections that are meaningful to them?	Learners are asked to determine what causes solar and lunar eclipses, and how the cause of eclipses is distinct from what causes the phases of the Moon.
Reflection • How will the task prompt learners to think back on the process for learning, to help reinforce their understandings and make them better learners in the future?	Learners revise their writing and compare and contrast their new entry with their previous ideas and explanations. They reflect on what they learned and want to be sure to remember, what specifically changed in their thinking about the concept, and what helped them come to a more accurate understanding.

Key Ideas from the Literature: How People Learn

(A) The Cognitive System & Prior Knowledge

1. **Working memory** is used to process information, such as organizing, contrasting, comparing, or manipulating information in some manner (*1*); this processing occurs primarily in our prefrontal cortex (the front top part of our brain). Such mental tasks may include language comprehension (e.g., keeping in mind ideas from a sentence to be combined with ideas later in that same reading), problem solving (e.g., remembering the price and number of items to calculate per unit cost of an item when comparison shopping), and planning (e.g., figuring out the order and route for picking up your children in timing with dinner and extracurricular activities). Our working memory capacity is limited. It can hold only about 5-9 items of information at one time (*2*), and that diminishes to about 2-3 items when we process information because any interactions between items held in working memory requires working memory capacity (*1*). Limitations in working memory only apply to new information obtained through sensory input, i.e., see, hear, feel etc. (*3*). There appears to be no known limitations on working memory capacity when dealing with information retrieved from long-term memory (*4, 5*).

2. *Long-term memory* consists of a large, relatively permanent collection of information. This information is organized in the form of mental models (schema) and can be retrieved and used automatically without much if any mental effort (automation) (*3*). For example, driving involves attending to and processing many mental and physical tasks simultaneously (e.g., checking mirrors, signaling, knowing rules of the road, etc.) that gets organized and automated over time and with practice. Mental models vary in their degree of complexity and automation. They reduce strain on working memory capacity because even highly complex mental models can be treated as one item in working memory, or not even require working memory at all (e.g., driving has become automated). Expertise comes from knowledge stored in these mental models, not from an ability to engage in reasoning with many elements that have not been organized in long-term memory (*3*). Expertise develops as learners mindfully organize and connect simple ideas into more complex ones. If nothing has been changed in long-term memory, nothing has been learned in an enduring way (*5*).

3. Mental models, which are constructed from *prior knowledge and experiences*, direct how new information is processed and organized in working memory (*3*). For example, when we do not have a mental model for a language foreign to us, we might not hear where one word ends and the other begins. Mental models enable an expert chess player to recognize a mid-game position at a single glance, while a novice player only sees an unstructured set of single chess pieces. If there is no prior knowledge (no existing schema), then we organize the new information randomly. That randomly organized information would need to be tested for soundness and effectiveness, as we begin to build a mental model. The processing requires working memory capacity. Prior knowledge may also have been incomplete or organized

(continues)

     This page may be duplicated for educational use.

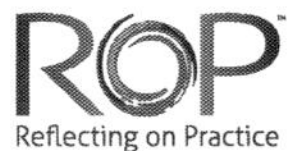

Key Ideas from the Literature: How People Learn *(continued)*

poorly, and thus new information may or may not fit (*6*). This situation, in turn, also strains working memory capacity to process and organize new information into existing mental models.

4. Prior knowledge does not refer just to "concepts," but also perceptions, social relationships, feelings and emotions, procedural skills, modes of reasoning, and beliefs about knowledge. Collectively, they are the neural connections that are built from past experiences over a person's lifetime. Learners' prior ideas, their "common sense" and "everyday thinking," are ***meaningful and useful*** to them. If those ideas are not engaged, the experience is not authentic to the learners. Consequently, learners often dismiss learning the new content as irrelevant (*7*) and the new information is less memorable (*8, 9*).

(B) The Social Cognitive System & Conversations

1. Humans are social organisms. Our brains have evolved towards social connections with others (*10*). At rest, when not engaged in a mental task, there are parts of our brain that become activated, which neuroscientists refer to as our brain's default mode (*11*). These same regions of our brain are turned on when we engage in social cognition—when we think of others, their relationships, and how they relate to us. In other words, when we're not tasked to think about anything, our brain automatically processes and consolidates our social relationships.

    a. Learning involves people, the things they use, the words they speak, the cultural context they're in, and the actions they take (*12, 13*). What we learn and how we learn are influenced by our environment and social interactions. In short, learning is as much a social activity as it is an individual endeavor.

    b. That humans are social organisms need not conflict with our introverted and extraverted personality traits. It does require educators to make appropriate space for learners. These traits are both grounded genetically in our temperament and shaped by our experiences (*14, 15*). Some people are highly sensitive to environmental stimulus (e.g., noise, light), and their brain triggers fear responses when in novel situations or among unfamiliar people (*16*). Over time, these individuals tend to have introverted personalities; they prefer an environment with less stimulation, intimate conversations with familiar people, and moments of quiet for introspection. Some people are uninhibited by environmental stimulus and novelty, and their brain triggers reward responses when being with others. As they grow up, these individuals tend to develop extraverted personalities; they often approach the unfamiliar and seek social activities (*17*). Regardless of personality traits, learners benefit from social interactions and moments for contemplation.

2. The opportunity to ***externalize*** and ***reflect on*** one's thinking and feelings facilitates learning, especially learning complex science concepts.

(continues)

Key Ideas from the Literature: How People Learn *(continued)*

a. Externalizing refers to expressing one's evolving understanding and emotions, whether by writing, talking, or drawing; it creates opportunities for learners to self-examine and manipulate their thinking and feelings, and obtain feedback from others (*18, 19*). Talking is particularly significant for learning. Students in K-12 to university settings show greater understanding when they engage in collaborative dialogue with peers where they provide explanations as part of arguments and justifications, and when they seek and provide help (*20-23*).

b. Reflection is the act of using one's mind to consider the process of learning. It's an opportunity to detect inconsistencies in one's thinking and emotional state, and to identify connections between areas of conceptual understanding (*24, 25*). This self-awareness is critical for learners to regulate their own learning (*19, 26*).

3. Families, friends, peer groups, and larger social networks are all units of learning, as well as significant contexts in which learning occurs (*12*). These units and contexts support social interactions that may occur in different, interdependent ways:

a. *Imitation*—learning from watching other people—is ubiquitous among humans across culture and lifespan (*12, 27*). Our ability to learn from imitation is attributed to transmitting cultural practices across generations and our evolutionary success beyond other primates (*10*).

b. *Collaboration*—learning from working with people—is a coordinated, synchronous activity that results from a continued attempt to build a common understanding of an idea or a problem (*28*), where the emergent understanding is a product of the group (*29*).

c. *Instruction*—learning through guidance from an adult, peer, or other conversant individual—is the process of more knowledgeable individuals helping less experienced learners make meaning of new experiences (*30, 31*).

(C) Motivation and Engagement in Learning

1. Learning complex ideas deeply involves expending considerable mental effort and persistence to organize and make connections, confront inconsistencies in one's own thinking, and overcome setbacks. Such commitment requires engagement and motivation from learners, both of which are situational and domain specific. They are also interrelated: motivation leads to achievement because it affects the quality of learners' engagement (*32*).

2. Engagement is multi-dimensional and is an interaction between the individual and their environment. It is presumed to be malleable, and has been studied in three ways (*33*):

a. *Behavioral engagement* refers to the ways in which learners participate in the learning experiences. The concept includes learners' conduct (e.g., showing up and adhering to rules of the environment) and levels of involvement in tasks (e.g., attention, concentration, effort, and contribution).

(continues)

          This page may be duplicated for educational use.

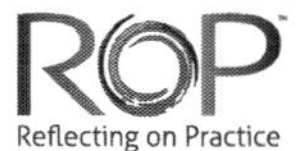

Key Ideas from the Literature: How People Learn *(continued)*

b. ***Emotional engagement*** refers to learners' affective reactions (their feelings and emotions) to the learning. It may be influenced by learners' interactions with the people and context involved; how they identify with the subject matter; their sense of belonging regarding the context and content; and how they value the subject matter.

c. ***Cognitive engagement*** refers to learners' investment in their own learning, which in turn affects the strategies they use for thinking and regulating their learning. It incorporates thoughtfulness and willingness to exert the effort necessary to comprehend complex ideas and master difficult skills.

3. Motivation is a dynamic, multifaceted phenomenon that influences a person's participation, persistence, and performance in learning (*34*). It is malleable. Learners can be motivated in many ways, and their motivation can change over time and vary depending on the circumstance and topic. Motivation to engage may be affected by several factors (*32*):

   a. ***Value*** refers to learners' regard that something has importance, worth, or usefulness (*35*). Value may be: intrinsic (e.g., interest in the topic), functional (e.g., perception of how tasks are related to future goals and life), or attainment oriented (e.g., personal importance placed on the task).

   b. ***Self-efficacy*** refers to learners' sense of their own ability to succeed in specific tasks (*26*). It is based on actual accomplishments and failures. It is influenced by learners' attitudes about their abilities (e.g., intelligence is developed and can be improved versus intelligence is innate and cannot be changed); and their attitudes towards goals (e.g., mastering the task and understanding versus for performance and task completion).

   c. ***Belonging*** involves having contact and interpersonal connections with other people (*36*). It refers to learners' feelings of fitting in and being welcomed. Sense of belonging is influenced by the relationships learners have with people (e.g., peers and instructors) and the content ("Others like me are engineers," "I feel welcomed in this lab") and, in turn, enhance learners' interest and engagement in the learning activities.

   d. ***Autonomy*** refers to learners' sense of agency. Feeling that they have choice and a role in directing their own activity can increase learners' interest and willingness to approach challenges (*19*).

4. Authentic means to be genuine, real, and trustworthy. Learning that is authentic to learners affects their motivation and engagement. The content and context may be authentic, in which case the learning is situated in **everyday experiences** that urge learners to determine how the new information is relevant and meaningful to them (*37, 38*). In short, the authentic context pushes learners to find a reason to understand—"I need to know this. I want to know this." As a result, learners put in the effort to form connections between new and old information, and develop deeper and more associated conceptual understanding (*32, 39, 40*). People involved

(continues)

 © 2021 Regents of the University of California ▪ **119**

Key Ideas from the Literature: How People Learn *(continued)*

in the learning activity may be authentic, meaning the **relationships are sincere**. That the instructor and peers, and the elicited interactions among them, are perceived to be genuine by learners cultivate trust and appeal to their affective responses to the experience that encourages deeper engagement in the learning activity (*19*).

References

1. J. Sweller, J. J. Van Merrienboer, F. G. Paas, Cognitive architecture and instructional design. *Educational Psychology Review* **10**, 251-296 (1998).

2. G. A. Miller, The magical number seven, plus or minus two: some limits on our capacity for processing information. *Psychological Review* **63**, 81 (1956).

3. J. J. van Merrienboer, J. Sweller, Cognitive load theory and complex learning: Recent developments and future directions. *Educational Psychology Review* **17**, 147-177 (2005).

4. J. Sweller, Evolution of human cognitive architecture. *Psychology of Learning and Motivation* **43**, 215-266 (2003).

5. J. Sweller, Instructional design consequences of an analogy between evolution by natural selection and human cognitive architecture. *Instructional Science* **32**, 9-31 (2004).

6. M. T. Chi, in *International Handbook of Research on Conceptual Change*, S. Vosniadou, Ed. (Routledge, New York and London, 2008), chap. 3, pp. 61-82.

7. D. Hammer, E. H. van Zee, *Seeing the Science in Children's Thinking: Case Studies of Student Inquiry in Physical Science*. (Heinemann, Portsmouth, NH, 2006).

8. M. T. van Kesteren, D. J. Ruiter, G. Fernández, R. N. Henson, How schema and novelty augment memory formation. *Trends in Neurosciences* **35**, 211-219 (2012).

9. D. A. Sousa, *How the Brain Learns*. (Corwin Press, Thousand Oaks, CA, 2016).

10. M. D. Lieberman, *Social: Why our Brains are Wired to Connect*. (Broadway Books, New York, NY, 2013).

11. R. L. Buckner, J. R. Andrews-Hanna, D. L. Schacter, The brain's default network. *Annals of the New York Academy of Sciences* **1124**, 1-38 (2008).

12. J. D. Bransford et al., in *Handbook of Educational Psychology*, P. A. Alexander, P. Winne, Eds. (Erlbaum, Mahwah, NJ, 2006), chap. 10, pp. 209-244.

13. B. Rogoff, in *Cognition, Perception and Language: Handbook of Child Psychology*, D. Kuhn, R. S. Siegler, Eds. (Wiley, New York, 1998), vol. 2, pp. 679-744.

14. C. E. Schwartz, C. I. Wright, L. M. Shin, J. Kagan, S. L. Rauch, Inhibited and uninhibited infants" grown up": adult amygdalar response to novelty. *Science* **300**, 1952-1953 (2003).

15. C. E. Schwartz, N. Snidman, J. Kagan, Early childhood temperament as a determinant of externalizing behavior in adolescence. *Development and Psychopathology* **8**, 527-537 (1996).

16. S. Cain, *Quiet: The Power of Introverts in a World that Can't Stop Talking*. (Broadway Paperbacks, New York, NY, 2013).

17. R. E. Lucas, E. Diener, A. Grob, E. M. Suh, L. Shao, Cross-cultural evidence for the fundamental features of extraversion. *J. Pers. Soc. Psychol.* **79**, 452 (2000).

18. R. K. Sawyer, in *The Cambridge Handbook of the Learning Sciences*, R. K. Sawyer, Ed. (Cambridge University Press, New York, NY, 2006), chap. 1, pp. 1-18.

19. L. Cozolino, *The Social Neuroscience of Education: Optimizing Attachment and Learning in the Classroom*. (Norton Inc., New York, NY, 2013).

This page may be duplicated for educational use.

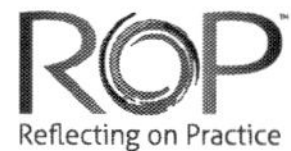

Key Ideas from the Literature: How People Learn *(continued)*

20. N. Mercer, L. Dawes, R. Wegerif, C. Sams, Reasoning as a scientist: Ways of helping children to use language to learn science. *British Educational Research Journal* **30**, 359–377 (2004).

21. G. J. Venville, V. M. Dawson, The impact of a classroom intervention on grade 10 students' argumentation skills, informal reasoning, and conceptual understanding of science. *J. Res. Sci. Teach.* **47**, 952–977 (2010).

22. F. M. van Blankenstein, D. H. J. M. Dolmans, C. P. M. van der Vleuten, H. G. Schmidt, Which cognitive process support learning during small-group discussion? The role of providing explanations and listening to others. *Instructional Science* **39**, 189–204 (2011).

23. S. Veenman, E. Denessen, A. van den Akker, J. van der Rijt, Effects of a cooperative learning program on the elaborations of students during help seeking and helping giving. *American Educational Research Journal* **42**, 115–151 (2005).

24. National Research Council, *Taking Science to School: Learning and Teaching in Science in Grades K-8.* R. A. Duschl, H. A. Schweingruber, A. W. Shouse, Eds. (National Academies Press, Washington, DC, 2007).

25. E. A. Davis, Prompting middle school science students for productive reflection: Generic and directed prompts. *Journal of the Learning Sciences* **12**, 91–142 (2003).

26. P. R. Pintrich, C. A. Wolters, G. P. Baxter, in *Issues in the Measurement of Cognition*, G. Schraw, J. C. Impara, Eds. (Buros Institute of Mental Measurements, Lincoln, NE, 2000), chap. 2, pp. 43–97.

27. A. N. Meltzoff, J. Decety, What imitation tells us about social cognition: A rapprochement between developmental psychology and cognitive neuroscience. *Philosophical Transactions of the Royal Society* **358**, 491–500 (2003).

28. J. Roschelle, S. D. Teasley, in *Computer Supported Collaborative Learning*, C. O'Malley, Ed. (Springer-Verlag, Heidelberg 1995), pp. 69–97.

29. P. Dillenbourg, M. Baker, A. Blaye, C. O'Malley, in *Learning in Humans and Machine: Towards an Interdisciplinary Learning Science*, E. Spada, P. Reiman, Eds. (Elsevier, Oxford, 1996), pp. 189–211.

30. L. Vygotsky, *Mind in Society: The Development of Higher Psychological Processes.* (Harvard University Press, Cambridge, MA, 1978).

31. D. Wood, J. S. Bruner, G. Ross, The role of tutoring in problem solving. *Journal of Child Psychology* **17**, 89–100 (1976).

32. P. C. Blumenfeld, T. M. Kempler, J. Krajcik, in *The Cambridge Handbook of the Learning Sciences*, R. K. Sawyer, Ed. (Cambridge University Press, New York, NY, 2006), chap. 28, pp. 475–488.

33. J. A. Fredricks, P. C. Blumenfeld, A. H. Paris, School engagement: Potential of the concept, state of the evidence. *Review of Educational Research* **74**, 59–109 (2004).

34. C. Senko, C. S. Hulleman, J. M. Harackiewicz, Achievement goal theory at the crossroads: Old controversies, current challenges, and new directions. *Educational Psychologist* **46**, 26–47 (2011).

35. Merriam-Webster (n.d.) Value. In *Merriam-Webster Dictionary.* Retrieved December 21, 2020, from https://www.merriam-webster.com/dictionary/value

36. R. F. Baumeister, M. R. Leary, The need to belong: desire for interpersonal attachments as a fundamental human motivation. *Psychological Bulletin* **117**, 497 (1995).

37. J. G. Greeno, in *The Cambridge Handbook of the Learning Sciences*, R. K. Sawyer, Ed. (Cambridge University Press, New York, NY, 2006), chap. 6, pp. 79–96.

38. J. L. Kolodner, in *The Cambridge Handbook of the Learning Sciences*, R. K. Sawyer, Ed. (Cambridge University Press, New York, NY, 2006), chap. 14, pp. 225–242.

39. R. C. Schank, *Dynamic Memory.* (Cambridge University Press, New York, NY, 1982).

40. J. L. Kolodner, *Case-Based Reasoning.* (Morgan Kaufmann Publishers, San Mateo, CA, 1993).

Module 2.
Session 2: Learners' Prior Knowledge

Session Overview

This session encourages participants to examine why it's necessary to access, activate, and connect to **learners' prior knowledge** in their practice. An investigation is used to invoke participants' prior knowledge and experiences with the subject matter in order to help them explain what they observe. Research on **prior knowledge and conceptual change** is discussed, to consider what can be learned from listening closely to learners' explanations and how these ideas can be applied in informal settings, where the interactions are unique and diverse.

Session Objectives

- Discuss the role of prior knowledge in learning.

- Engage educators in a learning experience and a reflection on how their prior knowledge influenced that experience.

- Discuss research perspectives regarding prior knowledge and conceptual change.

- Determine how educators in informal settings can activate and connect to learners' prior knowledge in their practice, given the unique and diverse nature of their interactions.

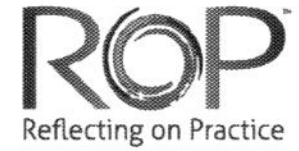

<table>
<tr><td colspan="3" align="center">SESSION AGENDA</td></tr>
<tr><th>Task
Routine</th><th>Description</th><th>Estimated Time
(in minutes)</th></tr>
<tr><td>Introduction
Session Objectives</td><td>The goals and objectives of the session are introduced.</td><td align="center">3</td></tr>
<tr><td>Retrieve & Connect
Think-Pair-Share: Learners' Prior Knowledge</td><td>Participants discuss their conception of prior knowledge, and how to identify prior knowledge in their learners. The discussion considers the benefits to learning when educators access and connect to ideas from prior knowledge, rather than merely assessing it.</td><td align="center">15</td></tr>
<tr><td>Hands-on
Ice Cubes Investigation</td><td>Participants engage in an ice cubes investigation to experience how their prior knowledge affects their explanations, process, and learning.
A. Explore the phenomenon
B. Gather more evidence
C. Construct an explanation</td><td align="center">50

(20)
(15)
(15)</td></tr>
<tr><td>Break option</td><td></td><td align="center">10</td></tr>
<tr><td>Research Discussion
Reading Partner Jigsaw: Prior Knowledge and Conceptual Change</td><td>The community is introduced to a framework for thinking about prior knowledge and misconceptions. The discussion pushes participants to consider the value and purpose of paying attention to what learners say.</td><td align="center">40</td></tr>
<tr><td>Let's Talk Practice
Think-Pair-Share: Connecting Ideas Across Experiences</td><td>Participants use ideas from the research discussion to frame a self-analysis of their learning in the Ice Cubes Investigation and then transfer to their practice.</td><td align="center">30</td></tr>
<tr><td>Continue the Learning</td><td>Participants are reminded of the upcoming Whole-Community Video Reflection in Session 3.</td><td align="center">2</td></tr>
<tr><td></td><td align="right">Total Estimated Time</td><td align="center">150 mins. (2.5 hrs.)</td></tr>
</table>

2.2

MATERIALS

Recurring for all Sessions

- See Module 1 Session 1

For this Session

- 1 small box kosher salt
- 1 long spoon or stirring stick
- 2 pitchers (one for salt water; one for fresh water)
- Small tub to dispose of used water
- A few sponges, cloth towels, or paper towels
- **Science Content Guide for Ice Cubes**

For each small group of 2–4 participants

- 2 identical, clear containers, e.g., glass or plastic cups, beakers; one labeled "salt water," the other labeled "fresh water."
- 1 small bottle of food coloring (any dark color—not yellow)
- 1 small cup to hold ice cubes
- 2 large ice cubes or 4 smaller ice cubes (See **Getting Ready** #4)
- Fresh water to fill freshwater container
- Salt water to fill saltwater container
- Colored pencils, at least two different colors
- Tray for materials
- (Optional) thermometer(s)

For each participant

- Handouts (1 copy each)
    - **Making Sense of Ice Cubes Investigation**
    - **Key Ideas from the Literature: Prior Knowledge and Conceptual Change**

The clear container needs to be at least 4.5 inches (12 cm) tall, in order for the convection current to move through it.

Getting Ready

1. Consider sending out the handout **Key Ideas from the Literature: Prior Knowledge and Conceptual Change** a few days before the session so that participants have the option to review the reading material ahead of time.

2. Review the handouts and slide deck for this session, and additional suggested readings.

    For the facilitator

    - *Taking Science to School*
      - Chapter 5 ("Generating and Evaluating Scientific Evidence and Explanations")
      - Chapter 6 ("Understanding How Scientific Knowledge is Constructed")

3. Duplicate handouts.

4. Review the design of the Ice Cubes Investigation and how it's used within this session.

    This entire session is designed using the Learning Cycle.

    a. The **Hands-on** Ice Cubes Investigation is the Exploration Phase for the focal content on prior knowledge. During the exploration, facilitators will need to listen closely to understand what prior knowledge participants recall, the **condition** of that prior knowledge, and the **grain size** of inaccuracies. (Refer to section A in **Key Ideas from the Literature: Prior Knowledge and Conceptual Change**.)

    b. These aspects of learners' prior knowledge are offered in the **Research Discussion** (Concept Invention Phase) for participants to consider. It will be helpful for facilitators to be attentive to these aspects and use the language as you facilitate the Ice Cubes Activity.

    c. This effort to model and make participants more aware of their prior knowledge will be helpful for them to reflect on and use what they learned in the **Let's Talk Practice**.

    The Hands-on part of the Ice Cubes Investigation doesn't include all the phases of the Learning Cycle because time is limited and the focus of this session is not to explore the science concept fully (no Application or Reflection Phases). The activity has three main parts.

The facilitation team should split up to listen and support the small groups. The whole team should be familiar with these details for the investigation.

a. <u>Explore the phenomenon</u>. Participants make predictions, then engage in exploring and making observations with the materials. Facilitator circulates and listens to understand what prior knowledge participants retrieve to help them explain the phenomenon. To aid your assessment and guidance as you eavesdrop, use the **Science Content Guide for Ice Cubes**.

b. <u>Gather more evidence</u>. Participants use food coloring and (optionally) thermometers to gather more evidence to help construct explanations.

c. <u>Construct an explanation</u>. Participants share their explanations and facilitator adds additional information and connections. The **Making Sense of Ice Cubes Investigation** handout is distributed.

5. Prepare saltwater pitchers as follows:

a. Add 11 tablespoons (~190 g) of kosher salt to 1 gallon (3.8 L) of fresh water. Stir and then allow the solution to sit until water clears. This ratio makes a solution of about 40 parts per thousand salinity (the ocean is ~35 parts per thousand).

b. Check the prepared salt water to make sure it works as you expect:

- Place one large or two small ice cube(s) in a few inches of the prepared salt water in a small cup.
- Let ice melt for about a minute or so and then add a drop of food coloring.
- If the salt water is salty enough, the food coloring will stay on the surface without drifting into the cup.
- If the food coloring sinks into the water in the cup, simply add more salt to your saltwater solution and test again.

6. Determine the number of small groups for the Ice Cubes Investigation, and set activity materials for each group on trays for ease in distribution. For each tray, include two labeled and water-filled cups (one salt and the other fresh) and a small cup to distribute and hold ice cubes. ***The food coloring is distributed later in the investigation.***

The density of this prepared saltwater solution ensures that the experiment will go as planned—water melting off the ice will layer on the surface of the salt water.

- If the saltwater solution was less salty, then the cold, fresh water melting from the ice cube might sink into the salt water. (Not what you want.)
- If the saltwater solution was made saltier, then the food coloring (dye plus fresh water) would float on the surface whether or not there is a freshwater layer from the melting ice overlying the salt water. (Not what you want.)

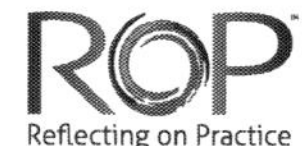

7. Use ice cubes approximately the size (~2 inches or 5 cm) you'd make in an ice-cube tray. Don't use bagged ice from the store; the chips are too small and melt too quickly.

8. (Optional) Decide whether you'll provide thermometers for groups that want to extend the investigation by gathering more evidence to support their explanations in the Ice Cubes Investigation. (See Step **(B) #3** in Hands-on: Ice Cubes Investigation.) Alternatively, you may decide to lead the "Factor in the temperature" step as a demonstration.

9. **Prepare for Whole-Community Video Reflection in Module 2 Session 3.** Determine who will be educator-presenter in the second opportunity for the whole-community to practice *Reflecting on Practice* Video Reflection during the next session. Ideally, the person would be another member of the facilitation team. If that isn't possible, then solicit a volunteer. A new tool will be introduced.

    - Have the presenter complete the **Reflection Exercise Worksheet to Prepare for Feedback** (Appendix A).

    - Check in with the presenter a few days prior to the session. Offer assistance as needed, including helping to select a video clip.

    - Ask the presenter to provide you the video clip at least one day before the video session.

Other variations of this activity involve making the ice cubes with food coloring mixed in. We don't use this variation because having the dye already in the ice cube shifts the emphasis of the engagement: waiting for something to happen rather than puzzling over what happened.

2.2

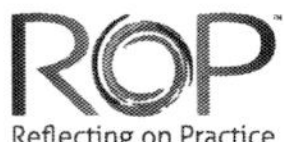

Session 2 Step by Step

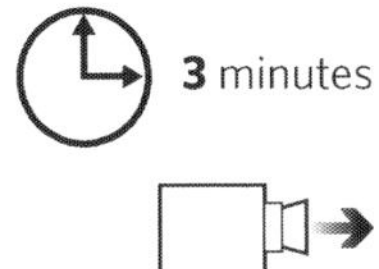

3 minutes

Introduction
Session Objectives

1. **Display the Ideas to Emphasize on learning so far.**
   Remind participants that, so far, their discussions about how people learn have emphasized the following displayed points:

   a. *Learning is <u>making connections</u> between past and new experiences, knowledge, skills, beliefs, etc., and building mental models to organize and understand.* Learners build knowledge through interactions, conversations, and experiences.

   b. *Learning is social.* It happens in social environments and is enhanced by social interactions. Learners need opportunities to express their thoughts and feelings with others as a part of making sense of these connections.

   c. *Learning requires time, effort, & persistence.* Learning isn't easy, most of the time. Learners need to be engaged and motivated to overcome setbacks, distractions, and even negative emotional responses.

2. **Introduce this session.**
   Share that this session will take a closer look at *what learners say,* to get a sense of how they may be organizing their thinking and making connections. The discussion will explore why this prior knowledge is important and necessary for supporting learning, what can be learned from it, and how to nudge learners' conceptual understanding.

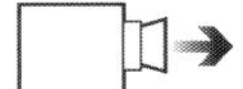

3. **Display and review the session objectives.**
   Share the following objectives:

   - Discuss the role of prior knowledge in learning.
   - Engage in a learning experience and reflect on how prior knowledge influenced that experience.
   - Discuss research perspectives regarding prior knowledge and conceptual change.
   - Determine how educators in informal settings can activate and connect to learners' prior knowledge in their practice, given the unique and diverse nature of their interactions.

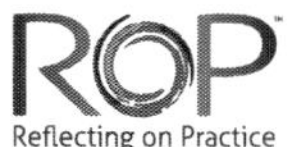

Think-Pair-Share: *Learners' Prior Knowledge*

15 minutes — Retrieve & Connect

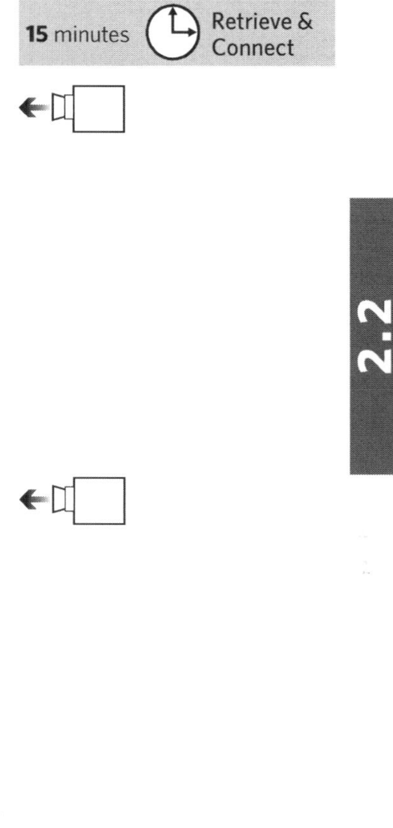

1. **Reminders about prior knowledge.**
   Display and review the ideas from the previous session about prior knowledge:

   - Prior knowledge exists not only as "concepts," but also as perception, procedural skills, modes of reasoning, and beliefs about knowledge.

   - Learners' common sense and everyday thinking are meaningful and useful for them. If those ideas aren't engaged when we teach, learners often dismiss new ideas from the learning experience as irrelevant.

2. **Display questions for thinking.**
   Ask participants to think about the following prompts:

   a. *What are your purposes for having learners retrieve their prior knowledge?*

   b. *What are your assumptions about learners' knowledge based on this information?*

   c. *What do you do with this information?*

3. **Form pairs.**
   After a minute or so, ask participants to pair up with someone and discuss their thoughts.

4. **Facilitate a whole-group share.**
   After about 5 minutes, invite participants to share their partner conversations. Facilitate a productive discussion using some of the following questions and prompts. Remember to stay neutral in your reaction to participants' comments and encourage them to listen deeply to what peers share.

   a. Ask participants to provide explanations, evidence, or clarifications to elaborate on their thinking. Suggested probing questions:

      - What makes you think that?

      - Please give an example from your experience.

      - What do you mean?

   b. Invite others to listen carefully to the ideas shared. Suggested probing questions:

      - Can anyone add to that comment?

As before, this facilitated discussion models the **Discussion Map** that will be formally introduced in Module 3: Learning Conversations. Continue to model the practice without calling attention to the Discussion Map.

□ If that were true, what assumptions are we making? What are we overlooking?

□ How can we consider that sentiment from a different perspective?

□ What do others think about that idea?

5. **Review discussion points.**

Offer these ideas where appropriate as the group conversation unfolds.

a. **Assess vs retrieval.** The most commonly offered reason for activating learners' prior knowledge is that it reveals to the educator what learners know, and allows the educator to adjust the depth and details of content to explore. This reasoning is sensible given the ephemeral nature of the interactions in informal learning environments. The educator is striving to support learners from where they are, which is helped if the educator knows what learners know. This reason makes assessment the purpose for learners to recall their prior knowledge, which serves the educator but not necessarily the learners. To invite learners to think about what they know or help them make connections to their prior knowledge, **the reason needs to be for learners to retrieve relevant memories** and consider how the new content is meaningful to what they already know and have experienced.

b. **Talking is for connecting with prior knowledge.** Driven by the need for assessment, educators often fail to offer learners the chance to share their prior knowledge simply because time doesn't permit getting to everyone's answers. It's important to remember that the educator doesn't have to be part of every conversation in order for the conversation to be of value for the learners. It may be more meaningful for learners to discuss their connections between new content and prior knowledge with people they know.

Becoming aware of their own prior knowledge is profoundly important for learners. Rather than emphasizing what they don't know, or how their understanding is wrong, retrieving prior knowledge draws attention to the value of their existing knowledge—experiences, ideas, skills, and opinions—for their learning. It also doesn't mean that flaws in learners' understanding are not addressed. However, their inaccuracies are more complicated than simply having the "right" or "wrong" answer to a question.

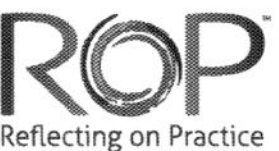

Ice Cubes Investigation

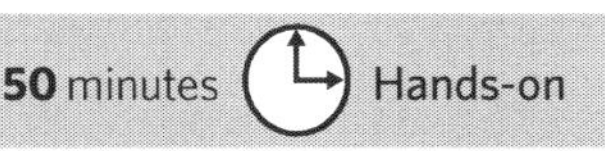

50 minutes   Hands-on

(A) Ice Cubes: Explore the Phenomenon (20 minutes)

1. **Introduce the purpose of the activity.**
   Let participants know they'll continue this conversation using a shared experience. Working in small groups of 2–4, they'll do an activity that puts them <u>in the role of the learner</u>. Remind them that, while doing the activity, they should keep mental notes on ideas about how prior knowledge is contributing along the way to their understanding and explanations.

2. **Display the guiding question.**
   Display and read aloud the guiding question: *Will ice cubes melt faster in fresh water or salt water?* Share the following:

   a. Ask them not to call out their answers or share their ideas with others yet.

   b. Explain that the same size and number of ice cubes were placed in each of two cups, one of fresh water and one of salt water.

   c. Invite and answer clarifying questions, such as "Is the water in both cups the same temperature?" [Yes, it was allowed to come to room temperature before adding in the ice cubes]; "How salty is the salt water?" [Quite salty, but not saturated]

3. **Participants record individual predictions and rationale.**
   Ask participants to think about the set-up and then record their <u>predictions</u> about which cup's ice will melt faster and a <u>rationale</u> for why they think that. Let them know they will be doing more recording so they should make sure there is sufficient space on their page. They don't need to write out a complete explanation (if they have one), but rather to record some words that describe the concept(s) or idea(s) they think would explain their predicted results.

4. **Exchange ideas with a partner or two.**
   Once participants have recorded their individual predictions and rationales, have them turn to people next to them to share a couple of words about the concept they think might explain their prediction, e.g. diffusion, solubility, keeping an ice chest colder, heat capacity, etc. Give them about one minute to exchange their ideas.

5. **Distribute materials to each small group.**
   Ask participants to form small groups of 2–4 people. As they're doing that, add ice cubes to the empty cups on each tray.

   a. Distribute a tray of materials to each group. Don't distribute the food coloring just yet.

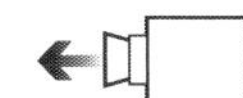

During discussions in the Ice Cubes Investigation, a range of comments will shed light on how prior knowledge contributes to participants' assumptions, ideas, and explanations. As the discussion proceeds, it may become apparent that participants tend to focus on prior knowledge that may NOT be helpful for explaining their observations in this activity. For some, this knowledge sidetracks them from considering other concepts.

Allow these distractions to happen naturally in the small-group discussions, and bring them up in the **Let's Talk Practice** discussion if they're not mentioned. They will be useful examples for thinking about the role of prior knowledge in learning.

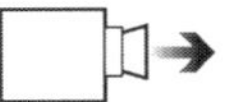

Participants have a total of 20 minutes for the whole investigation. Distribute the food coloring after about 10 minutes from the start.

The facilitation team should split up across the small groups to listen and guide them.

The **Science Content Guide for Ice Cubes** provides science explanations for the kinds of prior knowledge participants often retrieve for this investigation. What participants recall isn't always accurate or relevant. Additionally, probing questions are provided for use as needed.

b. Display and read out loud the following directions before starting the investigation:

Ice Cubes Investigation Instructions

1. Divide the ice cubes evenly between the two cups of water—and then be sure <u>NOT to disturb or stir the cups</u>.

2. Generate an explanation for what your group observes.

3. Continue to discuss your predictions and rationale as you make careful observations of the cups and melting ice.

4. Sketch a model (a diagram) to help explain what's happening. Revise as needed.

5. You have about 20 minutes for this investigation.

---

6. **Eavesdrop throughout the activity.**
While groups are working, circulate and listen with curiosity to understand what prior knowledge participants are using to help them make sense of this investigation. **Use the Science Content Guide for Ice Cubes** to assist your assessment and guidance. For this activity, the prior knowledge learners typically recall can be organized into four themes:

- Familiar experiences

- Equations and technical language

- Overlooking evidence from observations

- Distracted by emotional memories and social dynamics

Listen to understand the **condition** of their prior knowledge (missing, incomplete, misconceived), which can vary in their accuracy. The inaccuracies can be at different **grain sizes** (belief, mental model, category).

- **Missing.** Participants offer little to no prior knowledge from lived experiences, schooling, etc. to reference and/or consider for this activity.

- **Incomplete.** Participants recall some experiences or ideas, but don't remember what they mean or how they might be related to this investigation. Regardless of whether participants understood the concept before or not, they can't elaborate on it now. For instance, "I remember we studied heat capacity in college chemistry, but I don't remember what it means." "This set-up makes me think of salt on roads, but I'm not sure I know why salt is added to icy roads."

- **Misconceived.** Participants offer reasoning that is logical to them, but is scientifically inaccurate. The flaws can be at the grain size of individual belief, mental model, and/or category.

  ▫ Individual belief is a single idea, e.g., cold water is more dense than warm water. A common false belief in this activity is cold fresh water is more dense than warm salt water. However, the salt concentration in this activity is close to sea water, making it more dense than even ice cold fresh water.

  ▫ Mental model is an organized collection of individual beliefs. Individual beliefs could be accurate but connecting them inaccurately could make the mental model flawed. Some parts of mental models could be accurate while other parts incomplete and missing, which also results in a flawed mental model overall.

  ▫ Category is grouping ideas and concepts together; items in the same category share features and attributes. In this activity, participants may inaccurately categorize "salt dissolved in water" as a physical change because it can be reversed; when water is allowed to evaporate, the salt will return. However, it's actually a chemical change because a new substance and new bonds are formed as the salt dissolves. In addition, much like physical changes, some chemical changes are reversible.

7. **Encourage talking and sketching.**
   Help participants be attentive to the conditions of their prior knowledge (missing, incomplete, misconceived) and guide them to make connections. As you push on participants, there may be a tendency for some to give up on what they're thinking and choose another idea and term. Be sure to reassure them that you're seeking to understand and help them be more aware of how they're thinking about the ideas. It's unclear how they're "right or wrong" without more awareness of their understanding.

   a. Prompt participants to provide evidence to support their explanations—evidence can be from their observations, prior knowledge, etc.

   b. Challenge participants to describe how observations now either support or refute their prior knowledge and predictions.

   c. Insist that participants sketch models to depict what they think, particularly if connections are tentative. They can always refine their models and make multiple models to show different reasonings.

Participants may be reluctant to sketch models, but push on them to do so. Call it "sketching" to ease concerns that they aren't "good at drawing." Their sketch can be a combination of words and images, arrows, and question marks for what remains unclear.

In science and engineering, a model is a helpful tool for representing ideas and explanations. It can be diagrams, drawings, physical replicas, mathematical representations, analogies, and computer simulations. Making models is one of the Science and Engineering Practices in the Next Generation Science Standards.

For learning, making a model is a way to externalize one's thinking process and evolving understanding. Working with the model alleviates the cognitive load in our working memory to manipulate complex ideas. Also, combining illustrations and text involves multiple areas of the brain, which strengthens those memories.

    d. Give information, as needed.

        ◻ If it appears participants are missing or have incomplete prior knowledge, and they wouldn't be able to get that information from close observation of the investigation, then tell them. Either assess this from what they say or push on them to specify what they don't know.

        ◻ If there are inaccuracies in their explanations, encourage participants to articulate (in talk and drawing) their thinking so that both you and they can understand where are the flaws—individual belief, mental model, or category.

(B) Ice Cubes: Gather More Evidence (15 minutes)

1. Distribute food coloring and give additional directions.
As each small group is able to determine which ice cubes are melting the fastest, and is engaged in making explanations as to *why*, provide them with one bottle of food coloring. Instruct them to add two drops to the ice remaining in each cup (without stirring) and make careful observations of the results. Tell them to discuss these questions in their group:

a. How do these new observations with the food coloring help point to the cause for why ice cubes melted faster in *fresh water*?

b. What is the relationship between the results and your prior knowledge? What *knowledge*, *ideas*, *assumptions*, and *experiences* did you draw on during the investigation?

2. Encourage evidence-based explanations and revisions.
Encourage the groups to offer evidence-based explanations from observing the effects of the food coloring. Insist, again, that they sketch a new model or refine their existing one to reflect how their understanding is changing. As always, wait for participants to be ready to talk, especially the quieter ones, and stay neutral in your reactions to their comments.

Ask these follow-up questions to probe for elaboration, evidence, and clarification:

- What do you notice in the two cups?

- What explanation do you have for those results?

- What makes you think that?

- How is that explanation supported by the evidence?

- How does your sketch represent your understanding and/or explanation at this point?

Food coloring enhances this investigation because it becomes entrapped in the kind of water (i.e., salty or fresh) to which it was added, thus giving information about what's happening that wasn't visible previously. The dye entrapped in the cold, sinking water is carried throughout the cup in a convection current. In the saltwater cup, the dye is entrapped in the cold fresh water (from the melting ice) into which it was dropped. This provides evidence of currents forming in the freshwater cup.

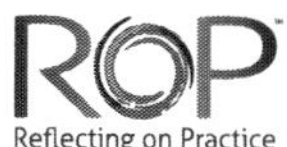

Draw others in the small group into the conversation by asking them to listen closely to what's shared.

- How does that comment help the group refine their explanation?

- What can be added to that explanation?

- What is an alternative explanation?

3. **(Optional) Factor in the water temperature.**
   If time permits and/or if water temperature came up in the group discussions, introduce the thermometers, either in a demonstration or by providing one or more to each group to explore with. To prevent accidental disruptions of the water layers, ask them to talk out what they plan to do with the thermometer before they insert it into their cups. (They should measure the temperature at the surface and at the bottom in each cup, without stirring.) Ask them to note the differences in temperature data and discuss how this additional information might help explain the results of the investigation.

(C) Ice Cubes: Construct an Explanation (15 minutes)

1. **Initiate whole-group share to construct an explanation together.**
   Clear away the materials. Begin a group discussion to generate an explanation for the phenomenon together. Use the Discussion Map to facilitate the discussion. Invite groups to share their observations and explanations, and encourage them to comment on each other's explanations. Ask these questions:

   a. Which ice cubes melted the fastest?

   b. Why do you think that was happening?

   How did food coloring add to the explanation?
   If these explanations don't come up, share them with participants:

   a. In the freshwater cup, the food coloring showed that the cold water melting off the ice cube sank to the bottom of the cup.

   b. The ice-cold fresh water was denser than the room-temperature fresh water in the cup. This set up a circulation pattern: cold water was continually sinking/moving away from the ice cube and warmer, room-temperature water from below was pushed upwards. As the rising, warmer water made continual contact with the ice cube, the ice melted. The food coloring was circulated through the cup.

   b. In the saltwater cup, cold fresh water from the melted ice cube remained at the surface, surrounding the ice. This cold fresh water insulated the ice, slowing the melt rate. Fresh water is less

After the ice melts in the saltwater cup, the water at the surface will be about 4°C and the water at the bottom of the cup will be about 16°C. In the freshwater cup, the water tends to be about 14°C throughout the water column.

Even without the thermometer, the differences in water temperature are noticeable when the outside of the cups are touched in different places, and also due to the location of water condensation on the outside of the cup.

2.2

dense than salt water, even though the fresh water was ice cold. The food coloring remained stuck in the melted ice water at the surface.

c. These results support the explanation that density is the concept at work in the activity.

How did temperature data add to the explanation?
If a thermometer was used, invite participants to share how temperature data supported their explanations. If these explanations don't come up, share them now:

a. In the freshwater cup, the temperature was uniform throughout, even after the ice melted. This showed that the water was circulating throughout the cup.

b. In the saltwater cup, the surface of the water was much colder than the rest of the cup, because very cold fresh water from the melting ice cube floated on top of the denser salt water beneath it. The fresh water couldn't sink into the salt water or circulate throughout the cup. As a result, the ice cube stayed bathed in very cold, melted ice water, which kept it from melting.

c. The temperature differences observed with the thermometer, like the visible effects of the food coloring, support the evidence that density played a major role in the results.

2. **Distribute the "Making Sense of Ice Cubes Investigation" handout.** Give participants a few minutes to read the explanation. Encourage them to discuss the handout and their reactions to its content in their small groups.

3. **Partners discuss.**
   Ask participants to turn to a partner to discuss their explanations of the investigation. Display the following prompt:

   What is your explanation <u>now</u> for why ice cubes melted faster in the fresh water?

4. **Provide quiet time to revise write.**
   Give participants a couple of minutes quiet time to revise their initial writing. Have them draw a line in their journals under their initial prediction and rationale, and record their new understanding. Remind them to review their models as a resource for this written explanation.

5. **Connect the investigation to the upcoming research discussion on prior knowledge.**
   Tell participants that the Ice Cubes Investigation shines light on a key idea—that prior knowledge influences how learners engage in a new experience and make explanations to demonstrate their

---

Actually, ice will start to melt faster in the salt water at the very outset of the investigation, because salt water has a lower freezing point than fresh water (meaning salt water has to be colder than fresh water for the ice to stay frozen/not melt). However since we don't allow stirring in this investigation, the less dense, icy cold fresh water (from the melting ice cube) was kept from mixing with the more dense salt water below. (Fresh water is less dense than salty ocean water, regardless of temperature differences.) These layers formed very quickly and the ice in the fresh water then started to melt faster than the ice in the salt water. Density differences rather than freezing point, became the more relevant and accurate concept to explain the phenomenon.

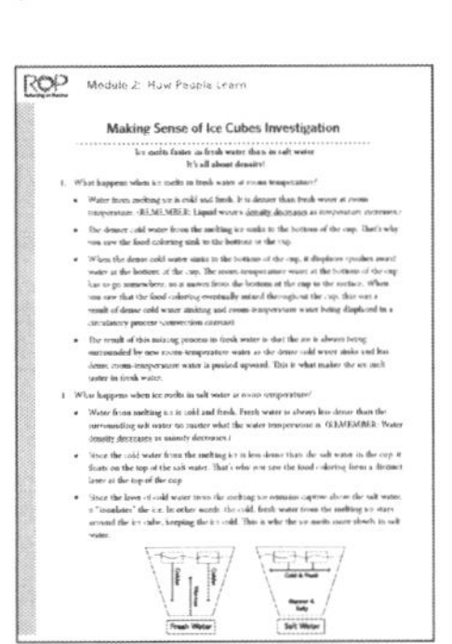
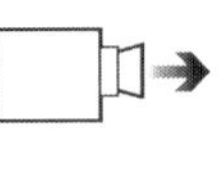

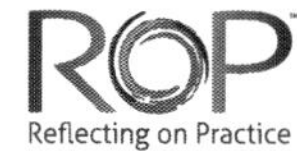

understanding of the concept. Only by listening carefully to what learners say, and pushing them to clarify and elaborate on their thinking, do we gain insight into how their prior knowledge is organized and successfully connect it to the learning experiences we offer.

Reading Partner Jigsaw: *Prior Knowledge and Conceptual Change*

40 minutes   Research Discussion

2.2

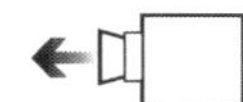

Remind participants that prior knowledge and experiences are stored in long-term memory, and they direct how new information is processed and organized in working memory.

The literature on prior knowledge, misconceptions, and conceptual change is extensive, dating back several decades and hundreds of studies. A 2006 paper by Michelene Chi (reference #1 in **Key Ideas from the Literature**) offers a useful way of thinking about prior knowledge and misconceptions without having to get lost in the details of the debates and controversies.

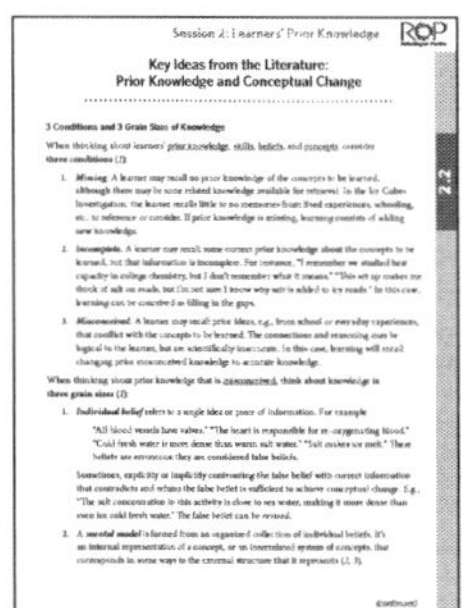

1. **Display and review the following information.**

   a. Humans are goal-directed agents who actively seek information. We have a range of prior knowledge, skills, beliefs, and concepts that significantly influence what we notice about the environment and how we organize and interpret it.

   b. This prior knowledge, in turn, affects our abilities to remember, reason, solve problems, and acquire new knowledge.

   c. Having errors in our understanding is an inescapable part of learning. So, is learning simply a matter of replacing those inaccuracies with the correct "answers"? We know from our own experiences that simply <u>telling</u> learners that their ideas, beliefs, and understandings are wrong doesn't work. Why not?

2. **Distribute "Key Ideas from the Literature: Prior Knowledge and Conceptual Change" handout.**
   Share with participants that they'll continue the conversation on prior knowledge, this time drawing on and considering more information from the research literature. Tell them that in this Research Discussion they'll read and discuss two big topics on prior knowledge and conceptual change, first with a reading partner and then in small groups of four. Those topics are:

   - Three Conditions and Three Grain Sizes of Knowledge

   - Addressing Misconceptions

3. **Form groups of four, composed of two pairs.**
   Direct formation of groups, or let participants form the groups on their own. Make sure there are at least four people in each group, so that everyone has a reading partner. Emphasize that both topics will be read and discussed within the groups after the partners have shared.

4. **Display and review the instructions.**

   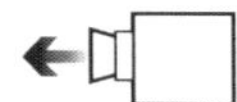

   a. Choose a partner with whom to read and (when ready) discuss one topic. Divide up the information so that each pair of reading partners in the group reads and discusses a different topic.

Remind them to find connections across all topics, and to use the discussion questions to push beyond rote descriptions of the content of their topic.

b. Make sure that all topics are addressed within the group.

c. After reading partners have a chance to share with each other, each pair will share their discussion with their small group.

d. Look for connections between the topics during the small-group discussion.

e. Be ready to share a couple of big ideas from the small-group discussion with everyone in the community.

5. **Circulate and monitor the pace of discussions.**
It will be quiet at first as everyone reads their assigned topic. As people finish reading, let them know partners can start discussing their topic whenever they're ready. Within each group, monitor to ensure that the partners are working at the same pace as others in their group, so that all partners finish at approximately the same time and the small-group discussion can begin. Across multiple groups, monitor to ensure that all groups are working at about the same pace, so that all will be ready for the whole-group share.

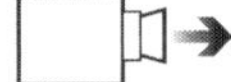

6. **Begin whole-group share.**
Call everyone back together. Use the Discussion Map to guide your facilitation and invite each group to share a couple of big ideas that emerged from their discussion. Suggested prompts to display to get groups to share:

a. What ideas generated interesting discussion within your groups?

b. What connections did you notice across the topics?

c. How can you use these ideas in your practice?

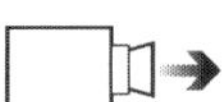

7. **Display the Ideas to Emphasize.**
Make connections between these ideas and what the community just discussed.

a. <u>**Three conditions of prior knowledge.**</u> We've emphasized that learning involves building on and making connections to/ between prior and new knowledge. This research discussion points out that sometimes learners' prior knowledge may be *incomplete, misconceived,* or *missing altogether.* Each condition has a different implication for teaching, so recognizing the condition can be helpful for both learners and their educators.

b. <u>**Three grain sizes of knowledge.**</u> Our knowledge can be conceived of as having three "grain sizes": *individual belief, mental model,* and *category.* But we don't advance from one grain size to the next as we learn. They exist simultaneously, and are broken into grain sizes for ease of talking about and studying learning—particularly misconceptions. This tells us that when learners offer a wrong answer, it's more complicated than that they just don't understand.

c. **<u>Wrong answers are complicated</u>.** Thinking about prior knowledge and misconception in terms of the three conditions and three grain sizes helps to explain why conceptual change sometimes happens readily (if misconceived at the grain size of *individual belief*), and at other times seems impossible to budge (when knowledge is misconceived at *mental model* or *category* level). Understanding where the inaccuracies may come from provides a reminder about the power of prior knowledge and that learning occurs everywhere, not just in schools or other designed spaces.

d. **<u>Nudging conceptual change</u>.** Despite the challenges, conceptual change is possible—but keep in mind that we can't eradicate misconceptions. Changing misconceptions is slow and takes time. **We nudge learners to forge new neural pathways while weakening existing elaborate connections that may have become automated.** It requires many opportunities to engage with the concepts in ways that get learners to articulate their thinking and identify what's missing, where their thinking is inconsistent, and how it may be flawed. For learners, knowing why the wrong answer is wrong is just as important as knowing why the right answer is right.

> This point will be explored further with information on making memories from the Research Discussion in Module 3 Session 1.

Think-Pair-Share: *Connecting Ideas Across Experiences*

1. **Explain the purpose and task.**
   Tell participants the purpose of this task is to explore ways to apply ideas from the reading into their work. First, they will **Think** about the ideas more concretely by analyzing their own experience as a learner in the Ice Cubes Activity. Then, they will **Pair** with someone to use their self-analyses to brainstorm actions they can put into practice. Finally, there will be a brief whole-group **Share** of their action steps.

2. **Review the instructions.**
   Ask everyone to review their explanations and models from the Ice Cubes Activity with the "three conditions and three grain sizes" framework in mind. Display and review the instructions.

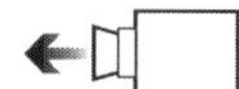

> **Think** on your own
>
> **a.** Review your model and explanation.
>
> **b.** Determine where and how you were "right" and "wrong."
> *It's also okay if you still don't fully understand the explanation.*
>
> **c.** Recall the experience and your conversations.
>
> **d.** Consider the prior knowledge you were relying on to make your explanation.
>
> **e.** Record your reflection.
>   - What prior knowledge did you recall and use?
>   - Was all the prior knowledge you recalled relevant to the investigation?
>   - How was your prior knowledge missing, incomplete, or misconceived?
>   - At what grain size were your inaccuracies?
>   - Where and how were you struggling to work with your prior knowledge?

> **Pair** up with someone.
>
> **a.** Exchange your self-analyses of your prior knowledge.
>
> **b.** Discuss what occurred in the experience that helped reveal your inaccuracies, make new connections, and transform understanding.
>   - What was happening?
>   - What were you and others doing?
>
> **c.** Determine what you can each take away into your own practice from your experience as a learner in this activity.
>
> **d.** Record your takeaways. (Sharing with whole-community, optional.)

Use the **Science Content Guide for Ice Cubes** for insights on participants' analysis of their prior knowledge. Gently push on participants to articulate how or when they realized there were gaps or flaws in their understanding. Invite them to consider how the realization felt and how it affected their engagement.

3. **Circulate and monitor.**
Give about 10 minutes for the self-analyses and then the remaining time to work with a partner. Insist on pairs so that everyone has to talk; allow trios only if there is an odd number. Support participants in specifying how their recalled prior knowledge was "right" or "wrong." Ideas to keep in mind as participants reflect and connect:

- It's possible their "wrong" explanations were not entirely wrong. It was just knowledge that turned out to be irrelevant to this specific investigation.

 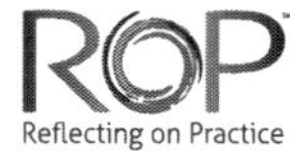

- They may struggle to explain connections because their understanding of a concept was incomplete or misconceived.

4. **Brief whole-community share.**
   This sharing can be done quickly with each pair calling out one thing or generating a group document afterwards.

Continue the Learning

 **2** minutes

1. **Reminder: second whole-community Video Reflection coming up.**
   Let participants know that, between now and the next session (Session 3), you'll be asking or assigning one facilitator (preferable) or volunteer participant to share a video clip with the whole community for the second *Reflecting on Practice* Video Reflection.

   Since this module focuses on How People Learn, that will be the focus of the whole-community Video Reflection—specifically, **Learners' Prior Knowledge**.

───────────────────────

Science Content Guide for Ice Cubes

In the Ice Cubes Investigation, the prior knowledge participants typically recall to help them explain what happened can be organized into four **themes**. The **conditions** of their prior knowledge within these themes can vary in their accuracy. The inaccuracies can be at different **grain size**. (Refer to **Key Ideas from the Literature: Prior Knowledge and Conceptual Change** for details on the conditions and grain size.)

Theme 1. Familiar Experiences

This activity involves salt and ice, which can trigger participants to retrieve memories of experiences that involve those items to make an explanation regardless of their prediction. The two most common lived experiences recalled by participants in the US are **"salt causes ice on roads to melt"** and **"salt is used in making ice cream."** Some people also recall adding salt to ice coolers to keep their beverages extra cold.

- Participants usually offer that *salt lowers the freezing point of water* as the causal mechanism in the familiar experiences described above. It's true that adding salt to water lowers the freezing point, but how that explains their experiences, or this investigation, is where their thinking tends to be unsteady. Moreover, the lower freezing point of salt water is not the causal mechanism here, so guidance from the facilitator is intended to help disassociate those experiences from this investigation.

Push on participants to articulate & sketch:	Probing questions:
1. the function of salt in those experiences	• How does that work?
2. how that mechanism would help them explain what is happening in this investigation	• If that were the case, what would you expect to see in the salt water cup?
	• What information do you feel is missing/incomplete from your understanding to help you explain?

"Salt lowers the freezing point of water." In this investigation, the water in both containers starts out at room temperature. At the very beginning, ice in salt water does melt faster than in fresh water because the lower freezing point of salt water means that the salt water would have to be at a much colder temperature than the fresh water for the ice to stay frozen, hence it would quickly melt in salt water at room temperature. But as the investigation progresses, the ice melts faster in the freshwater cup; thus, the lower freezing point of salt water is not the causal mechanism here. The "no mixing" rule of the investigation results in something else being the primary cause for what happens—ice melting slower in salt water.

(continues)

          This page may be duplicated for educational use.

Science Content Guide for Ice Cubes *(continued)*

Freezing point of water and heat flow

- Freshwater freezes at or below 32˚F (0˚C); saltwater freezes at or below 28.4˚F (-1.9˚C)

- Heat flows from warmer to cooler objects

"Salt causes ice on roads to melt." Salt is added to snow on roads to prevent snow melt during the day from freezing into ice at night, which makes the roads dangerous. Warm daytime temperature melts the snow and ice on the road. Adding salt to the road makes water from that snow melt salty, which lowers its freezing point. **The salt interferes with the freezing process of water.** After a warm day, the evening temperature would have to drop below 28.4˚F (-1.9˚C) for the salty slush to freeze into ice.

"Salt is used in making ice cream." Salt is used in ice cream makers to make the ice bath surrounding the container of sweet cream cold enough to cause ice crystals to form in the cream. The ice and salt are mixed constantly when making ice cream, resulting in a salt slushy. The salt lowers the freezing point of water, meaning the water temperature of the salt slushy can get below 32˚F (0˚C) before freezing begins. As a result, the extremely cold slushy can absorb more heat more quickly from the cream (energy flows from warmer to cooler objects), lowering the temperature of the cream enough to become ice cream. However, in this investigation, there's no mixing of ice and salt water that would lower the water temperature to a very low temperature.

Theme 2. Equations and Technical Language

This activity is an exploration into physical science concepts, which can trigger participants to retrieve memories from their science classes (chemistry and physics) for an explanation regardless of their prediction. The most common tendency is to offer formulas (**D=M/V**), terms (**freezing point, heat capacity, molecular bonds, molecular movement**), and rules and expressions (**"like dissolves like"**).

- Participants may recall concepts that might or might not be relevant to the investigation. Encourage them to elaborate on their understanding of the concepts to help them determine the relevance and connections to their actual observations. As they struggle to explain the meaning of the idea, help them recognize the condition of their prior knowledge. Is it missing altogether? incomplete? misconceived?

(continues)

Science Content Guide for Ice Cubes *(continued)*

Push on participants to articulate & sketch:	Probing questions:
1. what the formula, term, rule, or expression means 2. how the idea would help them explain what's happening in this investigation	• What does [that idea] mean? • What's your reasoning for why [that idea] would affect the melting rate? • If [that idea] were relevant here, what would you expect to see? • What are you missing or unsure of that would help you explain what's going on here?

"Dissolved salt causes molecules to move faster in salt water than molecules in fresh water. Faster molecule motion accelerates melting." Statements like this try to connect molecular composition of a substance with molecular motion and reflect flawed understanding of heat, temperature, chemical and physical change, and molecular motion. Participants' prior knowledge leads them to sense a connection between ideas, but they're unclear how all these ideas come together or how the ideas might be relevant. Their struggle may be due to not knowing (missing, incomplete) or misunderstanding (misconceived) any of the following: what causes molecules to move in a solution (thermal energy), and what causes a phase change from ice to liquid water (transfer of energy from a warmer to a cooler substance), what is a solution, and/or what constitutes a chemical change or reaction.

Heat affects molecular motion.

- Temperature is a measure of 1) how hot or cold something is (the internal thermal energy of a material), and 2) the average kinetic (moving) energy of molecules or atoms in a sample of matter.

- Heat is the transfer of energy from one substance to another; it's a process, not a thing.

Molecules are set in motion by the addition of energy. The greater the energy absorbed, the faster the molecules or other matter will move. As liquid water comes in contact with ice, heat energy is transferred from the water to the ice causing the liquid water to lose some energy (decrease in motion and temperature) while the ice gains energy and melts (increasing its molecular motion and temperature).

Participants may think that dissolving salt releases heat energy into the water in a chemical reaction, which could lead to an increase in molecular motion. However, not all chemical reactions cause an increase in temperature. You might remind participants that if there were a slight temperature difference between the fresh and salt water, it was accounted for as both cups were at room temperature when the investigation started.

- A chemical reaction is a process in which one or more substances are converted to one or more different substances. A change in temperature (increase or decrease) may occur.

    This page may be duplicated for educational use.

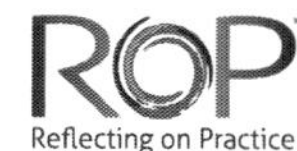

Science Content Guide for Ice Cubes *(continued)*

"Fresh water in the cup is the same as the ice cube(s), so the ice will melt faster in fresh water." Statements like this alludes to an expression in chemistry **"like dissolves like,"** which participants might also state to explain the investigation. Participants' (inaccurate) reasoning is that substances with the same molecular composition (ice cube in fresh water) will affect each other faster than substances with dissimilar composition (fresh ice cube and salt water). The expression is not relevant to this investigation. (The "like dissolves like" expression is used by chemists to remember how polar and nonpolar substances work.)

"Heat capacity." "Density." Offering up concepts without further elaboration doesn't mean participants understand the concept. Participants' reasoning is unclear without further explanations. As a learner, they may be habituated to respond with one-word or short answers, and then expect the educator to evaluate their response and explain for them.

- Heat capacity is the amount of heat energy that must be added to a gram of substance to raise its temperature by 1°C.

Water has a high heat capacity; thus, it can absorb a lot of heat before its temperature increases. Salt water has a slightly lower heat capacity than fresh water. This concept isn't relevant to this investigation.

- Density is the amount of mass in a given volume or space. It's how tightly packed molecules are in a substance or solution.

Salt water is more dense than fresh water because there's more matter (dissolved salt plus water molecules) in a given amount of space, which causes the matter to be more tightly packed together. This concept is the causal mechanism in this investigation and is made "visible" when food coloring is added and the circulation in the freshwater cup and layering in the saltwater cup become evident.

Theme 3. Overlooking Evidence from Observations

This activity is a science experiment, which can trigger participants' expectations of action. Instead, the investigation focuses on observations of ice melting.

- Participants might look closely, but not know what to look for. If they predicted correctly, they might ignore what they see while they rationalize their prior knowledge to explain what's happening. s

Science Content Guide for Ice Cubes *(continued)*

Push on participants to articulate & sketch:	Probing questions:
1. describe what they noticed 2. consider observations that you suggest 3. explain their predictions, using evidence from their observations	• Describe what you see, and what you think is happening. • What do you see that makes you think that? If that were the case, what would you expect to see? • What information are you getting from the food coloring, and how does that affect your explanation?

"Ice is more buoyant in salt water than in fresh water, so the ice cube will have less surface area in contact with liquid in the saltwater cup." Here participants use buoyancy for their explanation. Participants are striving to use observation to support their explanation, but perhaps aren't sure what to look for or pay attention to. They see ice float in water, so the concept of buoyancy is triggered. For this investigation, there is only a negligible difference in the amount of ice rising above the water from both cups. The results of the investigation, as evidenced by the food coloring, don't support the buoyancy concept to explain the phenomenon. Their reasonings are flawed. However, if their prediction is correct, then they might assume their reasoning must also be correct and pay little attention to new information provided when food coloring is added.

- Condensation is the process by which warmer water vapor in the air changes into liquid water on a cooler surface (e.g., portion of cup where water is cold).

Depending on the humidity level in the room, **condensation** may form on the outside of the cup. The condensation gives information about changes in water temperature that might help participants with their explanation. Participants usually don't notice condensation, so point it out and encourage them to consider how it can help them explain the situation.

- On the freshwater cup, condensation would be visible from the surface of the water to the bottom of the cup (the water temperature is uniformly colder than air temperature because ice cold water is sinking to the bottom).

- On the saltwater cup, condensation only forms near the top of the water level (water temperature in the upper layer only is colder than air temperature because ice cold fresh water is less dense than the salt water and stays at the surface).

(continues)

   This page may be duplicated for educational use.

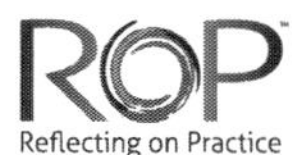

Science Content Guide for Ice Cubes *(continued)*

Theme 4. Distracted by Emotional Memories and Social Dynamics

This activity is an exploration into a physical science concept, which triggers participants to retrieve emotional memories from their science classes. Those memories might be negative, discouraging, or empowering for them. Those memories might also co-mingle with social dynamics among members in the community.

- Some who struggle to explain the meaning of the formula or term might quickly abandon it for another idea. Perhaps they didn't have positive experiences in chemistry or physics in high school or college, or don't remember the concepts. For the former, it's possible negative affective memories are recalled; for the latter, those memories aren't easily retrievable because they aren't used or connected to memories that are used regularly. Both of these can hinder participation. Moreover, the social dynamics among members in the community may affect behavioral engagement. Participants may be concerned about perceptions of who among them are "physical science experts," thus doubting themselves or deferring to those "experts."

Push on participants to	**Probing questions:**
1. describe what they see and what that makes them think, without emphasizing the need to use technical language 2. build their confidence and generate positive emotional memories in this investigation	• Describe what you see in your own words, and what you think is happening. • What do you see that makes you think that? • What are you missing or unsure of in your understanding that would help you explain what's going on here?

Making Sense of Ice Cubes Investigation

Ice melts faster in fresh water than in salt water.
It's all about density!

1. What happens when ice melts in fresh water at room temperature?

   - Water from melting ice is cold and fresh. It is denser than fresh water at room temperature. (REMEMBER: Liquid water's <u>density decreases</u> as *temperature* increases.)

   - The denser cold water from the melting ice sinks to the bottom of the cup. That's why you saw the food coloring sink to the bottom of the cup.

   - When the dense cold water sinks to the bottom of the cup, it displaces (pushes away) water at the bottom of the cup. The room-temperature water at the bottom of the cup has to go somewhere, so it moves from the bottom of the cup to the surface. When you saw that the food coloring eventually mixed throughout the cup, that was a result of dense cold water sinking and room-temperature water being displaced in a circulatory process (convection current).

   - The result of this mixing process in fresh water is that the ice is always being surrounded by new room-temperature water as the dense cold water sinks and less dense room-temperature water is pushed upward. This is what makes the ice melt faster in fresh water.

2. What happens when ice melts in salt water at room temperature?

   - Water from melting ice is cold and fresh. Fresh water is always less dense than the surrounding salt water no matter what the water temperature is. (REMEMBER: Water <u>density decreases</u> as *salinity* decreases.)

   - Since the cold water from the melting ice is less dense than the salt water in the cup, it floats on the top of the salt water. That's why you saw the food coloring form a distinct layer at the top of the cup.

   - Since the layer of cold water from the melting ice remains captive above the salt water, it "insulates" the ice. In other words, the cold, fresh water from the melting ice stays around the ice cube, keeping the ice cold. This is why the ice melts more slowly in salt water.

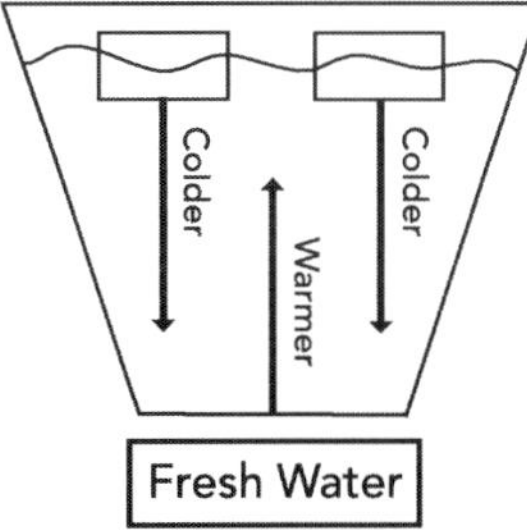

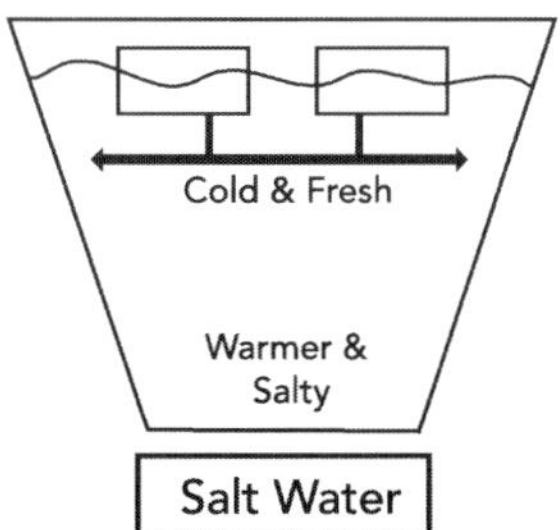

    This page may be duplicated for educational use.

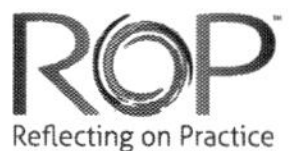

Key Ideas from the Literature:
Prior Knowledge and Conceptual Change

3 Conditions and 3 Grain Sizes of Knowledge

When thinking about learners' <u>prior knowledge</u>, <u>skills</u>, <u>beliefs</u>, and <u>concepts</u>, consider
three conditions (*1*):

1. *Missing.* A learner may recall no prior knowledge of the concepts to be learned,
   although there may be some related knowledge available for retrieval. In the Ice Cubes
   Investigation, the learner recalls little to no memories from lived experiences, schooling,
   etc. to reference or consider. If prior knowledge is missing, learning consists of adding
   new knowledge.

2. *Incomplete.* A learner may recall some correct prior knowledge about the concepts to be
   learned, but that information is incomplete. For instance, "I remember we studied heat
   capacity in college chemistry, but I don't remember what it means." "This set-up makes me
   think of salt on roads, but I'm not sure I know why salt is added to icy roads." In this case,
   learning can be conceived as filling in the gaps.

3. *Misconceived.* A learner may recall prior ideas, e.g., from school or everyday experiences,
   that conflict with the concepts to be learned. The connections and reasoning may be
   logical to the learner, but are scientifically inaccurate. In this case, learning will entail
   changing prior misconceived knowledge to accurate knowledge.

When thinking about prior knowledge that is <u>misconceived</u>, think about knowledge in
three grain sizes (*1*):

1. *Individual belief* refers to a single idea or piece of information. For example:

   "All blood vessels have valves." "The heart is responsible for re-oxygenating blood."
   "Cold fresh water is more dense than warm salt water." "Salt makes ice melt." These
   beliefs are erroneous; they are considered false beliefs.

   Sometimes, explicitly or implicitly confronting the false belief with correct information
   that contradicts and refutes the false belief is sufficient to achieve conceptual change. E.g.,
   "The salt concentration in this activity is close to sea water, making it more dense than
   even ice cold fresh water." The false belief can be *revised*.

2. A *mental model* is formed from an organized collection of individual beliefs. It's
   an internal representation of a concept, or an interrelated system of concepts, that
   corresponds in some ways to the external structure that it represents (*2, 3*).

(continues)

Key Ideas from the Literature:
Prior Knowledge and Conceptual Change *(continued)*

a. Learners may have missing, incomplete, or misconceived mental models. For example:

Perhaps a learner has no mental model for (no understanding of) the circulatory system, or else builds an incomplete or misconceived mental model of circulation as a single-loop system, in which the heart is the source of oxygenated blood. Examples from the Ice Cubes Investigation include: "Dissolved salt causes molecules to move faster in salt water than molecules in fresh water. Ice cubes melt faster in salt water because the faster molecule motion accelerates melting."

The learner's model is in conflict with a more accurate scientific view, but coherent in the sense that he can apply his mental model to arrive at similar (and consistently incorrect) explanations and predictions for a variety of questions. In the ideas learners put forth for Ice Cubes, some learners try to connect molecular composition of a substance with molecular motion. The inaccurate mental model includes the following assumptions and connections: (*1*) dissolving salt in water causes molecules to move faster, (*2*) molecules moving faster speeds up melting, and (*3*) therefore, ice melts faster in salt water. The statements reflect flawed understanding of heat, temperature, chemical change, and molecular motion.

b. Sometimes, when false beliefs within a flawed model are refuted by instruction and recognized by learners as contradictions, the learners can "self-repair" their flawed mental models (*1*). The flawed mental models can be *transformed* into the correct model. But knowing and learning many correct individual beliefs, or revising many false individual beliefs, doesn't guarantee successful transformation of a flawed mental model into the correct one. Critical individual false beliefs can undermine a correct mental model. For example, in understanding what causes moon phases, we saw that a common and critical false belief among learners is that Earth's shadow causes the phases of the Moon.

To counter the false belief's role in the mental model, in our moon phases activity we deliberately explored shadows as a part of the experience.

3. ***Categorizing*** is the process of identifying a new concept and assigning it to a known category to which you feel it belongs (*4*). We typically categorize everything we encounter, automatically—whether consciously or not (*5*). This process is an important learning mechanism, because in assigning something into a category, it inherits the features and attributes of that category. As a result, a learner can use knowledge of the category to make many inferences and attributions about the new concept, object, experience, or phenomenon.

(continues)

  This page may be duplicated for educational use.

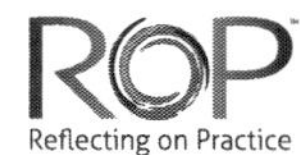

Key Ideas from the Literature:
Prior Knowledge and Conceptual Change *(continued)*

a. Categories can be organized by their properties and the relations between them, such as whether they are **entities** (e.g., can be contained, have weight, occupy space); **processes** (e.g., occur, take time); or **mental states** (e.g., emotion, intention). There can be several levels of sub-categories within a category, like branches on a tree.

b. Learners may have misconceptions because they have mis-categorized something:

- They may mis-categorize something *into one category rather than another*. For example, learners often mis-categorize heat as an *entity*. They think of heat as a physical object, such as "hot molecules," or a substance, such as "hot stuff" or "hotness" (6). But heat is actually a *process*, not an entity: it's the transfer of energy from a hot object to a colder object.

- Learners may mis-categorize something into different sub-categories *within a category*. For example, children who believe Earth is flat categorize it as a *physical* entity, while those who believe it is spherical categorize it as a *solar* entity in space (7). Learners may inaccurately sub-categorize "salt dissolved in water" as a physical change because it can be reversed; when water is allowed to evaporate, the salt crystals will return. However, it's actually a chemical change because a new substance and new bonds are formed as the salt dissolves. Some chemical changes are reversible.

Addressing Misconceptions

(A) Instructional approaches to building learners' conceptual understanding, which include addressing their misconceptions, need to consider how the misconception is occurring. Learners sometimes let go of their incorrect ideas readily. At other times, the misconceived understandings are robust and highly resistant to change, despite instructional interventions. Researchers and educators are still challenged with understanding why conceptual change can be so difficult.

Studies in cognitive neuroscience offers some interesting insights (8, 9). Functional MRI (fMRI) scans of students engaged in causal reasoning tasks showed that when students were given data consistent with their preferred theory, areas in the brain thought to be involved in learning were activated. When presented with data *in*consistent with their preferred theory, areas of the brain associated with error detection and conflict monitoring were activated.

Essentially, when people receive information inconsistent with their preferred theory, learning doesn't easily occur. So, what are the origins of their erroneous thinking?

(continues)

   © 2021 Regents of the University of California ▪ **151**

Key Ideas from the Literature:
Prior Knowledge and Conceptual Change *(continued)*

(B) Learners' inaccuracies arise in many different ways:

- *Personal experience*. Learners' misconceptions often arise from their experiences—that a heavy rock falls to the ground faster than a light piece of paper, or that it's hotter as they move closer to the fire. These observations support and strengthen their beliefs—that heavy things fall faster than light things (versus the physicist's view that all objects fall at the same rate), or that Earth is closer to the Sun (and therefore hotter) during summer (versus the astronomer's explanation that seasons are due to the tilt of Earth's axis in relation to the Sun). So while their everyday observations are sensible and useable for explaining many experiences, they can also be problematic.

- *Instruction*. Students may overgeneralize analogies, particularly if they're unfamiliar with a topic or don't understand the source example (*10*) or can't discriminate between good and faulty sources. The teaching materials may be inaccurate or misleading; science content in textbooks has been found to give erroneous explanations and incomplete information (*11, 12*), or the diagrams may be depicted in confusing and misleading ways. (For example, a nucleus is drawn large and the electron very small, with no indication in the text to clarify the scale representation; students are then surprised to learn that the nuclei themselves are very small, even the most massive ones (*11*).)

- *Diversity of information*. Topics like climate change, evolution, smoking, and vaccines have social, political, and economic implications. For some topics, such as climate change, the scientific concepts are complex and sometimes counterintuitive to personal experiences. For others, such as evolution, information is influenced by different belief systems. Moreover, the concepts may not be taught in schools at all, or what is taught competes with other, contradictory sources of information (*13*).

(C) Conceptual change does happen. It takes time and is a slow process. It involves learners forging new neural pathways while weakening existing elaborate connections that may have become automated. Conditions that facilitate conceptual change include:

- *Awareness of contradiction*. In order for learners to revise their false beliefs, transform their mental models, or even re-categorize concepts, they must recognize the contradiction between what they think and the scientific view (*1, 14*). Without this awareness, learners may simply assimilate the new information without clarifying for themselves how it fits, or they may ignore the new information altogether.

(continues)

    This page may be duplicated for educational use.

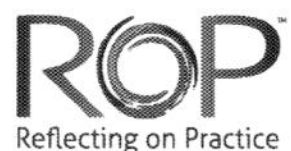

Key Ideas from the Literature:
Prior Knowledge and Conceptual Change *(continued)*

- ***Availability of alternative concept.*** Learners may revise a belief if they can think of (or are given) an alternative. If no alternative theory, model, or interpretation is available, they may stick to the old belief even when predictions from it are not supported (*14*).

(D) Instructional practices found to be especially effective in addressing learners' misconceptions include (*15*):

- Student participation, active engagement, and discourse

- Opportunities for reflection, metacognition (*16*)

- Interactive lecture, lecture tutorial (*17–19*)

- Bridging analogies (*10, 20*)

References

1. M. T. Chi, in *International Handbook of Research on Conceptual Change*, S. Vosniadou, Ed. (Routledge, New York and London, 2008), chap. 3, pp. 61–82.

2. J. Clement, in *International Handbook of Research on Conceptual Change*, S. Vosniadou, Ed. (Routledge, New York and London, 2008), chap. 16, pp. 417–452.

3. N. J. Nersessian, in *International Handbook of Research on Conceptual Change*, S. Vosniadou, Ed. (Routledge, New York and London, 2008), chap. 15, pp. 391–416.

4. J. D. Bransford, A. L. Brown, R. R. Cocking, Eds., *How People Learn: Brain, Mind, Experience, and School*, (National Academy Press, Washington, DC, 2000).

5. D.A. Sousa, *How the Brain Learns*. (Corwin Press, Thousand Oaks, CA, 2016).

6. M. Wiser, "Is heat hot?" inducing conceptual change by integrating everyday and scientific perspectives on thermal phenomena. *Learning and Instruction* **11**, 331–353 (2001).

7. M. Siegal, G. Nobes, G. Panagiotaki, Children's knowledge of the Earth. *Nature Geoscience* **4**, 130–132 (2011).

8. K. Dunbar, J. Fugelsang, C. Stein, in *Thinking with Data*, M. C. Lovett, P. Shah, Eds. (Lawrence Erlbaum Associates, Mahwah, NJ, 2007), pp. 193–206.

9. J. A. Fugelsang, K. N. Dunbar, Brain-based mechanisms underlying complex causal thinking. *Neuropsychologia* **43**, 1204–1213 (2005).

10. B. D. Jee *et al.*, Commentary: Analogical thinking in geoscience education. *Journal of Geoscience Education* **58**, 2–13 (2010).

11. J. L. Hubisz, Report on a study of middle school physical science texts. *The Physics Teacher* **39**, 304–309 (2001).

12. C. J. H. King, An analysis of misconceptions in science textbooks: Earth science in England and Wales. *Intl. J. Sci. Educ.* **32**, 565–601 (2010).

13. N. Oreskes, E. M. Conway, M. Shindell, From Chicken Little to Dr. Pangloss: William Nierenberg, global warming, and the social deconstruction of scientific knowledge. *Hist. Stud. Nat. Sci.* **38**, 109–152 (2008).

(continues)

Key Ideas from the Literature:
Prior Knowledge and Conceptual Change *(continued)*

14. K. Inagaki, G. Hatano, in *International Handbook of Research on Conceptual Change*, S. Vosniadou, Ed. (Routledge, New York & London, 2008), chap. 9, pp. 240-262.

15. S. R. Singer, N. R. Nielsen, H. A. Schweingruber, Eds., *Discipline-Based Education Research: Understanding and Improving Learning in Undergraduate Science and Engineering,* (National Academies Press, Washington, DC, 2012).

16. N. Zhao, J. G. Wardeska, S. Y. McGuire, E. Cook, Metacognition: An effective tool to promote success in college science learning. *Journal of College Science Teaching* **43**, 48-54 (2014).

17. R. Zimrot, G. Ashkenazi, Interactive lecture demonstrations: A tool for exploring and enhancing conceptual change. *Chemistry Education Research and Practice* **8**, 197-211 (2007).

18. D. R. Sokoloff, R. K. Thornton, Using Interactive Lecture Demonstrations to Create an Active Learning Environment. *AIP Conference Proceedings* **399**, 1061-1074 (1997).

19. K. M. Kortz, J. J. Smay, D. P. Murray, Increasing learning in introductory geoscience courses using lecture tutorials. *Journal of Geoscience Education* **56**, 280-290 (2008).

20. D. E. Brown, J. Clement, Overcoming misconceptions via analogical reasoning: Abstract transfer versus explanatory model construction. *Instructional Science* **18**, 237-261 (1989).

This page may be duplicated for educational use.

Module 2.
Session 3: Whole-Community Video Reflection on Prior Knowledge

Session Overview

This reflection session examines connecting to **learners' prior knowledge** in actual practice.

Participants revisit previous Research Discussions on reflective practice and professional learning to recall their **commitment to reflective practice and making their practice public**. They are introduced to additional **Tools for Reflective Practice**—which they soon practice using during a Video Reflection in whole group.

Finally, discussions of observation data and the Video Reflection experience reinforce how educators can **connect to learners' prior knowledge** in their work.

Session Objectives

- Revisit the rationale for professional learning and reflecting on practice.
- Introduce additional Tools for Reflective Practice.
- Practice doing a Video Reflection as a whole group, with a focus on the Educator Moves—Prior Knowledge.
- Discuss the video reflection experience.

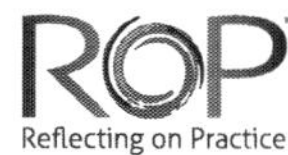

SESSION AGENDA		
Task Routine	Description	**Estimated Time** (in minutes)
Introduction *Session Objectives*	The goals and objectives of the session are introduced, and participants revisit reflective practice discussions from Module 1 Session 2.	5
Research Discussion Sentence-Phrase-Word: *Revisiting Reflective Practice and Professional Learning*	Participants revisit big ideas from a previous Research Discussion. They dig deeper into those ideas with a particular focus on capturing the essence of the text.	35
Let's Talk Practice Table Talk: *Educator Moves for Prior Knowledge*	Participants are reminded of the *Reflecting on Practice* video reflection experience and tools, and then introduced to another tool: Educator Moves.	15
Break option		8
Hands-on *Whole-Community Video Reflection*	Participants have a second opportunity to practice doing a Video Reflection session as a whole community. This time they will use all four Reflection Tools.	60
From Learners' Perspective Turn & Talk: *The Video Reflection Experience*	Participants pair up to discuss their experiences with this second Video Reflection.	7
Retrieve & Connect Minute Paper: *Personal Commitment to Reflective Practice*	Participants record in their Learning Journals their commitment to reflective practice, and their concerns about making their practice public.	5
Continue the Learning	Participants review preparation steps for their first video reflections in critical-colleague groups, including using the fifth tool, **Reflection Exercise Worksheet to Prepare for Feedback**. Critical colleagues set their schedule and assign roles for their small-group video reflections; to be completed before the next interactive session.	15
	Total Estimated Time	150 mins. (2.5 hrs.)

MATERIALS

Recurring for all Sessions

- See Module 1 Session 1

For this Session

- Video clip (from an educator-presenter, see **Getting Ready** #4)
- External speakers
- 2 Posters for **Roles** and **Agreements** (reuse from Module 1 Session 2, **Getting Ready** #4)

For each small group

- Chart paper and markers

For each participant

- Handouts (1 copy each) from Appendix A, *Tools for Reflective Practice*
  - **Educator Moves—Prior knowledge**
  - **Observation Instrument**
  - **Reflection Exercise Worksheet to Prepare for Feedback**
  - **Checklist for Reflection Sessions**
- Handouts (replacement copies as needed)
  - **Key Ideas from the Literature: Reflective Practice and Professional Learning** (Module 1 Session 2)
  - **Protocol for Video Reflection** (Appendix A)
  - **Feedback Chart** (Appendix A)

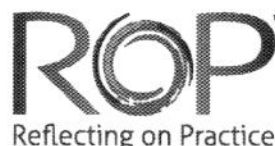

Getting Ready

1. Consider informing participants before the session that they will be revisiting **Key Ideas from the Literature: Reflective Practice and Professional Learning**, in case some individuals want to review the reading material ahead of time.

2. Review the handouts and readings from Module 1 Session 2 on Effective Reflective Practice. Review the **Hands-on** steps for *Practice a Video Reflection* (also from Module 1 Session 2) to recall how to be explicit about modeling the use of the protocol. Initially, participants may be reluctant to be assertive or do all the steps when it's their turn to guide their group. It's critical that they see how they can redirect comments and lead their colleagues.

3. Duplicate handouts.

4. Assign a presenter. Another member of the facilitation team should be the educator-presenter for this second practice. If that's not possible, then ask for a volunteer from the community.

   - Ask the presenter to complete the **Reflection Exercise Worksheet to Prepare for Feedback**.

   - Offer assistance as needed, such as selecting the clip, discussing the worksheet, or refining the focus question.

   - Ask the presenter to provide you the video clip at least one day before the video session.

5. Determine how examples for Educator Moves will be collected across Video Reflections. **Educator Moves** are categories for observable actions that educators take when they teach. How those Moves look can be different between educators and types of learning experiences. Members of the community negotiate the meaning of those categories as they use them, add details, and perhaps generate new Moves. Over time, those Moves can become part of the community's language to talk about its work. During Video Reflections, participants make observations and decide whether an observation is a Move or not; those decisions are informed by how the community thinks about and defines each Move.

   Collecting **Institutional Practices** was introduced in Module 1 Session 2. Determine if the process needs to be revised in preparation for Video Reflection in small groups. The processes for gathering Educator Moves and Institutional Practices across the

Another member of the facilitation team should be the educator-presenter. This is only the second time the community will have practiced Video Reflections and there may still be apprehension. Volunteering in this way shows to the community your commitment to using videos, and that everyone will have a turn.

2.3

small groups need to be available to everyone (e.g., participants add details to an online, collaborative document during each Video Reflection). Remember to review these community documents to consider how the community is thinking and talking about their work, and what issues require further attention.

6. Prepare the video clip. Be sure it's cued up and ready to play from the point where the presenter wants to begin sharing. Know where in the video to stop sharing. Test the sound before everyone arrives.

7. Be ready to give information during **Continue the Learning** about the processes for Video Reflections in small, critical-colleague groups.

   - Refer back to **Getting Ready** in Module 1 Session 2 *Preparing for Video Reflections*, for details on collecting and storing videos, suggestions for small group formation, and policies on video recordings.

   - Two handouts are used to help participants prepare and do this task (**Checklist for Reflection Sessions** and **Reflection Exercise Worksheet to Prepare for Feedback**).

   - Participants should decide the roles and rotation schedule in Step 1 in the **Checklist** before they depart from this session. Exact times and dates can be specified afterwards, if they don't have their calendars available.

   - Determine a way to collect the roles and rotation schedule for each group and make the information easily accessible to everyone. For instance, replicate the chart in the **Checklist** for each group onto chart paper (on the board, on an online collaborative document) for participants to add their details. Making this information public helps to ensure transparency and accountability, as well as ease your responsibility of keeping track of progress.

Groups of four will need to meet two times for everyone to have a turn (three meetings for groups of six). To give maximum flexibility and autonomy, groups determine when to meet. Consider announcing a date by which all Video Reflections need to be completed.

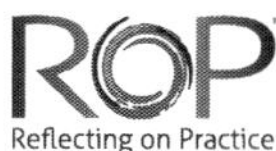

Session 3 Step by Step

Introduction
Session Objectives

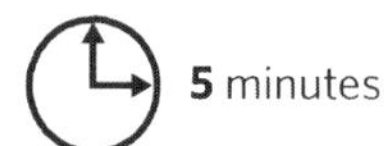
5 minutes

1. **Introduce the session.**
   Share that this session continues the discussion on <u>prior knowledge in learning</u> by looking more closely at actual practice. This session includes a second chance to practice Video Reflection as a whole community, in preparation for their small-group meetings. Remind participants they had their first practice in Module 1 Session 2 with Facilitator 1's video.

2. **Display and introduce the session objectives:**
   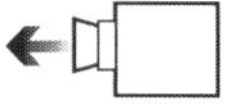

   - Revisit the rationale for professional learning and reflecting on practice.

   - Introduce additional Tools for Reflective Practice.

   - Practice doing a Video Reflection as a whole group, with a focus on the Educator Moves—Prior Knowledge.

   - Discuss the video reflection experience.

3. **Display and recap Session 2 from Module 1.**
   Remind participants that in Module 1 Session 2 they talked about reflecting on practice as an integral part of professional learning, and how reflection with colleagues is a way to cultivate a professional learning community.

Professional learning refers to practitioners' ongoing learning about their practice in order to increase their expertise and skills, and is valuable for improving practice regardless of the profession (Webster-Wright, 2009).

A professional learning community is a group of practitioners who share and critically examine their practice, in a way that is ongoing, reflective, collaborative, inclusive, learning-oriented, and growth-promoting. Members in the community: have shared values and visions on institutional priorities to support learning; take collective responsibility for their learners' learning; and promote individual and group learning (Stoll et al., 2006).

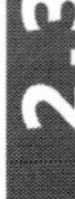

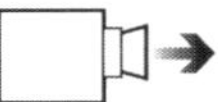

4. Revisit the "Reflective Practice and Professional Learning" Research Discussion.

Remind participants in that session they read and discussed four ideas on reflective practice. Display and review the Ideas to Emphasize on Reflective Practice:

- It is influenced by motivation.
- It is focused on understanding puzzling, troubling, or interesting situations.
- It is about making changes.
- It is deliberately learned over time within a community.

5. Retrieve copies of the handout.

Ask participants to retrieve their copies of the **Key Ideas from the Literature: Reflective Practice and Professional Learning** handout for use during the upcoming research discussion. If needed, distribute replacement copies.

Research Discussion 🕐 **35** minutes

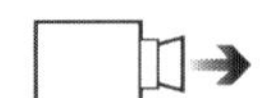

If participants are allowed to choose, make sure there are at least two people for each topic.

If your community is less than eight, select two or three topics to reread.

Sentence-Phrase-Word: *Revisiting Reflective Practice & Professional Learning*

1. Introduce Research Discussion and routine.

Tell participants that in this Research Discussion, they will use the Sentence-Phrase-Word routine to dig deeper into the big ideas from the literature by engaging with and making meaning from the text again. This time they focus on capturing the essence of the text: <u>what speaks to them.</u>

2. Display and call out the initial steps of the routine:

a. Choose and reread one of the big ideas. Tell participants to choose one of the ideas (or assign them one) from the **Key Ideas from the Literature: Reflective Practice and Professional Learning** handout. Ask them to read their chosen (or assigned) idea using the following "active reading" steps.

- Underline what's interesting
- Circle what's confusing
- Write questions in the margins

b. After rereading their idea, make three selections.

- Select a **sentence** <u>that captures the idea for them</u>. Everyone's selection will be different and a reflection of their experience.

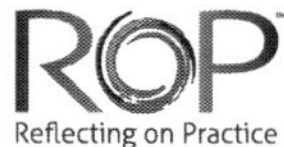

   ☐ Select a **phrase** <u>that helped them gain a deeper understanding of the idea</u>. The phrase should come from a different sentence.

   ☐ Select a **word** <u>that has either caught their attention or struck them as powerful or important</u>.

 c. Allow time and encourage silence for the active reading and selections.

3. **Form small groups and distribute chart paper.**
If groups haven't already been formed to reread a specific idea together, form them now based on participants' selections—all those who chose idea (A) form a group, (B) another, and so on. Keep the groups small so that everyone has an opportunity to talk. If many participants chose the same idea, make multiple small groups. Distribute chart paper for each group.

4. **Begin rotation to share, record, and discuss.**
Tell participants to share and record their ideas, explaining why they selected their sentences, phrases, and words. The discussion should emphasize their <u>reasoning</u> for selecting what's of interest; thus it's okay if the same excerpts were chosen.

Share that the purpose of this reread discussion is to think more deeply about these ideas on reflective practice and professional learning as a prelude to the Video Reflection. Display the steps to have them share in a rotation as follows:

 a. First participant shares a **sentence** and explains why it was chosen.

 b. Record the selection on the chart paper.

 c. The group comments on and discusses the selection.

 d. Then the next person shares, records, and discusses, and so on until everyone has shared a sentence.

 e. This process repeats with selected **phrases**, and finally, about selected **words**.

 f. Group prepares to share the essence of its idea with the whole community, drawing on the recorded selections.

5. **Circulate and monitor.**
Listen in on small-group conversations, paying attention to what participants are saying and not saying. Monitor to ensure that all groups are working at about the same pace.

This sharing structure keeps the discussion flowing and deepening with each round.

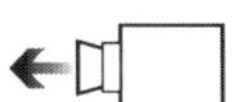

By now, participants should recognize that reflective practice is integral to professional learning. This discussion pushes them to understand that reflective practice is about making change; that change takes time and is driven by their own questions about their practice; and that it occurs within and because of this community.

6. **Call for a small-group review.**
   As groups finish their rounds, remind them to look at their recorded selections, think about their discussion, and determine the essence of their idea to share with the whole community.

7. **Initiate a whole-group share.**
   Call everyone back together and invite each small group to share the gist of its chosen or assigned idea. Suggested prompts:

   - What's the essence of your idea?

   - What's the powerful takeaway?

8. **Connect discussion to Video Reflection.**
   Tell participants to keep these ideas in mind as they practice another Video Reflection in a whole community. Display and remind them of the *Reflecting on Practice* process, and point out they are about to do another Reflection Session.

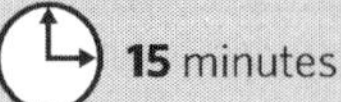

| Let's Talk Practice | 15 minutes |

Table Talk: *Educator Moves for Prior Knowledge*

1. **Reexamine the usefulness of videos.**
   Share the following:

   a. Observing teaching in real-time can be challenging because there's often so much going on.

   b. Observing videos of our teaching practice within a trusted professional learning community is helpful because everybody notices different details, for different reasons. This gives us multiple viewpoints to consider when reflecting on our practice.

   c. In the *Reflecting on Practice* program,

      - videos are used to <u>slow down</u> (and replay) the action;

      - an observation instrument is used to <u>focus attention</u> for collecting evidence; and

      - the Protocol for Video Reflection is used to keep the experience <u>transparent and supportive</u> for everyone.

2. **Display list of the Tools for Reflective Practice.**
   Tell participants they practiced their first Video Reflection as a whole community in Module 1 Session 2. At that time, they were introduced to three tools: **Protocol for Video Reflection, Observation Instrument**, and **Feedback Chart**. They will get the chance to use them again today. In this session, they will be introduced to three more tools.

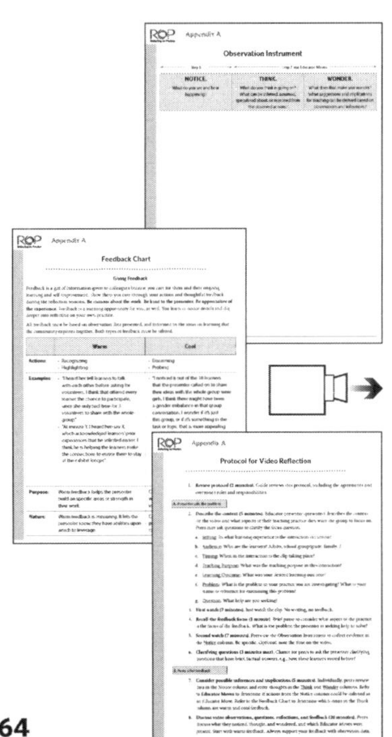

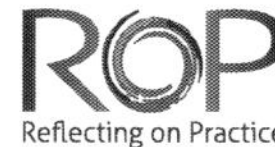

3. **Distribute "Educator Moves—Prior Knowledge" handout.**

   a. Explain that the **Educator Moves** is another tool to use during Video Reflections. This tool lists *actions* educators may take to support learning and possible reasons for each move.

   b. Let them know there is a different set of Educator Moves for each module; each set focuses attention on the aspect of practice explored in that module (prior knowledge, conversations, objects). For Educator Moves in Module 2, the focus is on activating, retrieving, and connecting learners' prior knowledge.

   c. Tell participants they reference the Educator Moves in Step 7 of the Protocol for Video Reflection when they <u>think</u> and <u>wonder</u> about what they noticed in the video clip, and consider whether the observed actions could be interpreted as any of these moves. During their discussion in Step 8, they can add examples and elaborate or revise the meaning to reflect the group's understanding of each Educator Move.

   These Educator Moves are generated from research on teaching science in formal and informal environments. This tool is both a reference for participants to consider the actions they make or want to make, and a document in which to record their evolving understanding of their practice. It isn't expected that an educator will use all of the Educator Moves in a learning experience. The moves collectively should become part of educators' practice, as they articulate what they notice and think about their own and one another's practice.

4. **Discuss with table group, and then whole group if needed.**
   Instruct participants to review and then discuss the Educator Moves —Prior Knowledge. They will be using the tool in the next task. Address their questions and concerns for how to use the tool now; more lengthy conversation about this tool should be saved until after they have used it.

Whole-Community Video Reflection

1. **Introduce the second opportunity to do a Video Reflection.**
   Tell participants it's now time to do another Video Reflection in whole community before they do it in their small groups. The *Reflecting on Practice* Facilitator 2 will be the Guide.

2. **Have tools and posters ready.**
   Affix the two posters (**Roles** and **Agreements**) in a prominent location. Remind participants to retrieve their copy of tools used previously as they will be used now. Distribute replacement copies as needed.

   - **Protocol for Video Reflection**

   - **Educator Moves—Prior Knowledge.** This tool is used in Steps 7 and 8 in the Protocol for Video Reflection.

   - **Feedback Chart.** This tool is used in Step 8 in the Protocol for Video Reflection.

3. **Distribute a new copy of "Observation Instrument."**
   This tool is used starting in Step 5 in the Protocol for Video Reflection.

4. **Call attention to the Reflection Session posters.**
   Direct everyone's attention to the two posters, **Roles & Responsibilities**, and **Community Agreements for Video Reflection**. Give them a moment to remind themselves of the details.

5. **Introduce presenter.**
   Introduce the educator-presenter whose video will be watched, and thank them for being willing to share.

6. **Start the practice Video Reflection.**
   Follow the steps in the protocol to guide the group through the Video Reflection.

   a. Remind participants that while the guide uses the Protocol for Video Reflection to lead the group through the Video Reflection, everyone should also use it to follow along with the process.

   b. Be sure to call out the Steps and the specific tools as they are used in the protocol. Refer to the Step by Step for Module 1 Session 2, if needed.

> Reiterate the problem the presenter wants help to solve, and the focus question for feedback. This emphasis will be helpful as participants prepare to present their own video in their upcoming critical-colleague groups.

> Inappropriate comments include those directed at the presenter *other* than when the presenter is setting the context or answering clarifying questions; and/or comments that are evaluative.

Tips for the Guide

Be assertive about keeping the Video Reflection centered on the presenter's problem and focus question. An easy way to re-direct the group is to use the Protocol.

- The presenter might stray too far in setting up the context, such as offering rationalizations and interpretations of what happened. Remind the presenter that the group will get the chance to gather observation data in Step 5 and interpret that information in Steps 7 and 8.

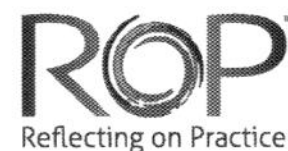

> ■ Peers might get caught up in the conversation, such as directing comments to the presenter or making comments beyond the focus question. As appropriate, remind peers that Step 8a requires them to leave the presenter out of the conversation, or that Step 8b insists they focus on the presenter's question and problem.

7. **Using the Educator Moves in Step 7.**
    As a part of their work completing the Think and Wonder columns in Step 7, instruct participants to use the Educator Moves. Ask them to consider whether they **think** any of the Moves showed up in the video clip, and if they **wonder** if any of the Moves would be helpful to address the presenter's question.

8. **Using the Educator Moves in Step 8.**
    If it doesn't come up, ask participants if they **think** any of the Moves showed up or **wonder** if any of the Moves would be helpful. Referring to the Educator Moves can occur as either warm or cool feedback. Encourage them to write down examples from the video clip for the Moves. What is meant by each of the Moves can also be debated and refined.

9. **Gather action items for changes in institutional practices in Step 11.**
    Let participants know how these items will be addressed. It's possible not every item will require massive change or will lead to change at all. It's important to have a process to consider all the items, and that this process is made transparent and available for everyone to participate.

> Share the process for gathering examples for Educator Moves across the community. Model how that can be done during a Video Reflection, e.g., details are added to a document at the end of Step 8.

2.3

> Remind the group how the action items regarding changes in institutional practice that come from their small-group conversations will be gathered.

Turn & Talk:
The Video Reflection Experience

1. After the Whole-Community Video Reflection, ask participants to talk with someone next to them about their experience. Display the suggested prompts:

    a. *What was interesting and challenging about the experience?*

    b. *What modifications would you suggest to make the experience better?*

2. **Facilitate whole-group share.**
    As participants share their thinking, listen to what people share and encourage the community to listen closely. Invite clarifications and elaborations, as well as different perspectives. Finally, summarize the ideas that have been shared.

7 minutes — From Learners' Perspective

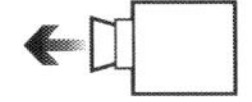

> Remember to follow the **Discussion Map**.

Retrieve & Connect 🕐 **5 minutes**

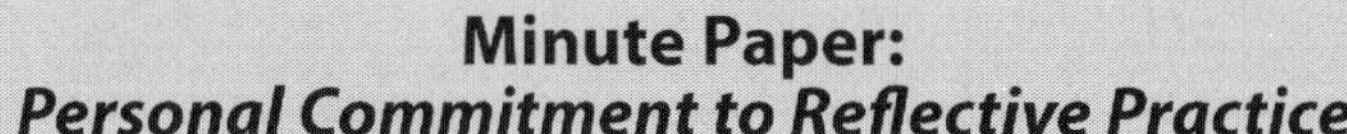

Display the following prompts and allow a few minutes for participants to gather their thoughts about this second video reflection session and make plans for themselves in their Learning Journal.

a. *What are you excited about, interested in, and/or curious to learn about your own practice?*

b. *What are your concerns about Video Reflections? What can you, the facilitators, or your peers do to help alleviate them?*

c. *What do you think will be the best part about sharing your practice in this way with your peers?*

🕐 **15** minutes

Continue the Learning

1. **Introduce critical-colleague video reflections.**
   Tell participants that between now and the next whole-community session, they will all do a Video Reflection in a small, critical-colleague group. Each critical-colleague group of 4–6 will convene amongst themselves to share videos and problem-solve with each other.

2. **Display and review processes for collecting and storing videos.**
   Let participants know the equipment available for their use and the procedures for sharing them. Describe the process for collecting and storing videos, including the organization's policy for video recordings. Invite their questions and input.

3. **Distribute "Checklist for Reflection Sessions" handout.**

   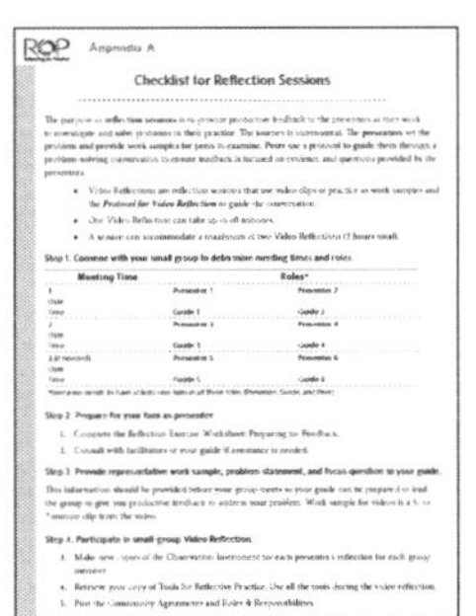

   a. Point out that the **Checklist** enumerates the steps they take from preparing for the Video Reflection to convening in their small groups. It also lists all the tools used in the Protocol for Video Reflection. Note that they will meet two times in their small groups for two hours each (or 3 times if there are more than 4 people in a group).

   b. Ask participants to convene in their small, critical-colleague groups to begin preparing for their Video Reflections. The critical colleagues should work together now to complete Step 1 in the Checklist—determining the rotation schedule and assigning roles. Each group should provide the facilitators with their rotation and roles by the end of this session.

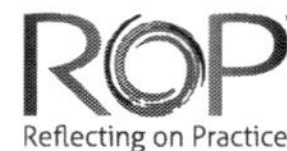

   c. In advance of the Video Reflection session, each presenter will complete Steps 2 and 3 on their own, or in consultation with the facilitator.

4. **Introduce and distribute "Reflection Exercise Worksheet to Prepare for Feedback."**

   a. Explain that this worksheet for Step 2 in the **Checklist** has two sets of questions. The first set of questions guides them through naming and framing a problem in their practice they want to investigate, and the second set focuses on generating a focus question that they seek to solve with help from their colleagues.

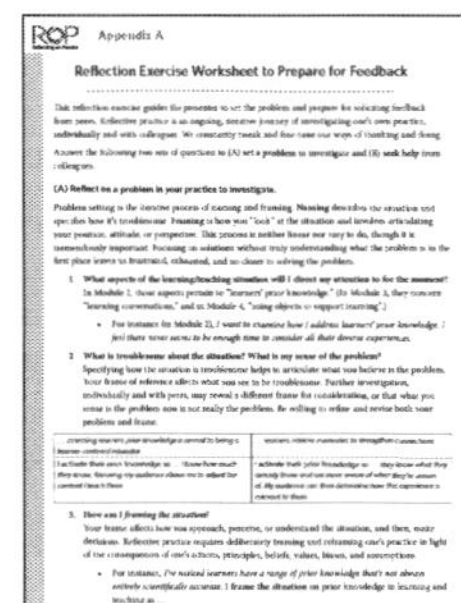

   b. Acknowledge that the reflection exercise will take time to do and might not be easy to complete; the facilitation team is available to help. It's necessary to do this task to make the Video Reflection experience meaningful and productive for everyone.

   c. Assure them that these routines and exercises will become familiar and easier to do. Emphasize that they shouldn't hesitate to ask for help if they struggle at any point.

5. Remind participants that the discussions in this module on how people learn have addressed activating, accessing, and connecting to learners' prior knowledge. For the coming small-group Video Reflection, they should think about <u>what they want to know about their own practice related to engaging with learners' prior knowledge</u>.

6. **Announce date for the next whole-community session.**
Inform participants of the date when all Video Reflections in small groups need to be completed.

❖

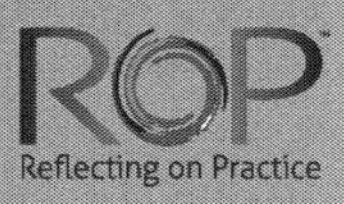

Overview

In this module, participants probe more deeply into the significance of *talk* in learning, and consider how to best facilitate conversations in their interactions, depending on their **teaching purpose**. One teaching purpose, Activate Prior Connections, was the focus of Module 2. In this Module, four more—**Give Information**, **Check for Understanding**, **Express Your Thoughts**, and **Transfer Understanding**—are examined in depth. The educators critically examine how they currently converse with learners in their practice, and explore methods for facilitating conversations that specifically support learning.

In Session 1, participants brainstorm **characteristics of quality learning conversations**. They take part in three scripted role-plays in which they analyze facilitated conversations, and discuss both the limitations and potentials inherent in these conversations for supporting specific **teaching purposes**. They read and discuss research on **the role of talk in learning**, including how we process information into long-term storage and how this relates to the importance of talk for supporting learning. After having seen it implicitly modeled since the program began, participants are now formally introduced to the **Discussion Map** for developing skills to facilitate conversations.

In Session 2, participants take part in a hands-on activity that offers a shared experience to probe more deeply into teaching purposes when facilitating conversations. They read and discuss different facilitation approaches, and learn how to use the **Facilitation Approaches Framework** to orchestrate ideas and voices in learning conversations.

In Session 3, participants do a close rereading of the Research Discussions on learning and talk from Modules 2 and 3. Following this priming, they do a **whole-community Video Reflection** on facilitating learning conversations strategically designed to achieve specific teaching purposes, including how to allow for the sharing of multiple ideas as learners are guided towards the scientific view.

Following Session 3, and on their own timetable, participants meet in their **critical-colleague groups to do their own Video Reflections** on the module's focus, the importance of talk to the learning process.

Module 3.
Session 1: Talking to Learn

Session Overview

This session engages participants in deeper discussion about learning and conversations, and specifically encourages educators to reflect on the ways in which they communicate with learners. Scripted role-plays are used to demonstrate facilitating conversations for three different purposes: to **give information,** to **check for understanding**, and to encourage learners to **express their thoughts**. Participants observe, analyze, and critique the scripts to understand and distinguish between conversations facilitated in different ways for different teaching purposes. Participants read and discuss key ideas from the literature on **how people process information** and **how our brains make memories**, and relate those findings to the critical importance of making opportunities for talk to support learning.

Session Objectives

- Examine the characteristics of conversations that support learning.
- Discuss how the brain processes information and why talk is important in this process.
- Examine facilitated conversations for three teaching purposes.
- Determine strategies for facilitating conversations that support learning.

Express your thoughts isn't limited to talking; expression can also take the form of writing, drawing, and movement. Talk is emphasized in this session and module.

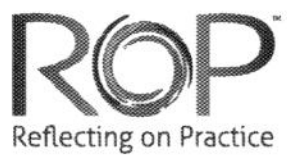

SESSION AGENDA

Task Routine	Description	Estimated Time (in minutes)
Introduction *Recap and Session Objectives*	This session probes more deeply into the role of conversations in constructing understanding and building knowledge. It explores ways in which educators in informal environments can facilitate conversations that support learning in their practice.	3
Retrieve & Connect Think-Pair-Share: *Characteristics of a Learning Conversation*	Participants discuss the key characteristics of a "learning conversation."	17
Hands-on *Role-Plays*	Role-plays are used to demonstrate facilitated conversations that can be used in practice to address three different teaching purposes.	50
Break option		10
Research Discussion Reading Partner Jigsaw: *Talking to Learn*	Participants engage in a *Reading Partner Jigsaw* as they read and discuss three big topics on how people process information and make memories, and relate that to why talking is important for learning.	40
Let's Talk Practice Connect, Extend, Challenge, & Apply: *Teaching Purpose & Discussion Map*	Participants are introduced to two Design and Teaching Tools from *Reflecting on Practice*. They are challenged to consider how these tools can be used in their practice.	28
Continue the Learning	Participants are assigned a Reflection Exercise in preparation for the next session.	2
Total Estimated Time		150 mins. (2.5 hrs.)

MATERIALS

Recurring for all Sessions

- See Module 1 Session 1

For this Session

- 3 copies of each **Role-Play Script** for the actors with their parts highlighted (See **Getting Ready** #4)
- 3 sheets of chart paper (or large section of whiteboard) for **Role-Play Debrief** discussions (See **Getting Ready** #6)
- Props for the role-plays:
    - Sea otter pelt (if available) or stuffed sea otter toy
    - 1 hand lens
    - 1-cm square piece of cardboard or cardstock
    - 1 large colored image of a sea otter

For each participant

- Handouts (1 copy each)
    - **Role-Play Scripts 1, 2, and 3**
    - **Key Ideas from the Literature: Talking to Learn**
    - **Teaching Purposes** (Appendix B)
    - **Discussion Map** (Appendix B)

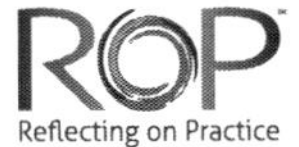

3.1

Getting Ready

1. Consider sending out the handout **Key Ideas from the Literature: Talking to Learn** a day or two before the session so that participants have the option to review the reading material ahead of time.

2. Review the handouts and slide deck for this session, and additional suggested readings.

   For the facilitator:

   - *Taking Science to School* (Chapter 5: "Generating and Evaluating Scientific Evidence and Explanations")
   - *How People Learn* (Chapter 5: "Mind and Brain")

   Continue the Learning suggestion for participants:

   - *Ready, Set, SCIENCE!* (Chapter 5: "Making Thinking Visible: Talk and Argument")

3. Duplicate handouts.

4. Prepare actors' scripts for each role-play:

   a. Highlight 3 copies of Role-Play #1.

      ○ Highlight the first actor's dialogue on the 1st copy.

      ○ Highlight the second actor's dialogue on the 2nd copy.

      ○ Highlight the third actor's dialogue on the 3rd copy.

      ○ Put this set aside for Role-Play #1.

   b. Follow the same steps for Role-Play #2 and Role-Play #3.

5. Prepare a space for recording the **Think-Pair-Share** discussion. Title a sheet of chart paper or section of the whiteboard, "Characteristics of Learning Conversations."

6. Prepare a space for recording **Role-Play Debrief.** Follow the directions and see the example chart below to draw a chart on the board (or on three chart papers) to record the debrief discussions:

   a. Draw <u>three columns</u> on the board (or use three chart papers), one for each role-play. Label the columns Role-Play #1, Role-Play #2, and Role-Play #3.

The **Sample Completed Role-Play Debrief Chart** at the end of this session shows how the chart might look by the end of the debrief.

b. Make <u>three rows in each column</u>, one for each Observation Question, and label as follows: Who is talking? What are they saying? and What is the pattern of talk?

Role-Play Debrief Chart

Observation Question	Role-Play #1	Role-Play #2	Role-Play #3
Who is talking?			
What are they saying?			
What is the pattern of talk?			

7. Advance preparation for Module 3 Session 3: Whole-Community Video Reflection. Determine who will be the educator-presenter. If there is another person on the facilitation team, then that person should go, otherwise seek a suitable volunteer from the community. Make one copy of **Reflection Exercise Worksheet to Prepare for Feedback** (Appendix A) to provide to the educator-presenter who will show their video during Session 3 of this module. Offer assistance as needed.

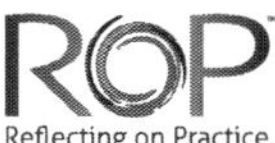

Session 1 Step by Step

Introduction
Recap and Session Objectives

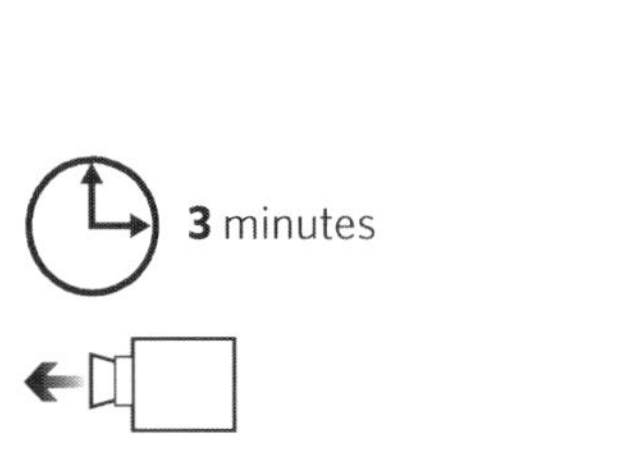

1. **Display reminders about learning and talk.**
   Remind participants about ideas on learning and talk introduced
   and discussed in previous sessions:

   a. *Learning is making connections between ideas and organizing
      these connections into mental models for storage in long-term
      memory.* Experiences and examples strengthen the connections
      because they offer details and context that can be useful for
      retrieval and application in other situations.

   b. *Learning is social.* The human brain has evolved towards social
      connections with others. Learning involves people, the things
      they use, the words they speak, the cultural context they're in,
      and the actions they take.

   c. *Talking is important for learning.* Students from K–12 to
      university learn best when they engage in collaborative
      dialogue in which they both provide explanations to bolster
      their reasoning, and seek and provide help.

2. **Introduce the session.**
   Let participants know that this session probes more deeply into
   what happens in the brain when we learn, and the significance of
   talk for learning. They'll consider different teaching purposes for
   facilitating conversations, and examine use of those purposes in
   their own practice.

3. **Display and introduce session objectives.**

   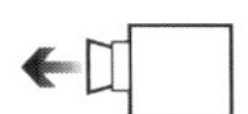

   • Examine the characteristics of conversations that support
     learning.
   • Discuss how the brain processes information and why talk is
     important in this process.
   • Examine facilitated conversations for three teaching purposes.
   • Determine strategies for facilitating conversations that support
     learning.

Retrieve & Connect 🕐 **17** minutes

Think-Pair-Share:
Characteristics of a Learning Conversation

1. **Introduce the topic of learning conversations.**
   In educational situations, conversations occur in two dynamics: between educators and learners, and between learners themselves. Both talking with another learner and talking with the educator support learning, each in its own way, and so serve different teaching purposes.

   For now, we focus on <u>facilitated conversations between the educator and learners</u>. Our big question in this module is this: how do we facilitate conversations that support learning?

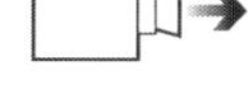

2. **Engage in Think-Pair-Share.**

   a. **Display the question** for <u>Think</u>. Ask participants to think about the following prompt:

      What are the characteristics of a conversation that would make it a "learning conversation"?

   b. **Form Pairs.** After a minute or so, ask participants to pair up with someone and discuss their thoughts.

3. **Facilitate a whole-group share.**
   After about 5 minutes, invite participants to share their partner conversations. Facilitate a productive discussion, making sure to push on participants to elaborate on their comments and listen closely to what's shared by others.

   a. Record their ideas on the board or chart paper under the heading of "Characteristics of Learning Conversations." Consider adding ideas from the list below if these items aren't mentioned.

      In *learning conversations*, participating speakers

      ☐ have mutual interest in the topic

      ☐ listen closely to understand each other

      ☐ ask questions to seek clarification and information they don't have

      ☐ provide information that identifies, describes, or explains objects, ideas, or phenomena

      ☐ generate explanations and produce justifications

      ☐ challenge current understandings

      ☐ are open to views that conflict with their own

Be sure to model the **Discussion Map** here and throughout this session. This tool will be introduced and distributed later in the session, with reference to how it's been used in this and previous sessions.

3.1

b. Explain that this list is helpful for developing a shared understanding of the key characteristics of learning conversations. Knowing what to look for is a critical step in observing and reflecting on facilitated conversations.

c. Make connections to previous discussions. As participants share their ideas, remind them to consider how specific ideas from the literature on *scaffolding, externalization and articulation*, and *reflection* occur in facilitated learning conversations.

Role-Plays

50 minutes   Hands-on

(A) Introduce the Activity

1. **Introduce role-play.**
    Tell participants that in this next routine they'll observe (or participate in) three scripted role-plays that simulate facilitated conversations in informal science learning environments. The role-plays are amalgamations of many different conversations, crafted to demonstrate key discussion points. The skits don't represent any one person's practice, though we all may see a little of ourselves in each of the skits.

2. **Display the purposes for this activity.**

    a. Examine and compare three educator-facilitated conversations.

    b. Discuss the limitations and potentials of each conversation in how it supports learning.

3. **Display Observation Questions.**
    Ask participants to observe the **characteristics** *of the conversation* that takes place in each of the three role-plays, focusing attention on three questions:

    a. *Who is talking?* refers to whose turn it is to talk (turn-taking), and is categorized according to the speaker, such as educator or learner(s).

    b. *What are they saying?* refers to what the speaker says when it's their turn to talk. Are they asking a question? What does the question seek? Are they providing information? What kind of information—an explanation, reasoning, observation, praise, opinion, directions?

    c. *What is the pattern?* refers to a repeated sequence of turn-taking (educator/learner/educator; learner/learner) and talk categories (e.g., question, information, praise, instruction).

Each of these role-plays is designed to illustrate one of three teaching purposes (not to be shared with participants yet): #1 demonstrates **Give Information**; #2, **Check for Understanding**; and #3, **Express Your Thoughts.**

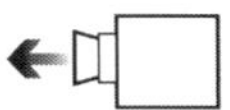

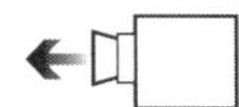

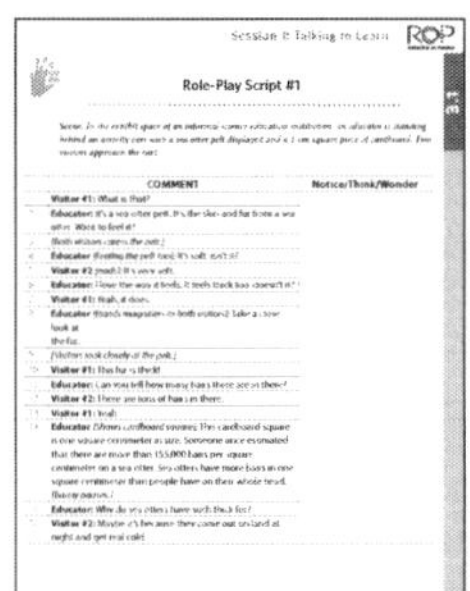

(B) Role-Play #1

1. **Cast Role-Play Script #1.**
   Ask for three volunteers to be in the first skit. Give each actor a highlighted **Role-Play Script #1**, and instruct them to read their characters' highlighted lines as written, not to ad-lib.

2. **Distribute Role-Play Script #1 to the community.**

   a. Remind everyone that these role-plays don't represent any one educator's teaching practice; they've been created to highlight specific ideas for discussion in this session.

   b. Ask participants to try not to focus on whether these facilitated conversations are "good" or "bad." The scripts are written to make the facilitated conversations distinct. Focus participants' attention on what the educator is doing and saying in each skit.

3. **Explain how to record observations on the script.**
   Tell participants they'll be given the script for each skit so they can record observation notes in the **Notice/Think/Wonder** column as they follow along.

4. **Begin the role-play.**
   Allow about 5 minutes to act out the script.

5. **Transition to small-group discussion.**
   When the volunteers have finished presenting, ask them to take their seats and make sure they get a clean copy of the scripts so they can participate in the small-group discussion.

   Give participants 3–5 minutes to review their notes and confer with others at their table about what they observed, and to discuss similarities and differences between their notes. Those who were in the skit can share what it felt like to be in each respective role.

6. **Post chart and initiate whole-group Role-Play Debrief.**

   a. Point out the role-play columns on the chart where you'll record the whole-community discussion about the skits. Instruct participants to answer the displayed Observation Questions, referring to their Notice/Think/Wonder notes, and to be sure to <u>provide evidence for their comments using examples from the script</u>.

   b. Use the probing questions below to facilitate discussion for each Observation Question, and record the community's comments in the appropriate section of the role-play chart.

If needed, refer to the **Sample Completed Role-Play Debrief Chart** for examples.

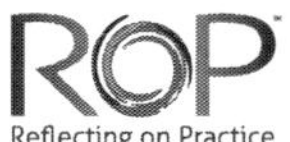

3.1

Who is talking?

- Do educator and learners contribute equally to the conversation?

  Discussion point:
  The educator is doing most of the talking. The educator or learner initiates a question or comment, and the educator responds with a long narrative, explanation, or summary of ideas.

What are they saying?

- What do you notice about the questions posed?

- Who is asking the questions?

- What's being asked or provided—question? information? instruction?

- What does the responder need to know in order to answer the question?

- Can the responder figure out the answer by examining the object?

- What do you notice about the information shared, whether in response to a question or not?

- Who shares the information?

- What kind of information is shared?

- How does the information contribute to learners making sense and meaning from the conversation?

  Discussion points:

  - *The questions ask learner(s) to provide information about what they know. Both the educator and learner(s) ask questions.*

  - *The kinds of information being shared are identification and description of objects and experiences, and explanations of scientific knowledge—typically as disseminated in textbooks and other reference materials.*

  - *Educator asks questions that the visitor either knows the answer to or not, and then the educator gives a narrative or explanation to answer the educator's own question (Rows 15–21 and Rows 23–30).*

As participants offer comments and interpretations, remind them to provide evidence from the skit, e.g., reference specific lines or quote from the text.

What is the pattern?

- What do you notice about the overall flow of the conversation?

- Is there a repeated sequence of talking between the educator and learners?

- What do you notice when it's the educator's turn to talk?

- What do you notice when it's the learners' turn to talk?

Discussion point:
Educator monologue. The turn-taking may or may not alternate between educator and learner(s), but the educator's turn to talk is noticeably longer than the learners'.

c. Probe for teaching purpose. Ask, **What is the educator's <u>purpose</u> for the conversation?**

Discussion point:
The educator's purpose is to <u>give information</u>. The educator tells learners what the educator knows, or believes learners need/want to know. The learners may or may not be interested in the information shared. It's tempting to infer that the learners are not interested, but it might also be the actors' interpretation of the script.

Nonetheless, focused on giving information, the educator cannot know what connections learners make or what is interesting to them. Since learners talk very minimally, the interaction is likely less memorable for them.

(C) Role-Play #2

Repeat Steps 1–6 from Role-Play #1.
Use the following discussion points for the second skit and record responses on the debrief chart.

Who is talking?

Discussion point:
Both educator and learners are talking.

What are they saying?

Discussion points:

- *The questions ask learner(s) to identify and describe what they see and tell what they know.*

- *The educator asks questions for which the educator already has an answer in mind, and determines the correctness of (evaluates) the responses.*

- *Learner's responses may be short answers.*

The point isn't for participants to recite the name of the purpose, specifically, but to elaborate on the educator's intentions for leading the conversations in the way observed.

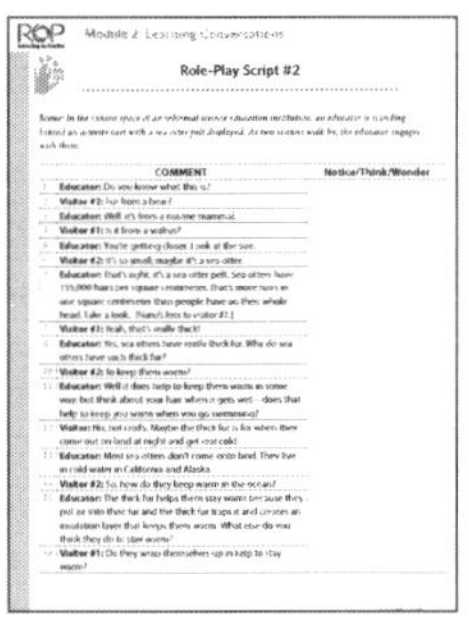

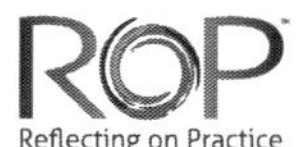

- *Educator's evaluation may be verbal or nonverbal, and educator elaborates on the learner's response.*

- *The kinds of information being shared are identification and descriptions of objects and experiences, and explanations of scientific knowledge—typically as disseminated in textbooks and other reference materials.*

- *The learners mostly identify and describe what they see.*

- *The educator mostly acknowledges the accuracy of the information.*

- *The educator explains the idea with scientific concepts.*

What is the pattern?

Discussion points:

- *The turn-taking regularly switches back and forth between educator and learner.*

- *The educator's turn to talk may or may not be longer than the learner's.*

- *There is a distinct 3-part pattern: the educator **initiates** the conversation with a question or comment, the learner **responds**, the educator **evaluates** the response and then provides an elaboration to the learner's response. The pattern repeats with another educator-initiated question or comment.*

What is the educator's purpose for the conversation?

Discussion point:
The educator's purpose is <u>checking for understanding,</u> specifically evaluating what the learner knows. There is more back-and-forth in this conversation than in Role-Play #1. As a result, the learners are more involved. The educator asks a question the educator knows the answer to (Rows 9–15 and Rows 19–26), as a means to check whether learners know the answer. The learners either know the answer or not. Then the educator provides an extended answer to the educator's own question. Learners aren't encouraged to explore the objects to figure out the answer, or may not be provided a way to do so.

(D) Role-Play #3

1. **Repeat Steps 1–6 from Role-Plays #1 and #2.**
   Use the following discussion points for the third skit and record responses on the chart.

   Who is talking?

   Discussion point:
   Both educator and learners are talking.

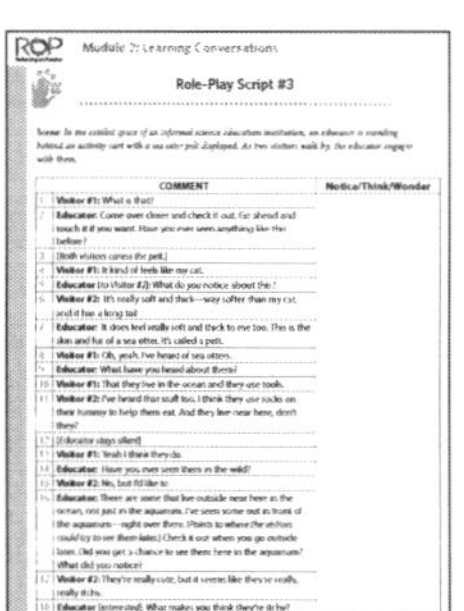

What are they saying?

Discussion points:

- *Educator and learners are asking and answering questions. The questions invite learners to explain and elaborate on what they think and respond to what others say.*

- *The kinds of information shared are identification and descriptions of objects and experiences, and statements and discussions of what both educator and learners consider to be possible explanations.*

- *Educator asks for additional viewpoints and probes learners to elaborate on their thinking with evidence, explanation, and clarification. Educator determines learners' understanding based on what they say, and gives relevant information for consideration.*

- *Learners elaborate on their thinking. Learners provide or consider multiple viewpoints.*

What is the pattern?

Discussion point:
The turn-taking switches back and forth between educator and learner.

- *Educator and learner initiate, respond to, and follow up on each other's comments.*

- *Learners are talking with each other, expressing their thinking out loud to each other.*

What is the educator's purpose for the conversation?

Discussion point:
The educator's purpose is for learners to <u>express their thoughts</u>.

In Role-Plays #1 and #2, learners identified and described what they saw (if they could) but didn't articulate what they were thinking, explain their understanding, or verbalize connections between ideas and observations.

In Role-Play #3, learners were invited to think aloud. The educator asked the learners to explain their answers (Row 18) and to examine the objects more closely to figure out the answer (Row 22). The educator did give information (Row 35), but it was given in smaller chunks, and only for things visitors couldn't determine on their own, and then visitors were prompted to use that information to figure out the answer.

The facilitation team has modeled aspects of the third skit during the Phases of the Moon and the Ice Cubes activities in the previous module. Facilitators gave relevant information when learners couldn't get the information from the object and it became apparent in the conversation that learners didn't already know the information. Information was given in small doses and just in time.

2. **Compare the three skits.**

Ask participants to compare and contrast the conversation in Role-Play #3 to Role-Plays #1 and #2. How are they similar and different in terms of what learners and educators are saying?

Be sure to raise the following points if they don't come up in discussion:

- The educator's questions and follow-ups in the first two skits didn't invite learners into the conversation. Learners either knew the answer or not; they weren't able to figure out the answers by exploring the objects for themselves. The questions may have led to sharing of interesting information about the topic, but they didn't take full advantage of the objects featured on the cart. This idea will be explored in more depth in Module 4.

- The facilitated conversation in each skit highlights a different teaching purpose: to give information, to check for understanding, to express thoughts. These purposes will be explored more in the Research Discussion. While only one teaching purpose dominated the conversation in each skit, in an extended interaction with learners, the educator likely has multiple teaching purposes and facilitates the conversation in a combination of ways.

Reading Partner Jigsaw: *Talking to Learn*

40 minutes — Research Discussion

1. **Transition to Research Discussion.**

Share the following:

a. The three skits took a closer look at facilitated conversations in order to provide an opportunity to analyze each conversation line by line.

   - The first two skits reflected many educator-facilitated conversations in informal environments, in K–16 classes, and even within families.

   - The facilitated conversation in the third skit is much less commonly practiced. It takes longer, and the information shared was not nearly as plentiful as in the other two skits; the emphasis was on getting learners to express their own thinking, out loud. This kind of conversation is distinctive because the learners are consistently invited to share what they think and what makes them think that.

b. The role-playing introduced an analytical approach to reflecting on conversations we facilitate, and demonstrated how we can facilitate conversations for different teaching purposes.

Now it's time to look beyond our personal experiences, opinions, and beliefs and dig deeper into why it's so important for learning that we make time for learners to share what they think and why they think that.

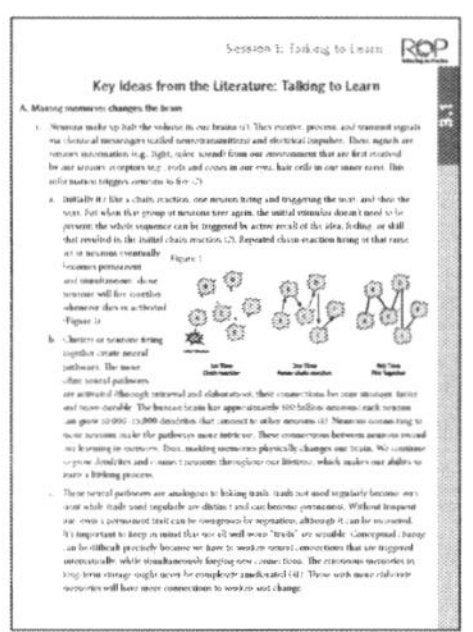

2. **Introduce Key Ideas from the Literature: Talking to Learn.**
   Distribute the handout. Tell participants that in this Research Discussion, as in the past, they'll read and discuss three big ideas, first in pairs and then in small groups.

   - Making memories changes the brain

   - Talking makes thinking visible

   - Teaching purpose informs how to facilitate conversations

3. **Form groups of six, composed of three pairs.**
   Direct formation of groups, or let participants form the groups on their own. Ask everyone in each group to pair up with a reading partner. Emphasize that all three topics will be read and discussed within each group.

4. **Display and review the instructions on the slide:**

   a. Work with a reading partner to read and (when ready) discuss one topic. Divide up the information so that each pair of reading partners in the group reads and discusses a different topic.

   b. Make sure that all the topics are addressed within the group.

   c. After reading partners have had a chance to share with each other, each pair will share their discussion with their small group.

   d. Look for connections between the topics during the small-group discussion.

   e. Be ready to share a couple of big ideas from the small-group discussion with the community.

5. **Circulate and monitor pace of discussions.**
   Ensure the room is quiet so everyone can read their assigned topic. As people finish their reading, let them know partners can start discussing their topic whenever they're ready.

   - Within each group, monitor to ensure that the partners are working at the same pace, so that all partners finish at about the same time and the small-group discussion can begin as soon as all the partners finish discussing their topics.

Remember to stay silent when participants are reading. Even whispering can be distracting.

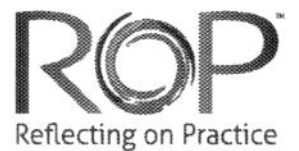

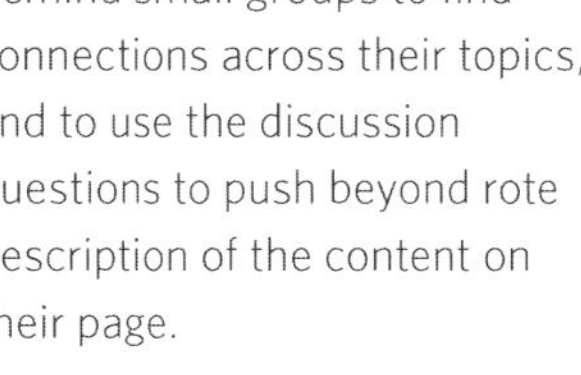

- Across multiple groups, monitor to ensure that all groups are working at about the same pace, so that all will be ready for the whole-group share. Move them along, if necessary.

6. **Begin whole-group share.**
   Call everyone back together and invite each group to share a couple of big ideas that emerged from their discussion. Model the **Discussion Map** as you facilitate. Suggested prompts:

   - What ideas generated interesting discussion within your groups?

   - What connections did you notice across the topics?

   - How can you use these ideas in your practice?

7. **Display the Ideas to Emphasize.**
   Make connections between these ideas and what the community shares.

   a. **Learning is making memories.**

      - Learning requires us to retrieve memories from long-term storage, and elaborate on them with new information or consideration.

      - Multiple experiences allow us to retrieve memories and consider them in new contexts through different modalities, which makes connections more elaborate.

   b. **Learning happens when learners externalize and articulate connections.** These actions allow learners to:

      - take ill-formed and incomplete ideas out of their heads to process and integrate.

      - get feedback from oneself (when they hear their own reasoning and connections) and from others (peers, educators).

   c. **Talking is learning. It is an integral part of the process of learning, not an indication of having learned.**

      - The educator needn't be part of every discussion for the conversation to be of value for the learners. Conversation between learners offers them the chance to retrieve prior knowledge and make connections without the influence of the educator.

      - It's important to assess learners' understanding throughout the learning experience, and to provide feedback to learners so they can regulate their own learning. However, the tendency to facilitate conversations for the purpose of

Remind small groups to find connections across their topics, and to use the discussion questions to push beyond rote description of the content on their page.

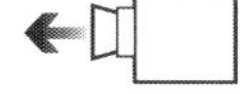

Remind the community that although externalizing, articulating, and elaborating are not limited to talking (expression can also take the form of writing, drawing, and gesturing), talk is the emphasis in this session and module.

checking for understanding, rather than inviting learners to *express their thoughts*, conditions learners to speak only when they know the answer.

Let's Talk Practice   28 minutes

Connect, Extend, Challenge, & Apply: *Teaching Purpose & Discussion Map*

1. **Introduce Teaching Purposes.**
   Distribute the **Teaching Purposes** handout. Tell participants this tool from *Reflecting on Practice* organizes the information from the Research Discussion into an instructional tool to keep our shifting teaching purposes in the forefront of our minds.

2. **Introduce the Discussion Map.**
   Distribute the **Discussion Map** handout. Tell participants that facilitating conversations to support learning—especially for learning by thinking out loud—is a learned skill that takes time and practice. The Discussion Map is a tool for developing that skill.

3. **Explain tools embedded in *Reflecting on Practice* program.**
   Point out that both tools are embedded in the *Reflecting on Practice* program, and that you've been modeling them in various ways in every session.

4. **Review the Discussion Map steps.**
   Go through each step of the Discussion Map, offering examples of how you've been using the steps to facilitate conversations throughout the program.

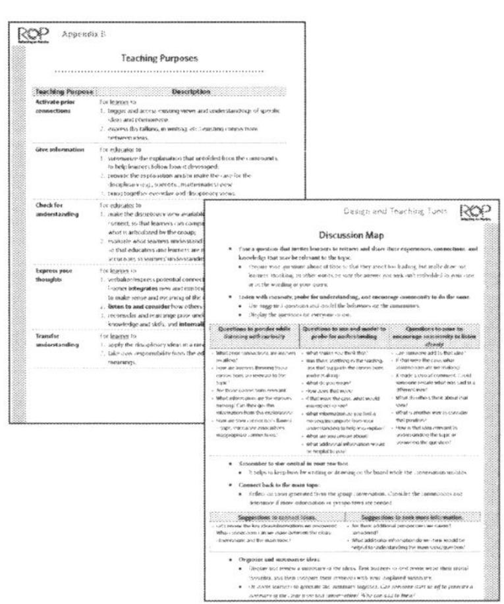

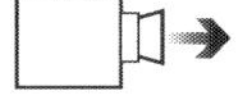

5. **Display the Let's Talk Practice chart.**
   As before, ask participants to use a blank page in their Learning Journals to make a **Let's Talk Practice** chart as shown.

CONNECT	EXTEND	CHALLENGE
APPLY		

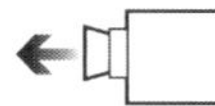

6. **Display the Let's Talk Practice questions.**
   Ask participants to think about the ideas discussed in today's session, and consider the two *Reflecting on Practice* tools.

   a. **Connect.** *How are the ideas and information discussed in this session connected to what you already know? What aspects of the tools are already in your work?*

   b. **Extend.** *What ideas or perspectives extended or broadened your thinking in new directions?*

   c. **Challenge.** *What still seems challenging or confusing? What questions or wonderings do you have?*

   d. **Apply.** *What one or two changes will you make in your practice to apply your new ideas or understandings? Where and how can you use the two tools in your practice?*

7. **Have participants share in small groups.**
   Ask participants to work in small groups of 3–4 to share their thoughts and determine how to integrate these new ideas into their practice.

8. **(Optional) Conduct a whole-group share.**
   If time permits, invite groups to share key points from their discussions with the rest of the community.

Continue the Learning

1. **Suggested reading.**
   If you've decided to assign the reading, share it now.

2. **Discussion Map & Teaching Purposes.**

   - Encourage participants to use the Discussion Map in their practice and share with colleagues how it went.

   - Invite participants to review and consider how they use each of the Teaching Purposes in their own practice.

2 minutes

Remind participants when the next whole-community Video Reflection will be. Let them know the focus for this round will be on learning conversations.

Sample for the Facilitator:
Completed Role-Play Debrief Chart

	Role-Play #1	Role-Play #2	Role-Play #3
Who is talking?	• Educator, mostly • Some learner	• Both educator & learner • Still more educator	• Educator and both learners
What are they saying?	*Educator* • Asks questions to which the educator knows the answer • Doesn't pause • Poses all questions • Assumes learners' prior knowledge, rather than asking what learners think • Provides information that *the educator* knows *Learners* • Share prior exposure or knowledge, but this isn't built upon by educator	*Educator* • Asks questions to which the educator knows the answer • Asks questions seeking what learners think • Evaluates what learners say • Elaborates on what learners say *Learners* • Answer questions • Share what they think, minimally	*Educator* • Asks learners to explain what they think & respond to what others say • Encourages learners to use multiple sources to get information • Builds from learners' observations, personal experience & prior knowledge *Learners* • Ask questions • Elaborate on and explain what they think • Respond to what others think
What is the pattern of talk?	• Information giving • Educator asks questions & sometimes learner responds • Kind of a lecture	• Guided dialogue; goal-oriented • Educator checks for what learners know	• Back-and-forth • Equal dialogue • Educator & learners share what they think, respond to what others say, and add to information provided
What is the teaching purpose?	• Give information	• Check for understanding	• Express your thoughts

     This page may be duplicated for educational use.

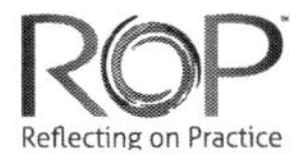

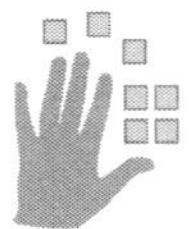

Role-Play Script #1

3.1

Scene: In the exhibit space of an informal science education institution, an educator is standing behind an activity cart with a sea otter pelt displayed and a 1-cm square piece of cardboard. Two visitors approach the cart.

	COMMENT	Notice/Think/Wonder
1	**Visitor #1:** What is that?	
2	**Educator:** It's a sea otter pelt. It's the skin and fur from a sea otter. Want to feel it?	
3	*[Both visitors caress the pelt.]*	
4	**Educator** *[Feeling the pelt too]:* It's soft, isn't it?	
5	**Visitor #2** *[nods]:* It's very soft.	
6	**Educator:** I love the way it feels. It feels thick too, doesn't it?	
7	**Visitor #1:** Yeah, it does.	
8	**Educator** *[Hands magnifiers to both visitors]:* Take a close look at the fur.	
9	*[Visitors look closely at the pelt.]*	
10	**Visitor #1:** This fur is thick!	
11	**Educator:** Can you tell how many hairs there are in there?	
12	**Visitor #2:** There are tons of hairs in there.	
13	**Visitor #1:** Yeah.	
14	**Educator** *[Shows cardboard square]:* This cardboard square is one square centimeter in size. Someone once estimated that there are more than 155,000 hairs per square centimeter on a sea otter. Sea otters have more hairs in one square centimeter than people have on their whole head. *[Barely pauses.]*	
15	**Educator:** Why do sea otters have such thick fur?	
16	**Visitor #2:** Maybe it's because they come out on land at night and get real cold.	

(continues)

This page may be duplicated for educational use.

Role-Play Script #1 *(continued)*

	COMMENT	Notice/Think/Wonder
17	**Educator:** It does feel cold when you get out of the water, doesn't it? Sea otters used to come out on land a lot, but now they hardly ever do. They spend almost all their time in the water, day and night. They live off the coast of northern California and in Alaska, and the water is very cold in both of those places. I've never heard of sea otters living in places with really warm water. Have you?	
18	**Visitor #1:** No.	
19	**Educator:** Thick fur is a structure that sea otters have that is an adaptation to survive in cold water. An adaptation is a structure or behavior that helps an organism survive. Their thick fur traps lots of air next to their skin. The air insulates their skin from the cold water. But that's not the only adaptation they have. Another adaptation they have is a behavior. They eat a lot of food. If you ever get to watch real sea otters—and I hope you do, 'cause they are so cool— you'll notice that they are constantly diving for food and eating the food they catch. Have you ever seen a real live sea otter?	
20	**Visitor #2:** No, but I did see one on the *Nature Channel* and I want to see the sea otters here in the aquarium too.	
21	**Educator:** The sea otters are inside this gallery and they are very cool. Did you know they have to eat about one-quarter of their weight every day? That's like if an average 10-year-old ate about 40 to 50 sandwiches every day. Have you ever known someone who is always eating and eating, but they don't gain weight? That's what sea otters are like. They're eating all the time but they burn off the calories in the cold water.	
22	**Visitor #2:** Interesting.	
23	**Educator:** Do you know why they would have to eat so much?	
24	**Visitor #2:** I guess they're really hungry? Are they like… really fat?	

(continues)

     This page may be duplicated for educational use.

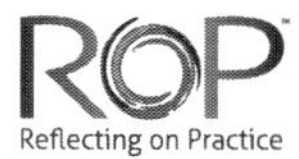

Role-Play Script #1 *(continued)*

		COMMENT	Notice/Think/Wonder
25		**Educator** *[shows photo of sea otter]:* Here's a picture of one. They only grow to be about the size of a large dog, which is small compared to other mammals that live in the ocean.	
26		**Visitor #2:** Why would living in cold water make them eat more?	
27		**Educator:** Why do you think they'd need to eat more?	
28		**Visitor #2:** When I'm cold I get really hungry. They probably get really hungry being in cold water all the time.	
29		**Visitor #1:** Yeah, and it probably makes them hungry if they're swimming around all the time.	
30		**Educator:** Well a whale, which is very big, has a low surface-to-volume ratio, right? A shrew, which is tiny, has a high surface-to-volume ratio, so it has to run around all the time so it can stay warm. And as far as marine mammals go, sea otters are small. And since sea otters are relatively small, they have a higher surface-to-volume ratio. So that's why they have to eat a lot of food.	
31		**Visitor #1:** Really? Interesting.	
32		**Visitor #2** *[to visitor #1]:* Let's go check out the sea otters.	
33		**Educator:** Okay. Come back after you check them out, and I'll tell you more about them.	
34		**Visitor #1:** OK, we'll try.	

© 2021 Regents of the University of California

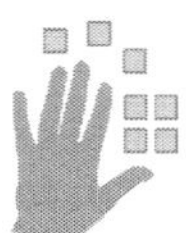

Role-Play Script #2

Scene: In the exhibit space of an informal science education institution, an educator is standing behind an activity cart with a sea otter pelt displayed. As two visitors walk by, the educator engages with them.

	COMMENT	Notice/Think/Wonder
1	**Educator:** Do you know what this is?	
2	**Visitor #2:** Fur from a bear?	
3	**Educator:** Well, it's from a marine mammal.	
4	**Visitor #1:** Is it from a walrus?	
5	**Educator:** You're getting closer. Look at the size.	
6	**Visitor #2:** It's so small; maybe it's a sea otter.	
7	**Educator:** That's right, it's a sea otter pelt. Sea otters have 155,000 hairs per square centimeter. That's more hairs in one square centimeter than people have on their whole head. Take a look. [*Hands lens to visitor #1.*]	
8	**Visitor #1:** Yeah, that's really thick!	
9	**Educator:** Yes, sea otters have really thick fur. Why do sea otters have such thick fur?	
10	**Visitor #2:** To keep them warm?	
11	**Educator:** Well it does help to keep them warm in some way, but think about your hair when it gets wet—does that help to keep you warm when you go swimming?	
12	**Visitor:** No, not really. Maybe the thick fur is for when they come out on land at night and get real cold.	
13	**Educator:** Most sea otters don't come onto land. They live in cold water in California and Alaska.	
14	**Visitor #2:** So, how do they keep warm in the ocean?	
15	**Educator:** The thick fur helps them stay warm because they put air into their fur and the thick fur traps it and creates an insulation layer that keeps them warm. What else do you think they do to stay warm?	
16	**Visitor #1:** Do they wrap themselves up in kelp to stay warm?	

(continues)

This page may be duplicated for educational use.

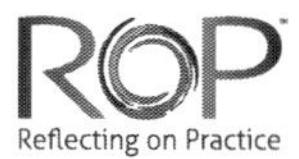

Role-Play Script #2 *(continued)*

	COMMENT	Notice/Think/Wonder
17	**Educator:** Well that's how they keep their babies from floating away. To stay warm, sea otters have to eat a lot of…. [*waits for visitor to fill in the word*]	
18	**Visitor #1:** Food.	
19	**Educator:** That's right. They eat a lot of food to stay warm. Why do they have to eat more if they live in cold water?	
20	**Visitor #2:** Maybe it has something to do with them being so small?	
21	**Educator:** That's right. As far as marine mammals go, sea otters aren't big. And since sea otters are relatively small, they have a hard time staying warm in cold water even though they have such thick fur. So they have to eat a lot of food to get more energy to stay warm. Right?	
22	**Visitor #1:** Yeah.	
23	**Educator:** In fact, they have to eat about one-quarter of their weight every day. That's like if a person weighed 100 pounds, and they ate...how much food per day?	
24	**Visitor #1:** I don't know.	
25	**Visitor #2:** 25 pounds.	
26	**Educator:** That's right. That's a lot of food to have to eat every day. They live where the water is cold, so that's why they have all these adaptations.	
27	**Visitor #1:** OK.	
28	**Educator:** Another thing to know about sea otters is that they almost went extinct. Do you know why?	
29	**Visitor #2:** Ummm, no.	
30	**Educator:** It's because sea otters have such thick fur and people wanted the nice thick fur, so they hunted sea otters almost to extinction.	
31	**Visitor #2:** Wow.	
32	**Educator:** Go take a look at the sea otters in the gallery. If you have more questions about sea otters, come and ask me.	
33	**Visitor #1:** OK. Thanks.	

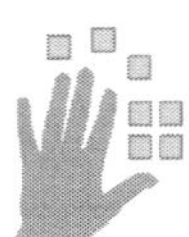

Role-Play Script #3

Scene: In the exhibit space of an informal science education institution, an educator is standing behind an activity cart with a sea otter pelt displayed. As two visitors walk by, the educator engages with them.

	COMMENT	Notice/Think/Wonder
1	**Visitor #1:** What is that?	
2	**Educator:** Come over closer and check it out. Go ahead and touch it if you want. Have you ever seen anything like this before?	
3	[*Both visitors caress the pelt.*]	
4	**Visitor #1:** It kind of feels like my cat.	
5	**Educator** [*to Visitor #2*]: What do *you* notice about this?	
6	**Visitor #2:** It's really soft and thick—way softer than my cat, and it has a long tail.	
7	**Educator:** It does feel really soft and thick to me too. This is the skin and fur of a sea otter. It's called a pelt.	
8	**Visitor #1:** Oh, yeah. I've heard of sea otters.	
9	**Educator:** What have you heard about them?	
10	**Visitor #1:** That they live in the ocean and they use tools.	
11	**Visitor #2:** I've heard that stuff too. I think they use rocks on their tummy to help them eat. And they live near here, don't they?	
12	[*Educator stays silent*]	
13	**Visitor #1:** Yeah I think they do.	
14	**Educator:** Have you ever seen them in the wild?	
15	**Visitor #2:** No, but I'd like to.	
16	**Educator:** There are some that live outside near here in the ocean, not just in the aquarium. I've seen some out in front of the aquarium—right over there. [*Points to where the visitors could try to see them later.*] Check it out when you go outside later. Did you get a chance to see them here in the aquarium? What did you notice?	
17	**Visitor #2:** They're really cute, but it seems like they're really, really itchy.	
18	**Educator** [*interested*]: What makes you think they're itchy?	

(continues)

     This page may be duplicated for educational use.

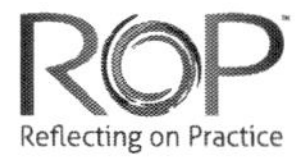

Role-Play Script #3 *(continued)*

	COMMENT	Notice/Think/Wonder
19	**Visitor #1:** They were always scratching and rolling around.	
20	**Educator:** I know what you mean, and scientists have noticed that too. Scientists have figured out a bit about why the sea otters scratch themselves all the time. Hey, would you like to try to figure out the mystery of the itchy sea otters too?	
21	**Visitors #1 and #2:** [*together*] Sure.	
22	**Educator:** OK, great. Let's try it. Here's a magnifying glass so you can take a closer look.	
23	**Visitor #1:** Whoa! There are a lot of hairs packed together.	
24	**Educator:** I know! If you look at it really closely, you can see that it's made up of a bunch of smaller hairs.	
25	[*Visitor #1 continues to look closely at the pelt.*]	
26	**Visitor #2:** Yeah, and it looks like there are different *kinds* of hairs.	
27	**Educator:** Their fur is thicker than almost any other animal.	
28	**Visitor #1:** I wonder why?	
29	**Visitor #2:** To keep them warm, maybe.	
30	**Educator** [*to Visitor 1*]: Do you agree?	
31	**Visitor #1:** Yes. Lots of other animals have thick fur to keep warm too.	
32	**Educator:** Have you been in the water near here?	
33	**Visitor #1:** Yeah, it's pretty cold.	
34	**Visitor #2:** And they're out there in the water all the time.	
35	**Educator:** One thing I can tell you about fur is that it doesn't work very well to keep an animal warm in water. Think about your hair when it gets wet—it loses the ability to keep you warm anymore, right? Same thing with a sea otter. So, here's the mystery: how do you think their thick fur helps to keep them warm? Remember the things you noticed before about the sea otters.	
36	**Visitor #2:** When the otter was diving, I saw lots of air bubbles.	
37	**Visitor #1:** …and remember the way it was always scratching.	
38	**Visitor #2:** But how can scratching and air bubbles keep the otter warm?	

(continues)

This page may be duplicated for educational use.   © 2021 Regents of the University of California ■ **197**

Role-Play Script #3 *(continued)*

	COMMENT	Notice/Think/Wonder
39	**Educator:** Let's think about it together. When the otter scratches itself, it creates a layer of air next to its skin and the thick fur helps to keep the air in there. The layer of air keeps its skin nice and dry and warm, even in the cold water.	
40	**Visitor #2:** Hey, could those air bubbles we saw coming off the sea otter come from the bubbles that are caught in between the skin and the hair?	
41	**Visitor #1:** Ooh, I bet yeah. And I bet the sea otters have to scratch all the time because it looks like the air bubbles escape a lot.	
42	**Educator:** Hm. Pretty good thinking. I think you may be on to something. Let's see what else we can figure out about sea otters. You mentioned earlier that they use rocks on their tummy to help them eat.	
43	**Visitor #1:** I saw in a book once where an otter was using rocks on a clam or something.	
44	**Visitor #2:** Yeah, they would need to use something to get through that shell.	
45	**Educator:** Let's take a look at the shells of some things they eat. Do you think they would need rocks for all of them?	
46	**Visitor #2:** Not this little mussel.	
47	**Visitor #1:** Oh, but look at this. [*Picks up abalone shell and shows it to #1 and Educator.*] This one is really thick. They would definitely need a rock for this one.	
48	**Educator:** Why don't you go take a look at the otters in the tank and see if you notice any behaviors that could give you some clues about their feeding behavior, like you did before about how they keep warm. There are a few books over here that you could look through too. Be sure to come back and let me know if you find out anything interesting.	
49	**Visitor #1:** OK.	

 This page may be duplicated for educational use.

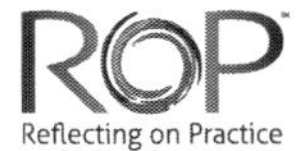

Key Ideas from the Literature: Talking to Learn

(A) Making memories changes the brain

1. Neurons make up half the volume in our brains (*1*). They receive, process, and transmit signals via chemical messengers (called neurotransmitters) and electrical impulses. These signals are sensory information (e.g., light, color, sound) from our environment that are first received by our sensory receptors (e.g., rods and cones in our eyes, hair cells in our inner ears). This information triggers neurons to fire (*2*).

   a. Initially it's like a chain reaction, one neuron firing and triggering the next, and then the next. But when that group of neurons fires again, the initial stimulus doesn't need to be present; the whole sequence can be triggered by active recall of the idea, feeling, or skill that resulted in the initial chain reaction (*2*). Repeated chain-reaction firing of that same set of neurons eventually becomes permanent and simultaneous; those neurons will fire together whenever they're activated (Figure 1).

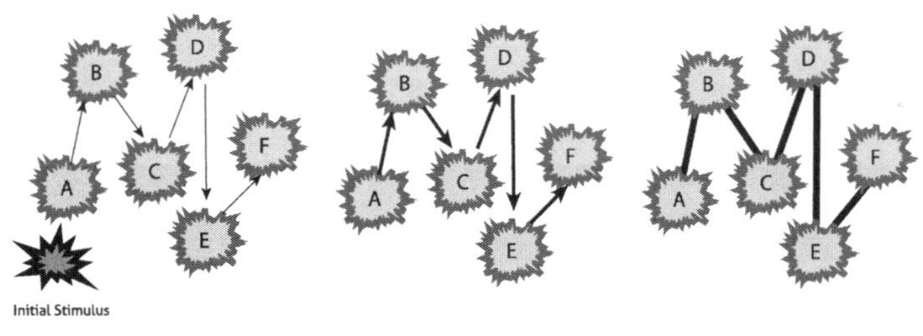

Figure 1. Repeated chain reaction firing of neurons forms memory

   b. Clusters of neurons firing together create neural pathways. The more often neural pathways are activated (through retrieval and elaboration), their connections become stronger, faster, and more durable. The human brain has approximately 100 billion neurons; each neuron can grow 10,000–15,000 dendrites that connect to other neurons (*1*). Neurons connecting to more neurons make the pathways more intricate. These connections between neurons record our learning in memory. Thus, making memories physically changes our brain. We continue to grow dendrites and connect neurons throughout our lifetime, which makes our ability to learn a lifelong process.

   c. These neural pathways are analogous to hiking trails: trails not used regularly become very faint while trails used regularly are distinct and can become permanent. Without frequent use, even a permanent trail can be overgrown by vegetation, although it can be recovered. It's important to keep in mind that not all well-worn "trails" are sensible. Conceptual change can be difficult precisely because we have to weaken neural connections that are triggered automatically, while simultaneously forging new connections. The erroneous memories in long-term storage might never be completely ameliorated (3). Those with more elaborate memories will have more connections that need to weaken and change.

(continues)

Key Ideas from the Literature: Talking to Learn *(continued)*

2. The model in Figure 2 offers a big-picture overview of how new sensory information is processed and integrated with what a person already knows (*2*). Though this model focuses on cognition, memories from social interactions and emotions can be triggered and affect how new information is processed.

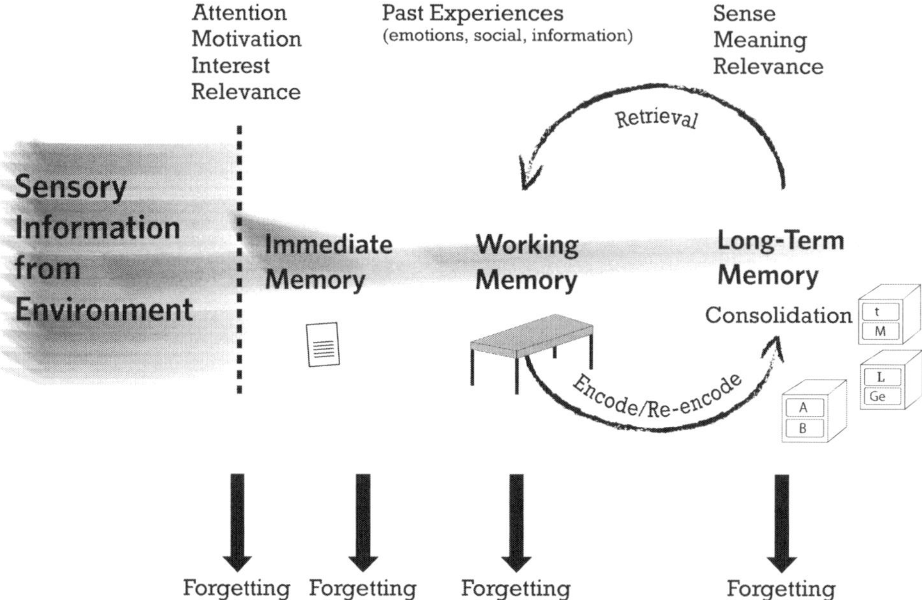

Figure 2. Big picture overview of processing sensory information with existing memory.

a. **Sensory input.** Our sensory receptors receive hundreds of inputs from our environment at any one time. While we may receive all these bits of data at once, our brain doesn't ascribe equal importance to all the information. Within milliseconds of receiving these data, parts of the brain involved in filtering sensory information (e.g., thalamus) screens them depending on their importance for the individual; that is, the extent to which the information is relevant and/or meaningful, as informed by past experiences. In the model, the dashed line represents the filtering of sensory information such that some, but not all, information passes through.

b. **Immediate memory to Working memory.** Sensory data that make it through are passed to the processing areas of the cortex. This information is immediate memory that lingers for up to 30 seconds. In the model, a note page represents immediate memory—a place where we put information briefly until we do something with it. If nothing is done with it, the information fades away. If the information captures our focus and demands our attention, it has made it to the table in the model, which represents working memory.

(continues)

  © 2021 Regents of the University of California   This page may be duplicated for educational use.

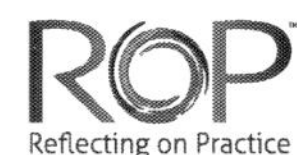

3.1

Key Ideas from the Literature: Talking to Learn *(continued)*

c. **Working memory to Long-term storage.** Working memory handles data received from immediate memory and retrieved from long-term memory. At the table of working memory, we build, take apart, and rework existing connections and integrate them with new information for eventual storage in different parts of the brain. There is, however, limited "space" on the table. Our working memory has the capacity to hold about 5–9 items at a time, which diminishes to about 2–3 when we process the items (*4*). These numbers vary by individual, age, and type of information (visual/spatial, verbal/auditory) (*2, 5*). Chunking the information (e.g., FBI-NSA-JFK-FDR versus F-B-I-N-S-A-J-F-K-F-D-R) takes advantage of preexisting information in long-term memory so that we can more efficiently hold new information (*6*). For cognitive performance, there are many additional factors to consider in addition to storage space (*7*). We must be able to control our attention in order to hold and process relevant items in the face of internal and external distractions, irrelevant information, and lapses of attention (*8, 9*). We need to be able to search and retrieve pertinent information stored in long-term memory in the presence of irrelevant or unfamiliar information (*9, 10*). Not everything processed in working memory becomes long-term memory—that is, encoded, consolidated, and stored long-term for future retrieval and use. Information that has high survival value and/or strong emotions (positive or negative) is highly memorable (*2*).

d. **Long-term storage.** We experience the greatest loss of newly acquired information and skills within the first 18–24 hours of receiving the data in working memory. Connections between neurons are strengthened with activation. The implication for learning is that, without dedicated time for retrieval and reconsolidation (repetition, use, elaboration) of memories, information and skills aren't retained in long-term storage.

(B) Talking makes thinking visible

1. Human thinking is shaped by the social activities, such as discourse, and use of the materials and symbols invented by our cultures, which in turn are temporally and geographically influenced (*11*). Vygotsky (*12*) further detailed the importance of discourse by arguing that higher mental functions have social origins that are first expressed between individuals before they are internalized within the individual. In other words, meanings are rehearsed and made explicit as a result of conversations and interactions between people before transforming into the individual's thoughts. From this sociocultural viewpoint, learning relies on conversation. For learners, engaging in conversations can foster more generative thinking and enable them to practice dialogic skills, such as asking questions and communicating ideas in an effective manner (*13, 14*). These thinking and dialogic skills form the basis of active, analytic, individual thought, and allow learners to develop their ability to communicate their ideas. For educators,

(continues)

 © 2021 Regents of the University of California ▪ **201**

Key Ideas from the Literature: Talking to Learn *(continued)*

talk from learners offers valuable information into their learners' minds, offering a glimpse of how their learners make sense of the new information in light of what learners already know (*15*).

2. The evidence for the value of *talking* on learning is overwhelming. Students show greater understanding when they engage in collaborative dialogue with peers where they provide explanations as part of arguments and justifications, and seeking and providing help (*16, 17*). Students who are given the opportunity to talk, argue, and defend their ideas in small groups have shown positive change in their understanding of difficult and complex concepts, like evaporation (*18*) and climate change (*19*). In a study of university students in a genetics class, researchers found that peer discussion enhanced understanding, even for those students who started the discussion being wrong (*20*). Laboratory studies with university students also supported this finding (*21*). Students demonstrated deeper understanding of the concept themselves, including by having to make inferences about the concept to solve a problem when tasked to explain the concept to someone else, compared to those who heard or read the explanation. Thus, it is the act of speaking one's understanding out loud to someone (and to oneself) that produces the greatest learning gains, long-term retention, and transfer of learning.

3. Externalizing and articulating are *joint processes of expressing one's unformed ideas* while developing understanding (*22*), and is necessary throughout the process of learning. **Externalizing** is taking the ideas out of one's head and communicating them in any form (e.g., talking, writing, gesturing, drawing, etc.), and **articulating** is putting those ideas into words. These expressions push learners to put forward how they're thinking about ideas (the relationships) for themselves and others to consider, regardless of whether those connections are ill-formed or well-organized. The construction of an explanation (such as through reflective writing and talking out loud) requires activating existing neural connections (retrieval) and then rearranging and connecting them with new information (elaboration). This articulation of what they think, reason, predict, and assume about concepts and ideas is helpful for learners because it requires a great deal of active, in-depth processing of the material to formulate and refine their understanding. Limitations in working memory processing capacity (*4, 5*) suggest that externalizing alleviates cognitive load on that capacity, allowing learners to manipulate and connect complex ideas more easily.

4. Talking is a way for learners to make neural connections more memorable. *Talking out loud* obliges learners to engage the motor, auditory, and **self-referential** processing areas of their brain, which in turn, makes what is said more memorable (*23*). The memory of what is voiced by oneself ("I said that") is more distinct than reading it silently or hearing someone speak the words. *Talking with others* in a conversation has a **social interaction** component to the act of voicing one's thinking that is of value to learning. When engaging in tasks that involve

(continues)

     This page may be duplicated for educational use.

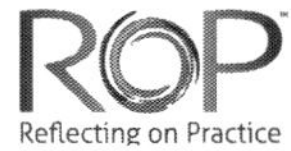

Key Ideas from the Literature: Talking to Learn *(continued)*

interacting with others, regions of our brain associated with reward, attention, and social cognition (thinking about others) become activated (*24, 25*). Thus, it's possible that giving learners the opportunity to converse with their peers as a part of the learning activity triggers reward responses, cues them to be attentive, and prompts them to envision and anticipate what the other person is thinking. All of which would make the learning activity more meaningful and memorable.

(C) Teaching purpose informs how to facilitate conversation

The idea that "talking is learning" necessitates a shift in thinking from both the learner's and educator's perspectives. It requires us to reconsider statements like "I know when I understand something because I can explain it to someone else," or "if you can't explain something, it means you didn't understand it." There is truth to these comments. However, their underlying sentiment is that talking out loud is reserved for when the ideas are fully understood, when in fact, talking with others helps us unravel the uncertainties in our understanding. Equating explanations to understanding leads educators to listen for right answers and focus on correcting wrong ones; in turn, this makes learners reluctant to speak unless they have "figured it out" in their heads. **Talking is learning** means that talking out loud is not just a representation of having learned, but importantly, learning is happening in that moment. The opportunity to express their ill-formed and incomplete understanding enables learners to retrieve a variety of memories, hear their own thinking, and consider different connections (*23*). Meanwhile, the social interaction triggers reward responses that can spark motivation and attentiveness (*24*). Thus, being learner-driven in teaching and when designing learning experiences requires educators to be purposeful in making discourse opportunities available for learners.

Teaching purpose refers to the instructional reasons for different segments (or parts) in a designed learning experience. The teaching purpose drives the educator's moves as the educator guides learners along their journey towards the learning goal. The following are five possible teaching purposes that inform how to facilitate learning conversations towards science understanding. Learning experiences are more effective when they comprise multiple teaching purposes.

1. **Activate Prior Connections**
   The purpose is for learners to retrieve memories that might be relevant to the ideas and phenomena about to be explored, bringing them forward for consideration and consolidation with new memories (*26*). **Retrieval** is deliberately recalling information from learners' prior knowledge and experiences (*27*), which urges them to pull out knowledge from long-term memory for closer examination in working memory. The more often learners summon their previous memories, the more often those neural connections are activated. In turn, those neural connections are strengthened, and retrieved more easily next time. Retrieval strategies include quick write, self-test, exit ticket, and think-pair-share. These are moments for productive

(continues)

Key Ideas from the Literature: Talking to Learn *(continued)*

struggle. Learners are forced to stop and consider, "what do I know about this" before they confront the flaws and gaps in what they have stored in long-term memory. Providing feedback with detailed examples following retrieval leads to greater learning gains because it offers learners the chance to examine and adjust the memories they just retrieved (*28*).

In learning conversations, learners are invited to call to mind past experiences triggered by the new situation. In recalling what they know, learners have the opportunity to draw from their strengths, which can contribute to their comfort and confidence in approaching new content (*29*, *30*). Activating learners' prior connections gives learners the chance to determine how the new learning can be meaningful to them.

2. **Give Information**
The purpose is for the educator to convey details, such as the scientific argument or explanation, that are relevant to the experience for learners to consider. Learners may not have all the information they need to understand what is to be learned, or they may not know how the knowledge and ways of knowing they already possess are pertinent. The educator pulls these details to the forefront for learners in this moment of learning.

In learning conversations with this purpose, the educator typically tells learners these details. For instance, the educator communicates a series of logical connections between various ideas, terms, and tasks, or gives a narrative account of a set of events and actions that establishes chronological and causal relations among them (*31*, *32*). Giving information by telling is an efficient way of broadcasting scientific knowledge, but it doesn't ensure learners actually "get" all the information given. Space needs to be made for learners to consider and articulate their own understanding of the ideas. Information can be given in ways that encourage learners to obtain it for themselves, such as through written text (like this reading), close observation of objects, or manipulating materials.

3. **Check for Understanding**
The purpose is for the educator to extend the learner's answer, draw on its significance, or make connections with other parts of the learner's total learning experience (*14*). The educator evaluates learners' understanding about the scientific idea being taught, and makes the flaws and accuracies available for the community to consider (*15*). Equally important, though often overlooked, are helping learners use the errors to adjust their own thinking (*33*, *34*) and cultivating habits of self-monitoring and -assessment among learners (*35*).

Conversations facilitated for this purpose have a distinct 3-part pattern: the educator **initiates** the discussion with a question, learners **respond**, and then the educator **evaluates** learners' responses (*32*, *36*). This teaching purpose for facilitating conversations is a popular way to make the scientific normative view available for all learners through social interaction. It fails to

(continues)

     This page may be duplicated for educational use.

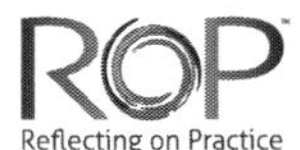

Key Ideas from the Literature: Talking to Learn *(continued)*

provide learners with an opportunity to articulate their understanding and express themselves in the language of the discipline (*13, 37*), and risks habituating learners to only talk when they know the answer. Understanding can be checked in ways that shares the responsibility with learners, such as reflective writing, opportunities for revisions, asking learners what is unclear to them, and listening in on conversations between learners.

4. **Express Your Thoughts**
    The purpose is for learners to externalize and articulate their thinking as they try to make sense of new ideas; in so doing, they are elaborating on relevant memories. **Elaboration** involves clarifying and specifying the relationship between information-to-be-learned and related information from a learner's prior knowledge and experience (*38*). It is "encoding the original content in a different but related way" (*38*). Elaboration strategies include paraphrasing and summarizing the material to be learned, creating analogies, adding details, reorganizing and making connections between ideas while note taking, explaining ideas to someone else, and asking and answering questions (*39*). "Elaboration leads to deep levels of information processing, and is assumed to inhibit forgetting, because it produces a richer, more redundant memory structure" (*17*).

    In conversations, learners can be prompted to verbalize their own thoughts, questions, and connections, with peers and educators. The educator acknowledges and restates learners' comments; encourages learners to think about and elaborate on their ideas; urges learners to contemplate one another's thinking; and remains silent to offer learners time to think about what was said (*40, 41*). Expressing learners' thoughts enables them to voice their scientific views of the world in everyday language, with the educator available to make connections between these personal and scientifically normative views (*23, 41*). The educator, also, needn't be a part of the conversation for talking to be of value to learners. It takes time, persistence, and patience for learners to reflect on their thinking and speak their thoughts aloud, and both educator and learners need to listen carefully to understand what's said.

5. **Transfer Understanding**
    The purpose is for learners to use their new knowledge to strengthen new neural connections (2). Without further retrieval and use, newly acquired information and skills are easily lost.

    Learners are tasked to consider the new ideas in novel situations, or to apply new skills on their own. The educator is available to reassure learners and offer assistance as needed; the emphasis is for the learners to use the new language and associations for themselves. Learners take ownership of the new knowledge introduced by the educator. Transferring understanding gives learners the chance to make the new learning relevant to them.

(continues)

Key Ideas from the Literature: Talking to Learn *(continued)*

References

1. L. Cozolino, *The Social Neuroscience of Education: Optimizing Attachment and Learning in the Classroom*. (Norton, Inc., New York, NY, 2013).

2. D. A. Sousa, *How the Brain Learns*. (Corwin Press, Thousand Oaks, CA, 2016).

3. A. R. Vaughn, R. D. Brown, and M. L. Johnson, Understanding conceptual change and science learning through educational neuroscience. *Mind, Brain, and Education* **14**, 82-93 (2020).

4. G. A. Miller, The magical number seven, plus or minus two: Some limits on our capacity for processing information. *Psychological Review* **63**, 81 (1956).

5. A. Baddeley, Working memory. *Current Biology* **20**, R136-R140 (2010).

6. T. Drew, E. Vogel, in *Encyclopedia of Neuroscience*, L. R. Squire, Ed. (Elsevier Ltd, ebook, 2009), pp. 523-531.

7. N. Unsworth, G. J. Spillers, Working memory capacity: Attention control, secondary memory, or both? A direct test of the dual-component model. *Journal of Memory and Language* **62**, 392-406 (2010).

8. R. W. Engle, M. J. Kane, Executive attention, working memory capacity, and a two-factor theory of cognitive control. *Psychology of Learning and Motivation* **44**, 145-199 (2004).

9. N. Unsworth, R. W. Engle, The nature of individual differences in working memory capacity: active maintenance in primary memory and controlled search from secondary memory. *Psychological Review* **114**, 104 (2007).

10. J. A. Mogle, B. J. Lovett, R. S. Stawski, M. J. Sliwinski, What's so special about working memory? An examination of the relationships among working memory, secondary memory, and fluid intelligence. *Psychological Science* **19**, 1071-1077 (2008).

11. L. Schauble, G. Leinhardt, L. M. W. Martin, A framework for organizing a cumulative research agenda in informal learning contexts. *Journal of Museum Education* **22**, 3-8 (1997).

12. L. Vygotsky, *Mind in Society: The Development of Higher Psychological Processes*. (Harvard University Press, Cambridge, MA, 1978).

13. R. Alexander, *Towards Dialogic Teaching*. (Dialogos, York, 2005).

14. G. Wells, *Dialogic Inquiry: Towards a Sociocultural Practice and Theory of Education*. (Cambridge University Press, New York, 1999).

15. P. H. Scott, Teacher talk and meaning making in science classrooms: A Vygotskian analysis and review. *Studies in Science Education* **32**, 45-80 (1998).

16. N. Mercer, L. Dawes, R. Wegerif, C. Sams, Reasoning as a scientist: Ways of helping children to use language to learn science. *British Educational Research Journal* **30**, 359-377 (2004).

17. F. M. van Blankenstein, D. H. J. M. Dolmans, C. P. M. van der Vleuten, H. G. Schmidt, Which cognitive processes support learning during small-group discussion? The role of providing explanations and listening to others. *Instructional Science* **39**, 189-204 (2011).

18. R. Tytler, S. Peterson, Deconstructing learning in science: Young children's responses to a classroom sequence on evaporation. *Research in Science Education* **30**, 339-355 (2000).

19. L. Mason, M. Santi, Discussing the greenhouse effect: Children's collaborative discourse reasoning and conceptual change. *Environmental Education Research* **4**, 67-85 (1998).

20. M. K. Smith et al., Why peer discussion improves student performance on in-class concept questions. *Science* **323**, 122-124 (2009).

     This page may be duplicated for educational use.

21. E. B. Coleman, A. L. Brown, I. D. Rivkin, The effect of instructional explanations on learning from scientific texts. *The Journal of the Learning Sciences* **6**, 347-365 (1997).

22. R. K. Sawyer, in *The Cambridge Handbook of the Learning Sciences*, R. K. Sawyer, Ed. (Cambridge University Press, New York, NY, 2006), chap. 1, pp. 1-18.

23. N. D. Forrin, C. M. MacLeod, This time it's personal: the memory benefit of hearing oneself. *Memory* **26**, 574-579 (2018).

24. M. D. Lieberman, Social: *Why our Brains are Wired to Connect.* (Broadway Books, New York, NY, 2013).

25. R. Hari, L. Henriksson, S. Malinen, L. Parkkonen, Centrality of social interaction in human brain function. *Neuron* **88**, 181-193 (2015).

26. J. W. Antony, C. S. Ferreira, K. A. Norman, M. Wimber, Retrieval as a fast route to memory consolidation. *Trends in Cognitive Sciences* **21**, 573-576 (2017).

27. P. K. Agarwal, H. L. Roediger III, M. A. McDaniel, K. B. McDermott, in *http://pdf.retrievalpractice.org/ RetrievalPracticeGuide.pdf.* (Washington University in St. Louis, 2018).

28. B. Finn, R. Thomas, K. A. Rawson, Learning more from feedback: Elaborating feedback with examples enhances concept learning. *Learning and Instruction* **54**, 104-113 (2018).

29. L. C. Moll, C. Amanti, D. Neff, N. Gonzalez, Funds of knowledge for teaching: Using a qualitative approach to connect homes and classrooms. *Theory into Practice* **31**, 132-141 (1992).

30. J. Andrews, W. C. Yee, Children's 'funds of knowledge' and their real life activities: Two minority ethnic children learning in out-of-school contexts in the UK. *Educational Review* **58**, 435-449 (2006).

31. D. Edwards, N. Mercer, *Common Knowledge: The Development of Understanding in the Classroom.* (Methuen, London, UK, 1987).

32. J. L. Lemke, *Talking Science: Language, Learning, and Values.* (Ablex Publishing, Westport, CT, 1990).

33. P. Black, D. Wiliam, Inside the black box: Raising standards through classroom assessment. *Phi Delta Kappan* **8**, 139-144 (1998).

34. B. Bell, B. Cowie, The characteristics of formative assessment in science education. *Science Education* **85**, 536-553 (2001).

35. P. R. Pintrich, C. A. Wolters, G. P. Baxter, in *Issues in the Measurement of Cognition*, G. Schraw, J. C. Impara, Eds. (Buros Institute of Mental Measurements, Lincoln, NE, 2000), chap. 2, pp. 43-97.

36. H. Mehan, *Learning Lessons: Social Organizations in the Classroom.* (Harvard University Press, Cambridge, MA, 1979).

37. J. J. Wellington, J. F. Osborne, *Language and Literacy in Science Education.* (Open University Press, Buckingham, 2001).

38. R. J. Hamilton, Effects of three types of elaboration on learning concepts from text. *Contemporary Educational Psychology* **22**, 299-318 (1997).

39. C. E. Weinstein, R. E. Mayer, in *Handbook of Research on Teaching*, M. C. Wittrock, Ed. (Macmillan, New York, 1986), pp. 315-327.

40. E. H. van Zee, J. Minstrell, Reflective discourse: Developing shared understandings in a physics classroom. *Intl. J. Sci. Educ.* **19**, 209-228 (1997).

41. P. H. Scott, E. F. Mortimer, O. G. Aguiar, The tension between authoritative and dialogic discourse: A fundamental characteristic of meaning making interactions in high school science lessons. *Science Education* **90**, 605-631 (2006).

© 2021 Regents of the University of California

Module 3.
Session 2: Facilitating Conversations

Session Overview

In this session, educators consider the nuances in facilitating conversations, allowing learners to express their everyday ideas while learning about the scientific views. Participants read and discuss ways to approach the **tension between everyday ideas and scientific views, how power relations affect learning**, and **the nature of family conversations in informal science environments**.

Participants engage in a hands-on demonstration designed to set up the opportunity to analyze Educator Moves focused on **teaching purposes** and **approaches to facilitating discussions**. They apply their understanding of facilitation approaches to their practice, thinking about their teaching purposes and how they'll allow for the sharing of multiple ideas while guiding learners to a more accurate understanding of the scientific idea.

Session Objectives

- Understand the tension between everyday thinking and scientific views, and how educators can guide discussion to help learners connect their ideas with the scientific view.

- Discuss how explicit teaching purposes and consequent facilitation approaches influence how educators teach and communicate science to their learners.

- Consider the influence of power dynamics and family conversations in informal learning environments.

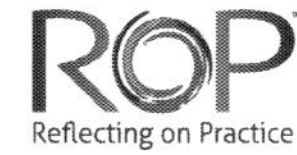

SESSION AGENDA		
Task Routine	Description	**Estimated Time** (in minutes)
Introduction *Session Objectives*	An overview of Session 2 and its objectives is presented.	2
Retrieve & Connect Quick Write: *Using the Discussion Map*	Participants respond to two prompts in a Quick Write about applying the Discussion Map to make room for learners to share diverse ideas. The quick write is followed by a whole-group sharing of ideas and experiences.	10
Research Discussion Expert Jigsaw: *Tension, Power, and Family*	Participants read and discuss three key aspects of facilitating discussions: Tension between Everyday and Scientific Views, Power & Scaffolding, and Family Conversations in Informal Environments.	30

Hands-on
Two Balloons
	Participants engage in a hands-on activity designed to model how educators can strategically address the key ideas on facilitating discussions so that learners are guided to a deeper understanding of a phenomenon. The Facilitation Approaches Design and Teaching Tool is introduced.
	    A. *Invitation Phase*
	    B. *Exploration Phase*
	    C. *Concept Invention Phase*
	    D. *Application Phase*
	    E. *Reflection Phase*
		55
		
		(5)
		(8)
		(15)
		(15)
		(7)

Break option		10
From Learners' Perspective Turn & Talk: *Learning from Two Balloons*	Pairs use the *Two Balloons* activity as a common reference point to understand facilitation approaches, and how they are interrelated with Teaching Purposes and the Discussion Map.	20
Let's Talk Practice Micro Lab: *Using New Tools*	Participants write about and work in groups of three to discuss application of ideas from this module to their practice.	20
Continue the Learning	Participants are assigned a Reflection Exercise in preparation for the next session.	3
	Total Estimated Time	**150 mins. (2.5 hrs.)**

3.2

MATERIALS

Recurring for all Sessions

- See Module 1 Session 1

For this Session

- 6 balloons (See **Getting Ready** #5)
- Basin to place under the water balloon
- Access to water faucet to fill balloons
- 1 lighter
- **Science Content Guide for Two Balloons**
- Access to the Internet for *Heat Reservoir* animation, and classroom response platform (if using)
- (Optional) Online classroom response platform that has open-response option (e.g., www.polleverywhere.com, padlet.com, www.socrative.com)

For each small group

- 1 set of **Key Ideas from the Literature: Tension, Power, and Family** (See **Getting Ready** #3a.)

For each participant

- Handouts (1 copy each)
    - **The Ocean: A Giant Heat Reservoir**
    - **Teaching Two Balloons**
    - **Key Ideas from the Literature: Tension, Power, and Family**
    - **Facilitation Approaches** (Appendix B)
    - **Teaching Purposes** (Appendix B) for replacement copies

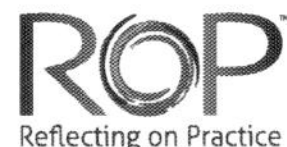

Getting Ready

1. Consider sending out the handout **Key Ideas from the Literature: Tension, Power, and Family** a day or two before the session so that participants have the option to review the reading material ahead of time.

2. Review the handouts and slide deck for this session, and additional suggested readings.

   For the facilitator:

   - *Taking Science to School*
     - Chapter 6: "Understanding How Scientific Knowledge Is Constructed"
     - Chapter 7: "Participation in Scientific Practices and Discourse"

   Continue the Learning suggestion for participants:

   - *Learning Science in Informal Environments* (Chapter 4: "Everyday Settings and Family Activities")

3. Duplicate handouts.
   For the Research Discussion, **Key Ideas from the Literature: Tension, Power, and Family**.

   a. Make a single-sided copy of the entire research discussion for each team of three to distribute among the topic experts. Don't staple this copy as the team will distribute the pages between them. (Reference list doesn't need to be included; these sets can be reused.)

   b. Duplicate and staple one copy of the entire research discussion for participants to take away.

4. Review how Sessions 1 and 2 of this module are connected through design with the Learning Cycle.

   a. This session digs deeper into content on facilitating learning conversations that began in Session 1 of this module. That first session serves as the *Invitation* and *Exploration Phases* for this session.

   b. The **Retrieve & Connect** task focusing on the Discussion Map invites participants to recall their experiences applying concepts from Session 1, as a means to prime them for new information

This is an example of how learning cycles can be interconnected across learning experiences. Consider sharing this information with the community, as appropriate.

in this session. The **Research Discussion** (*Concept Invention Phase*) offers concepts to expand on their experiences.

c. The **Hands-on** Two Balloons activity is the *Application Phase* for the focal content on facilitating conversations. During the activity, the facilitator will need to listen closely to understand what prior knowledge participants recall (the condition and grain sizes). Use the **Science Content Guide for Two Balloons** to aid your assessment of their prior knowledge and to facilitate the conversations to ease *tension* between everyday and science language, while being conscious of the participant's *power* in the interactions.

d. Deliberate facilitation of Two Balloons with these aspects (teaching purposes, facilitation approaches, tension, and power) in mind can help participants recognize them. **From Learners' Perspective** is an opportunity to reflect on how the concepts "feel" as the learner, which prepares participants to consider how to use what they learned in the **Let's Talk Practice**.

5. Prepare and practice the balloon demonstration.

a. Prepare balloons:

- Make two balloons (one filled with air and other with water) to about the same size; be sure the balloons are stretched out, but not distended.
- Test and practice with the air and water balloons (b. below).
- Make two more sets of air and water balloons, based on results of your test, and save them for use during the session.
- Place the lighter nearby.

b. Practice the balloon demonstration, using the trial balloons to become familiar with the procedure and test the material:

- Hold the air balloon in one hand and the lighter in the other hand.
- Touch the flame to the balloon *directly under the bottom of the balloon*, not on the side of the balloon. It should quickly pop. (If the balloon isn't inflated enough, it will take a few seconds to pop; consider inflating the balloon more for a more dramatic and straightforward result.)
- Do the same with the water balloon, touching the flame to the bottom of the balloon for about 30 seconds (or even more—we've done this for up to 3 minutes!). It will *not* pop.

To visualize the demonstration, watch ~1 minute of the video, *Oceans of Climate Change*, beginning at 1 minute, 13 seconds at climate.nasa.gov/climate_resources/40/.

Optional. Have extra balloons available, in case there's interest to test particular conditions from questions participants raise. Alternatively, have multiple set-ups for participants to do the activity in small groups.

Regardless, try not to get so distracted by the investigations that participants run out of time to make sense of the concepts.

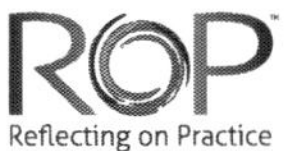

(If water drips from the balloon, the balloon material may be old or compromised; consider replacing the balloons for the demonstration.)

6. Prepare a poster-sized version of the **Teaching Two Balloons** handout, either on chart paper or on the board.

   a. One facilitator will lead the activity while the other facilitator will fill in this poster-sized chart as the actions take place.

   b. (Optional) Make printed (or hand write) labels of each teaching purpose and facilitation approach so that the appropriate item can be affixed into the chart according to the details in the handout. Be sure the font size is large enough to be read from the back of the room.

   c. Both facilitators should review and discuss the activity beforehand to agree upon when to write or affix which purpose and approach on the chart.

7. (Optional) Follow the instructions from your chosen classroom response platform to make two spaces for the community to enter their thoughts online (in the *Exploration Phase*):

   a. Share a couple of words that captures the concepts or ideas you discussed.

   b. What additional information would be helpful?

8. Prepare for Whole-Community Video Reflection in Module 3 Session 3.

   • Determine who will be the educator-presenter.

   • Make one copy of **Reflection Exercise Worksheet to Prepare for Feedback** (Appendix A) for presenter to complete in preparation for the Video Reflection. Offer assistance as needed.

9. Have a web browser open for the Heat Reservoir animation (http://static.lawrencehallofscience.org/mare-sims/1-2-heat-reservoirs.html)

An online classroom response platform is a convenient way to gather a lot of information from many participants in a short amount of time. Everyone can enter their ideas simultaneously, and the facilitator can quickly choose the entries that are appropriate for the activity.

However, participants will need to have web-enabled devices, e.g., smart phone, mobile tablet, computer, in order to participate. PollEv offers participants the option to text in their entry, though cellular charges will apply.

Session 2 Step by Step

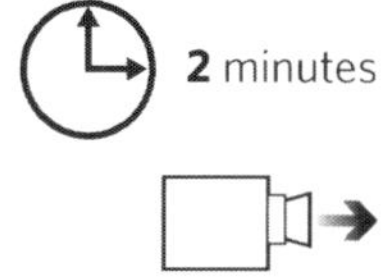
2 minutes

Introduction
Session Recall and Objectives

1. **Harken back to learning and talk.**
   Remind participants about ideas on learning and talk discussed in previous sessions. Display the Ideas to Emphasize:

   a. *Learning requires making new memories and retrieving relevant existing memories to connect with new information.* Making memories physically changes the brain on a cellular level.

   b. *Talking is critical for learning because it makes thinking "visible" to the learner (and educator).* We all have limitations in our ability to process and focus on information. Talking alleviates the cognitive load, pushes us to test out new connections, and makes the thinking available for feedback.

   c. *Awareness and deliberateness of our teaching purposes for facilitating conversations makes a difference in the learning experience.* Facilitating conversations for the purposes of giving information and checking for understanding can be helpful and necessary at times—but they don't make learners' thinking visible to them or the educator.

2. **Introduce the session.**
   Share that this session will dig deeper into facilitating conversations—specifically, how the different teaching purposes introduced in the last session call for the use of different facilitation approaches to support learning.

3. **Introduce and display session objectives.**

   - Understand the tension between everyday thinking and scientific views, and how educators can guide discussion to help learners connect their ideas with the scientific view.

   - Discuss how explicit teaching purposes and consequent facilitation approaches influence how educators teach and communicate science to their learners.

   - Consider the influence of power dynamics and family conversations in informal learning environments.

Quick Write: *Experiences Using the Discussion Map*

10 minutes   Retrieve & Connect

1. **Recall the Discussion Map.**
   Recall the Discussion Map from Module 3 Session 1, and that it's a tool for facilitating conversations so that learners **express their thoughts**. Remind participants that they were all urged to try the Discussion Map in their practice prior to this session.

2. **Take time for Learning Journal entries.**
   Ask participants to take three minutes to reflect on their experience using the Discussion Map and write an entry in their Learning Journals. Display the questions below.

   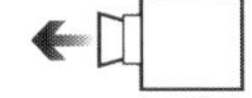

   - *How do you feel it went? What's your evidence and what did you notice?*

   - *How did your learners respond to your use of the Discussion Map?*

   If they haven't or don't use the Discussion Map in their practice, have them respond to the following prompts:

   - *How do you encourage and allow for learners to articulate their thinking?*

   - *How do you encourage and invite learners to listen closely to what others share and consider multiple viewpoints?*

3. **Hold a brief whole-group share.**
   Allow a couple of minutes for a few volunteers to share their reflections. Here are a few challenges participants may voice after first trying to use the Discussion Map:

   - Learners are reluctant to talk.

   - It takes so much time.

   - Not sure we got the correct science explanation communicated.

   Reassure them that their sentiments are understandable, and that they'll be provided even more tools and opportunities to gain confidence in guiding discussions.

It takes practice to get more skilled with facilitating learning conversations. Tools like the Discussion Map offer guidance to educators for gaining comfort and confidence in their abilities.

<table>
<tr><td>Research Discussion   30 minutes</td><td>**Expert Jigsaw:**
Tension, Power, and Family</td></tr>
</table>

1. **Introduce research discussion.**
   Acknowledge that facilitating learning conversations isn't easy. It's definitely a skill that's learned and practiced over time. Point out that the discussions so far have focused on why we need to make time for learners to talk and express their thinking, and listen and consider different viewpoints. The Discussion Map is a tool for doing this.

   Let them know they're about to do a Research Discussion that will give them more information to consider regarding the dynamics of their conversations.

Each team should have at least one topic expert. Make a team of four if the community cannot be divided evenly into teams of three.

2. **Set up Research Discussion as an Expert Jigsaw.**
   Tell participants they'll be divided up into teams of three. Each team member will take on the task of becoming more expert on a topic for their team.

3. **Divide participants into teams of three and distribute the handout.**
   Once the participants are in their teams, distribute one set of the handout **Key Ideas from the Literature: Tension, Power, and Family** per team. Explain that each person will become the topic expert for a different section of the handout for their team. Tell them to determine the topic experts and distribute that part of the reading to them, but not to start reading yet.

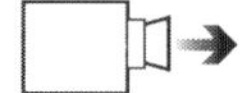

4. **Experts from teams convene to gather information.**

   a. Tell participants that topic experts from each team will now convene for 15 minutes to read and discuss their topics. After that, topic experts will return to their teams to report back what they found out. There will be no whole-group share, so team members are reliant on one another to piece together all the information.

Each big idea includes one or two discussion questions to guide topic expert groups to think and talk about the implication of the topic. Facilitators should join the topic conversations to address questions and concerns. If there are fewer than three facilitators, then circulate among the three topics. Power is a topic that may need support from a facilitator.

   b. Display instructions and discussion prompts:

   ❑ *Discuss the meanings of the ideas and discussion question(s) included in your topic.*

   ❑ *Recall instances from your experience.*

   ❑ *Generate implications and applications for your own practice.*

   ❑ *Be ready to share these discussion points with your team.*

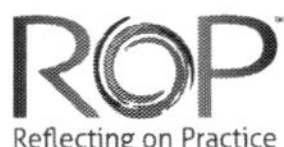

5. **Display discussion prompts for team conversations.**
   Have topic experts return to their teams for reporting out and group conversation. Tell them they have another 15 minutes. Display the prompts as a guide.

   - *What is the idea about?*

   - *What are some instances from your experience?*

   - *What are implications and applications for our practice?*

   Remind participants that, *as they listen,* they should pay attention to and call out connections between ideas, questions that emerge, and challenges and concerns for practice.

6. **Circulate and listen during team conversations.**
   Pay particular attention to corollary ideas voiced by participants during their team conversations.

   - **Tension** comes up as participants talk about heat and temperature during the animation (Part (C), Step 4 of Two Balloons). Participants are asked to describe what they see, and the educator directs and guides their everyday language towards the scientific terms and explanations.

   - **Power** comes up when the facilitator joins or eavesdrops on a small-group conversation or leads a whole-group conversation. The facilitator's presence inevitably affects the dynamics of the situation.

   - **Family** comes up throughout as the facilitator insists that participants talk with each other, rather than have an extensive exchange with the facilitator. While participants are not family groups, the point is that more learners are able and likely to express their thoughts when they're talking with familiar people.

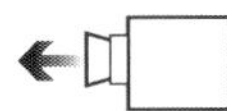

There is no whole-group discussion. The facilitation team should circulate and eavesdrop on team conversations. Take note of ideas and themes in these conversations to mention in the **From Learners' Perspective** conversation as you connect the **Research Discussion** here with the **Hands-on** task coming up next.

Two Balloons

55 minutes Hands-on

1. **Introduce the activity.**
   Explain that this next hands-on activity will provide a shared experience from which to reflect further about facilitating learning conversations. Alert participants that as they engage in the activity, they should be conscious of two aspects of the facilitation.

   a. Instances in which the ideas just brought up in the Research Discussion emerge and are attended to.

   b. How the teaching purpose shifts as the activity progresses, and as a result, the facilitation approach changes as the educator uses the Discussion Map.

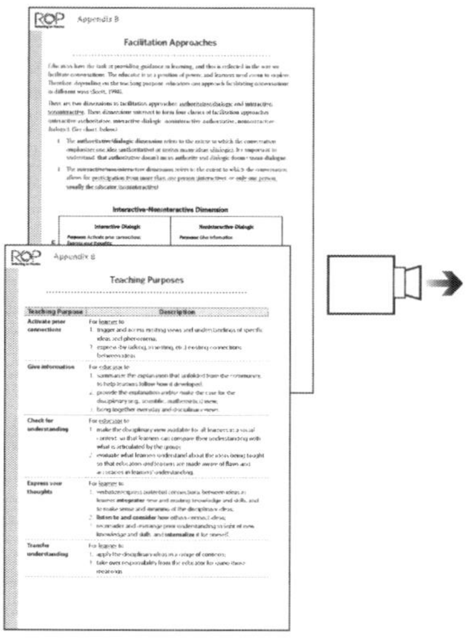

The second facilitator should use the **Teaching Two Balloons** handout as a "cheat sheet" for completing the chart. The facilitator will have to pay close attention to what's happening in order to fill in the chart in real-time.

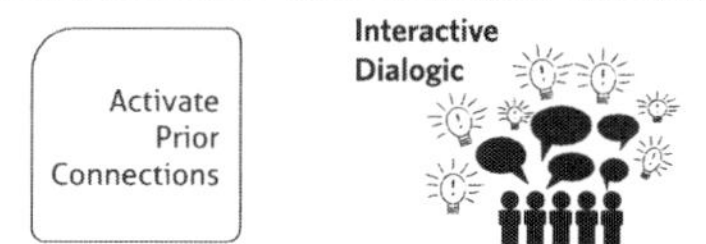

There will be recurring opportunities for participants to recall and express their thinking, and to articulate and test connections privately in writing and in small group discussions. This back and forth is intended to alleviate anxiety of "not knowing the answer" by giving participants space to reflect and prepare to contribute their ideas.

2. **Introduce "Facilitation Approaches" handout and revisit "Teaching Purposes."**

   a. Distribute the **Facilitation Approaches** and **Teaching Purposes** handouts. Remind participants that the **Teaching Purposes** were introduced in Module 3 Session 1.

   b. Display and review aloud the **Facilitation Approaches**. Tell participants that you'll be modeling these approaches in this activity. For now, they should skim the handout to get a sense of the four approaches.

   c. Give participants 2-3 minutes to skim the two handouts; allow them to talk with a partner if they desire.

3. **Being mindful of the Purpose and Approach.**
   Direct participants' attention to the Teaching Two Balloons chart posted on the wall. Explain the following:

   a. The chart has three columns (Action, Teaching Purpose, and Facilitation Approach) that will document particular events in the activity.

   b. The other facilitator will fill in the chart as the activity progresses, and will call attention to it when appropriate. The participant's task is to be mindful of what's happening for them from the learners' perspective as the teaching purpose and facilitation approach are called out.

(A) Two Balloons: Invitation Phase (5 minutes)

1. **Introduce balloon demonstration.**
   Show participants one air-filled and one water-filled balloon and describe how you'll use a lighter to add heat energy to the balloons by applying a flame directly to the balloon. Share the following:

   a. The two balloons are made of the same material.

   b. The water balloon is filled with water from the tap earlier in the day. The air balloon is filled with air blown into it by one of the facilitators. Both balloons are at room temperature.

2. **Display Thought Experiment: What do you think will happen to each of the balloons?**
   Ask them to first think about their predictions and then record their ideas in their Learning Journals, being sure to include the reasoning for their prediction. Share that you anticipate that some people may know the explanation well, and others will have bits and pieces of an explanation. Assure them that they won't show anyone what they

write down, so they should write freely and their reasoning doesn't need to be in complete sentences. Tell them they'll have about a minute. Don't do a whole-group share-out of their ideas yet.

3. **Begin the demonstration.**

   a. Hold up the air-filled balloon and ask for a show of hands—how many think it will pop? How many think it won't? Apply the lighter flame directly to the underside of the balloon. [The balloon will pop immediately]

   b. Hold up a water-filled balloon and place the basin underneath. Urge participants to move back a little.

   c. Tell them you'll use a lighter to add heat energy to the balloon as before. With a show of hands, how many think it will pop? How many think it won't? Hold the lighter flame directly to the underside of the balloon for 30 seconds or more. [The balloon won't pop]

4. **Initiate Turn & Talk.**
   Display the question prompt:

   Why did one balloon pop and the other one not?

   Have participants share their ideas and reasoning with a partner for a couple of minutes.

5. **Eavesdrop throughout the activity.**
   When participants talk amongst themselves, circulate and listen with curiosity to understand what prior knowledge participants are using to help them make sense of this investigation. Use the **Science Content Guide for Two Balloons** to assist your assessment and guidance. Similar to the Ice Cubes Investigation, the prior knowledge participants typically recall can be organized into four themes:

   - Familiar experiences

   - Equations and technical language

   - Overlooking evidence from observations

   - Distracted by emotional memories and social dynamics

   Listen to understand the **condition** of their prior knowledge (missing, incomplete, misconceived), which can vary in their accuracy. The inaccuracies can be at different **grain sizes** (belief, mental model, category).

Using a basin and encouraging people to move back adds drama and keeps the explosion a viable option, even though the water balloon will not pop.

Participants with expertise in this content might share a detailed explanation of the phenomenon with their partner. Don't worry; it will set up a discussion about how the expert's explanation influenced their partners' sense-making.

3.2

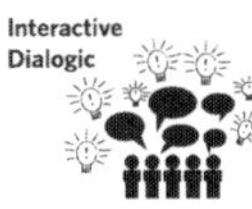
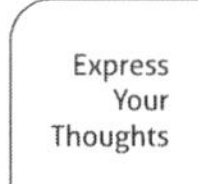

Encourage many people to voice the many ideas they may be thinking for the community to consider. Constraining the sharing to "a couple of words" prevents any one idea or voice from dominating the conversation at this point.

Asking participants for additional information urges them to retrieve relevant memories and recognize the gaps, and signals to them that they aren't expected to know everything already.

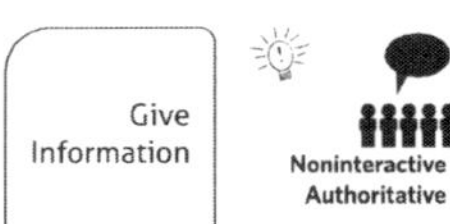

Participants are now primed to receive information. There will be a lot of information to give, with learners having varying levels of conceptual understanding.

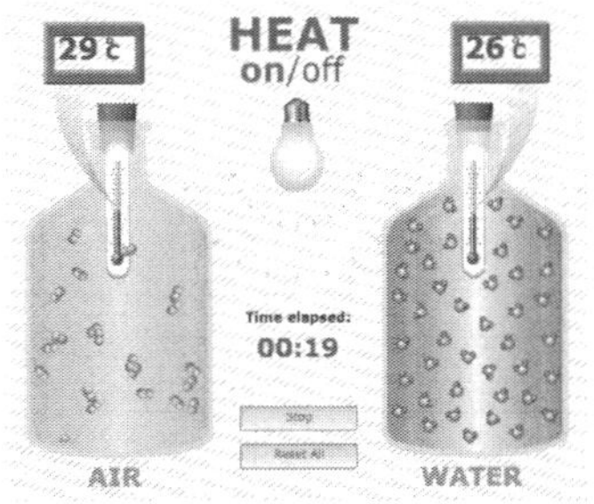

(B) Two Balloons: Exploration Phase (8 minutes)

1. **Briefly elicit more ideas.**

   Invite participants to share a couple of words that captures the concepts or ideas they discussed, either verbally and record their contributions on the board, or using the classroom response platform your team chose. Stay neutral as ideas are shared. Don't provide an explanation, but rather encourage participants to share ideas.

2. **Turn & Talk with a different partner.**

   Display the following prompts:

   - *What are you unsure about in your understanding to explain what happened?*

   - *What additional information would be helpful to make an explanation for the balloon phenomenon?*

   Have participants turn to a different partner to consider the ideas brought up so far. They should share their initial prediction and reasoning, and discuss how their thinking may be changing based on what others think. If no change, then explain why not. Perhaps they disagree with ideas put forth, or don't understand the reasoning or explanation. They should also discuss what additional information would be helpful to explain the phenomenon.

3. **Gather requests for more information.**

   Invite participants to share what information they would like to know, posed in question form, that would help them refine their explanation. Either record their questions on chart paper, or have participants submit them to the classroom response platform. Consider whether there are relevant ideas you overheard that should be shared with the group. As they request information, ask them to clarify how they think the additional information will help them explain what happened. (See **Science Content Guide for Two Balloons** for potential probing questions.)

(C) Two Balloons: Concept Invention Phase (15 minutes)

1. **Introduce the Heat Reservoir animation.**

   Tell participants the following animation will provide information that may answer some of their questions. If possible, attribute this information to members of the community to connect the animation to ideas they're discussing, e.g., "Person A and B wondered about how the molecular movement is related to heat. Let's see how this animation can offer some useful information to address their question."

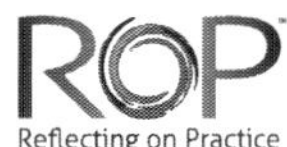

2. **Show the animation and describe what's displayed:**

   a. Two sealed bottles, one filled with air and the other filled with water.

   b. Air and water molecules are visible in each bottle.

   c. A light bulb provides the heat energy.

   d. A thermometer above each bottle displays the temperature inside the bottle.

   Click the light bulb "on."

3. **Elaborate on observations.**
   Invite participants to call out what they notice in the animation. These observations provide evidence for their explanations, hence when participants call out what they think is happening (the bottle of air is getting hotter), urge them to provide evidence from their observations, e.g., "what do you see to make you think that." Observations that should come up.

   In everyday language:

   a. Air molecules bounce around all over the bottles. They bounce around faster and faster over time. Temperature in the bottle increases.

   b. Water molecules slowly begin to wiggle. They continue to wiggle over the 5 minutes. Temperature in the bottle doesn't change much.

4. **Provide concise explanation.**
   Display and tell participants the following information:

   In the language of science:

   a. Heat is the <u>spontaneous transfer</u> of (thermal) energy from hotter materials to cooler materials. (Remember, heat is a *process*, not an entity.)

   b. Temperature is a <u>measure</u> of a) how hot or cold something is (the internal thermal energy of a material), and b) the average kinetic (moving) energy of molecules or atoms in a sample of matter.

   c. The animation shows how water and air molecules behave differently as they absorb thermal energy. The model provides evidence that water was able to absorb a lot of thermal energy before the temperature in the bottle (or balloon) increases.

The steps in this phase (C) modulate between interactive (many voices) and noninteractive (one voice), as learners are given the scientific idea that explains the phenomenon. They move from considering their many ideas (dialogic) towards the one scientific normative idea the facilitator presents (authoritative). This modulation is aimed at easing the tension between everyday and disciplinary views, and gives learners space to integrate the new ideas.

3.2

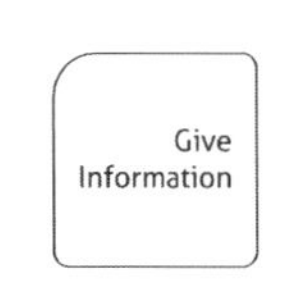

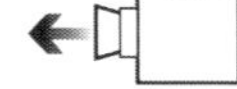

Temperature is more than how hot or cold a substance is. If you do an Internet search for definitions of temperature, be sure to look for scientific definitions. This situation highlights the tension between everyday language and scientific views.

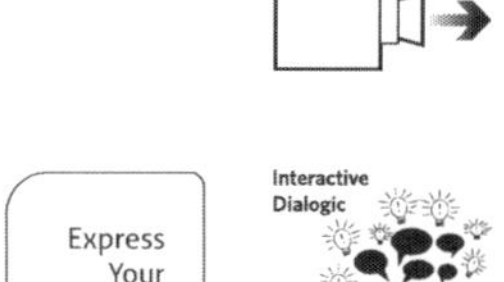

5. **Initiate small-group discussions.**
   Ask participants to discuss the following displayed questions in pairs or with their table group:

   - *How can you use this information to explain the phenomenon?*

   - *What is still unclear to you?*

   - *What other information do you still want or need to know?*

   Wander around to eavesdrop on their conversation, but don't join in. Be as unobtrusive as possible. Pay attention to how participants are making sense of the concepts—their explanations, connections, struggles, and errors.

(D) Two Balloons: Application Phase (15 minutes)

1. **Provide reading for further evidence.**
   Tell participants that, to support their explanations and answer some of their remaining or new questions, they may want or need additional information. Distribute the **Ocean: A Giant Heat Reservoir** handout. Ask them to answer the following questions.

   - *Why did one balloon pop and the other one not?*

   - *How might this balloon activity be similar to what happens on Earth?*

   Tell them they'll have about five minutes to read parts of the article they think will help them gain a deeper understanding and answer the focus questions. Share that they might want to use the following Active Reading strategy to find evidence from the text to support their explanations.

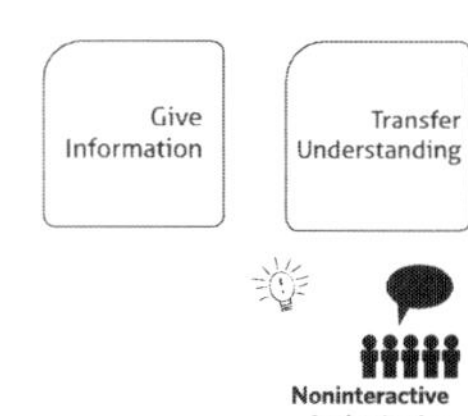

There is additional information in the reading to explain the phenomenon. The second question directs them to transfer their understanding to another context as a way to make the information more meaningful and relevant.

> **Active Reading reminder:**
> - Underline ideas you find interesting or that seem important.
> - Circle ideas you find confusing or have questions about.
> - Write your questions in the margins.

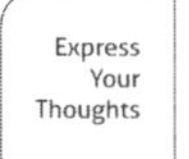

2. **Have participants discuss with others.**
   Invite participants to discuss their ideas with others (in a small group or with someone sitting near them) for about 2-3 minutes. Remind them to explain their reasoning and to support their explanations with evidence.

   Again, wander and eavesdrop unobtrusively. How are their comments changing? Where is there confusion and clarity?

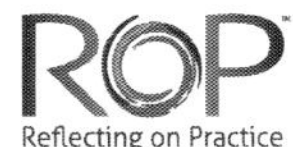

3. **Ask participants to elaborate on their writing.**
    Have participants return to their Learning Journal and draw a line under their first entry. Display the following questions and ask each one aloud, one at a time.

    a. *Why did the air balloon pop?*

    b. *Why didn't the water balloon pop?*

    c. *What role does the concept of heat reservoir play in the larger Earth system?*

    Allow participants about 3-4 minutes to revise and add to their initial explanation. Encourage them to compare and contrast their new entry with their previous ideas and explanations.

(E) Two Balloons: Reflection Phase (7 minutes)

1. **Lead a whole-group share.**
    With the previous questions displayed, facilitate a whole-community discussion using the information you overheard as a guide to know how much you need to push, and also who to call upon to provide more accurate explanations.

    Use the Discussion Map—probe for evidence (e.g., *What makes you think that? Where's your evidence?*) and encourage participants to listen closely to what's shared (e.g., *What do others think about that idea? Can someone rephrase that statement?*).

2. **Display and review Ideas to Emphasize about the balloons and Earth system.**

    a. The **balloon filled with air** popped because when matter inside the balloon (air) absorbed heat energy, it quickly increased in temperature and couldn't transfer the heat energy away from the rubber of the balloon. As soon as the rubber heated, it lost its integrity and failed.

    b. The **balloon filled with water** didn't pop because the matter inside (water) was able to absorb a lot of heat energy before increasing in temperature. The water molecules were able to continually transfer the heat away from the rubber of the balloon. The transfer of heat away from the rubber kept it from melting and failing.

    c. In the **Earth system**, the vast ocean constitutes one large source of matter (water) into which heat energy is absorbed. Just like the water in the balloon, that oceanic body of water absorbing heat energy keeps the other things around it relatively cool.

        ▫ The ocean covers 70% of Earth's surface, and all this water has a huge impact on temperatures on Earth.

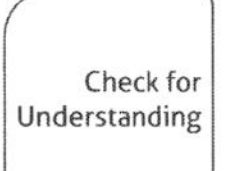

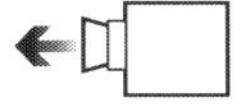
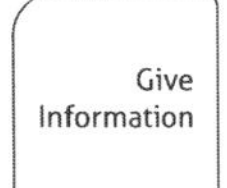

      □ Water acts as a heat reservoir. Water absorbs a lot of heat energy before it changes temperature, and it holds onto the heat for a relatively long time before releasing it to the atmosphere as the water cools down.

      □ In absorbing and releasing heat energy, the ocean warms cold air and cools warm air. The heat reservoir of the ocean keeps air temperatures moderate all over the planet, not just in coastal places.

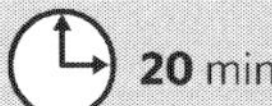 **From Learners' Perspective**   **20** minutes

Turn & Talk: *Learning from Two Balloons*

1. **Introduce task.**

   Explain that, like the Phases of the Moon activity, the Two Balloons activity modeled how a community of learners can work together to generate explanations for a phenomenon and build understanding of a concept. Acknowledge that the context for the learning experiences they facilitate and design is different from this Two Balloons activity. They might be at a cart of objects, large group demonstration, etc.

   Emphasize that the purpose of this task is to use their experience as a learner in this context to consider how to apply the tools and ideas into their own teaching situation.

2. **Provide summary of three tools for Two Balloons**

   Share the following details to remind participants about the *Reflecting on Practice* tools:

   a. **Teaching Purposes.** Harken back to the Role Plays activity in Module 3 Session 1. Recall with participants that one of the points discussed is that there are typically multiple teaching purposes throughout interactions with learners. The skits featured three of the five teaching purposes described in the tool.

   b. **Discussion Map.** Listening with curiosity and probing for understanding are fundamental parts in the Discussion Map. They emphasize that the point of group discussions is to listen and understand what others are saying and thinking. The Research Discussion highlights there are issues like language and power that affect (and perhaps complicate) the dynamic of facilitating conversations, and that the chance to talk with people they know makes the experience more memorable.

   c. **Facilitation Approaches.** The third tool introduced at the beginning of the Two Balloons activity offers a way to think about orchestrating the voices and ideas to co-construct understanding when they facilitate conversations.

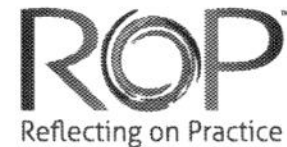

- ◻ Interactive/Noninteractive refers to the number of people speaking.

- ◻ Authoritative/Dialogic refers to the number of ideas being considered.

3. **Summarize relationship between the three tools.**
Display the following details and review each individually.

   a. The four classes of facilitation approaches remind educators that when they facilitate learning conversations, the number of voices and ideas can vary.

   b. The educator can make space for learning conversations among learners, and not be an active participant in the conversation.

   c. The educator's teaching purpose, which shifts throughout an interaction with learners, prompts the facilitation approach.

   d. Shifting the teaching purpose and facilitation approach can address the ideas discussed in the Research Discussion.

4. **Ask participants to think about the learning experience.**
Distribute the **Teaching Two Balloons** handout and have participants review the details. Explain that it details the various actions they took during the activity, organized within the Learning Cycle. The handout lays out what the facilitator was intending during each teaching moment. Display and review the instructions:

   a. Individually, recall what it was like for you as a learner in each of those teaching moments.

   b. Share your experiences with one partner.

   c. Discuss the educator moves and design.

   d. Jot down what you learned about teaching and learning in this experience that you want to bring into your practice.

   e. Be ready to share one thing from your list with the community.

5. **Begin whole-group share.**
After about 10 minutes of partner talk, call time and invite each pair to share one implication for practice for the community to consider.

   Share the facilitation notes in the sidebar that were included throughout the Step by Step for the activity, as well as the information below, as appropriate in this discussion.

   a. **Information can be given in many different ways.** Most common is the educator *telling* learners all the information—in a lecture following a demonstration, for example. In Phases of the Moon, learners manipulate the model (moon ball and shadow) to get information to answer their questions. The educator tells

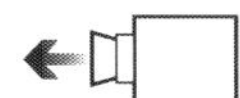

The power dynamic is typically viewed between educator and learner, but can also exist between learners (e.g., those more knowledgeable and/or outspoken).

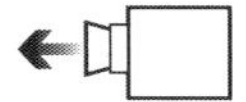

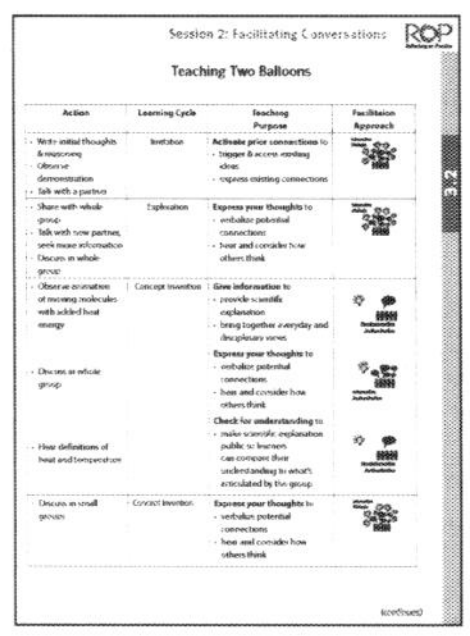

information that learners wouldn't be able to figure out using the model. In Two Balloons, an animation and text are used to give information learners wouldn't be able to figure out just by messing with the balloons and lighter. There were several ways participants were invited to *Express your thoughts* to consider the information given.

b. **The animation gives information visually about how air and water molecules behave.** Learners are guided to make their thinking visible (whether they already know the concept or not), both to themselves and to their peers—by first being prompted to share what they see, and then encouraged to articulate what the observation makes them think and how their thoughts might help explain the demonstration. This is actually the same routine (Notice-Think-Wonder) participants use in the **Observation Instrument** to articulate their thinking.

c. **The reading gives even more information, this time through text and at the reader's pace.** Learners are given questions to focus their reading, an active reading strategy to guide how to extract information, time to talk with others to process what they've read, and finally to regulate their learning by writing down how their thinking has evolved even further.

Let's Talk Practice 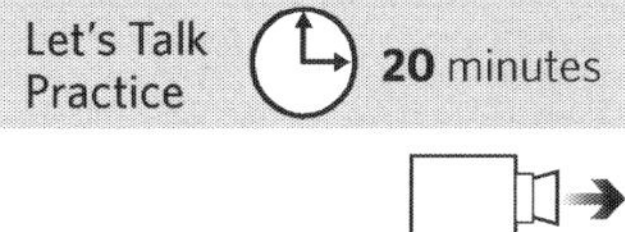 20 minutes

Micro Lab: *Using New Tools*

1. **Begin with individual reflection in Learning Journals.**
Ask participants to use a blank page in their Learning Journals to reflect on this session. First, select a learning experience they typically teach or recently taught. Then, respond in writing to the following displayed reflection prompts. They'll have about five minutes to write.

   a. *What teaching purpose(s) and facilitation approach(es) are present in that learning experience?*

   b. *How would you like to modify it to incorporate the ideas discussed in this session?*

2. **Begin Micro Lab**
Tell participants that in this routine, the **Micro Lab**, they'll share their reflections in groups of three. Have them number off 1-2-3 to form groups of three. (If the community doesn't divide evenly, have the remaining individuals work in pairs.) Tell them you'll serve as timekeeper, and begin the protocol:

If your community members work at a diverse range of tasks and programs, consider sorting them into small groups based on the work they do or the audience with whom they most interact (i.e., by types of programs or activities, subject matter, age group, etc.).

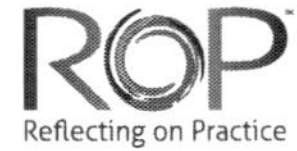

a. **Share**: Ask *Number 1s* to share their ideas with their groups for two minutes. No one speaks except the speaker. Other group members listen and take notes.

b. **Silence**: Allow 20–30 seconds of silence for everyone to take in and mentally review what was heard.

c. **Share for rounds 2 and 3**: Repeat a. and b. above until all members of each group have shared their thinking.

d. **Begin small-group discussions**. Groups may now have an open discussion for five minutes. Encourage groups to make connections between what other members have said or to ask questions for clarification.

3. **(Optional) Conduct a whole-group share.**
   If time permits, invite groups to share key points from their discussions with the rest of the community.

Continue the Learning

3 minutes

1. **Additional suggested reading for this session.**
   If you've decided to assign the reading, share it now.

2. **(Optional) Extra copies of reading.**
   Have additional copies of the **Key Ideas from the Literature: Tension, Power, and Family** available for participants to take topics that they didn't read.

❖

Science Content Guide for Two Balloons

In Two Balloons, the prior knowledge learners typically recall and draw on to help them explain what happened can be organized into four **themes**. The **conditions** of their prior knowledge within these themes can vary in their accuracy.

Listen carefully and probe:	**Potential probing questions:**
1. Ask learners to clarify how they think the additional information they seek will help them explain what happened. 2. Listen to understand what memories they're retrieving (theme) and the condition (missing, incomplete, misconceived) of their prior knowledge.	• How does [that information] help you explain what happened? • What concept (or big idea) is this information you seek related to? • If [that idea] were relevant here, what would you expect to see? • How is that memory you're referencing related to this experience? • What specifically are you wondering about?

Theme 1. Familiar Experiences

This activity involves balloons, water, and heat from a flame, which can trigger learners to retrieve memories of experiences that involve those items for an explanation.

Familiar experiences recalled:	**Learners usually ask:**
• Balloons pop when they're blown up too large; and hot air balloons expand and rise due to heated air pushed into them. • Water changes from a liquid to a gas when heated to a high enough temperature. • Fire burns/melts certain types of material.	• "What happens to water/air when heated?" • "Does the air inside the balloon expand with heat?" • "What is the internal temperature of each balloon?" • "What happens to the rubber/latex of the water balloon?" • "Does the air balloon ultimately pop because of something happening on the outside of the balloon (i.e., does the flame act somewhat like a pin prick), or because of a change inside the balloon?"

■ Learners may offer that **"heat causes molecules to spread apart from one another and take up more volume," "heat energy causes molecules to move faster and with more energy,"** as the causal mechanism. These are both accurate statements, but how that explains their experiences, or this experiment is where their thinking tends to be unsteady. They often suggest that the heated air molecules hit the latex with enough force to pop it, or that the air molecules expand so much that the latex is pushed beyond its capacity.

(continues)

 This page may be duplicated for educational use.

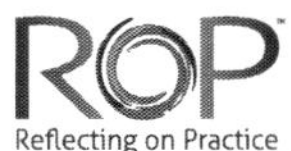

Science Content Guide for Two Balloons *(continued)*

However, there isn't evidence that the air balloon expands. (Some have suggested that the expansion happens too quickly to see it, so want a slow-motion video of the experiment.) Moreover, this explanation doesn't explain why the water balloon didn't pop or why the air and water balloon are reacting differently to added heat energy.

Theme 2. Science Terms and Technical Language

This activity is an exploration into physical science concepts, which can trigger learners to retrieve memories from their science classes (chemistry and physics) for an explanation.

Terms and ideas recalled:	Learners usually ask:
• heat capacity • molecular movement • phase change • latent heat • kinetic energy • melting point of material (latex/rubber), boiling point of water • breaking bonds, gas laws • heat transfer	• "What is the melting temperature of the balloon?" • "What about the molecules and movement?" • "What is actually going on with the molecules in the two balloons?" • "What is the heat capacity of water/air?" • "O_2 molecules are light … water molecules are tightly bonded and heavy. How does that change their popping potential?" • "Does density of water vs air have something to do with the results?"

- Learners recall concepts that might or might not be relevant to the experiment. It's necessary for them to elaborate on their understanding to determine the relevance and connections. Learners might struggle to explain the meaning of the idea for many reasons due to different levels of prior knowledge—missing, incomplete, misconceived.

- As learners share their ideas, encourage them to elaborate, and focus on those ideas that involve **molecules**, **molecular motion**, **density**, **heat** and **temperature**, and **heat capacity**. The Concept Invention Phase will provide more information to help fill in the holes in their explanations.

(continues)

Science Content Guide for Two Balloons *(continued)*

Theme 3. Overlooking Evidence from Observations

This activity is a science experiment, and as such requires learners to gather evidence based on their observations. The balloon and flame part of the investigation happens quickly and is done as a demonstration.

Learners are likely to seek	Learners ask:
• information to clarify the set-up • information to enable them to "see" what they missed	• "Are the balloons made of the same material?" • "Is the flame put directly on both balloons?" • "Is the flame placed on the same spot on both balloons?" • "Are the temperatures of the water and air the same?" • "Can we get a slow-motion video?"

- Some ask **what if** … the water added to the water balloon was at the boiling point? …the liquid used wasn't water? …we could somehow capture that the air balloon was expanding? …the flame was kept on the water balloon longer? These questions are imaginative and aim to invite more exploration, but can also be distracting. In each instance, be sure to ask, "how would that information help to explain what happened to these two balloons?"

- Be cautious that if learners predicted correctly, they might rationalize their prior knowledge to explain what happened, rather than seek information and evidence that would help them to support and be more sure of their explanation.

Theme 4. Distracted by Emotional Memories

This activity is an exploration into a physical science concept, which triggers retrieval of emotional memories from their science classes. Those memories might be negative, discouraging, or empowering for them.

- Learners might not engage in the discussion of ideas, or they might defer to more vocal peers. Watch their behaviors, offer encouragement, and connect to their ideas. There will be multiple opportunities coming up in the activity to gather additional information and build their confidence.

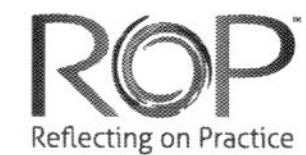

Teaching Two Balloons

Action	Learning Cycle	Teaching Purpose	Facilitation Approach
• Write initial thoughts & reasoning • Observe demonstration • Talk with a partner	Invitation	**Activate prior connections** to • trigger & access existing ideas • express existing connections	Interactive Dialogic
• Share with whole group • Talk with new partner, seek more information • Discuss in whole group	Exploration	**Express your thoughts** to • verbalize potential connections • hear and consider how others think	Interactive Dialogic
• Observe animation of moving molecules with added heat energy	Concept Invention	**Give information** to • provide scientific explanation • bring together everyday and disciplinary views	Noninteractive Authoritative
• Discuss in whole group		**Express your thoughts** to • verbalize potential connections • hear and consider how others think	Interactive Authoritative
• Hear definitions of heat and temperature		**Check for understanding** to • make scientific explanation public so learners can compare their understanding to what's articulated by the group	Noninteractive Authoritative
• Discuss in small groups		**Express your thoughts** to • verbalize potential connections • hear and consider how others think	Interactive Dialogic

(continues)

This page may be duplicated for educational use.

Teaching Two Balloons *(continued)*

Action	Learning Cycle	Teaching Purpose	Facilitation Approach
• Read *Ocean: A Giant Heat Reservoir*	Application	**Give information** to • Provide scientific explanation **Transfer understanding** to • Apply the disciplinary ideas in a range of contexts	Noninteractive Authoritative
• Discuss in small group • Revise writing: what was learned, what changed, & what questions arose		**Express your thoughts** to • verbalize potential connections • hear and consider how others think • reconsider and rearrange connections	Interactive Authoritative
• Discuss in whole group • Listen to summary of key concepts explored • Revise writing, as needed	Reflection	**Check for understanding** to • make scientific explanation public so learners can compare their understanding to what is articulated by the group **Give information** to • provide scientific explanation • bring together everyday and scientific views	Interactive Authoritative Noninteractive Authoritative

This page may be duplicated for educational use.

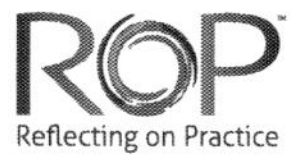

The Ocean: A Giant Heat Reservoir *(1 of 2)*

Solar Energy Is Absorbed by Earth

A vast amount of energy makes its way from the Sun to Earth. Much of that energy gets absorbed by matter on Earth: air, water, and land. As these different kinds of matter absorb heat energy, their molecules begin to move faster, causing their temperature to rise. An increase in temperature is evidence that molecular particles are speeding up.

> **Temperature** is a *measure* of 1) how hot or cold something is (the internal thermal energy of a material), and 2) the average kinetic (moving) energy of molecules or atoms in a sample of matter. **Heat** is the *transfer of energy* from one substance to another; it's a process, not an entity.

Water Absorbs Heat Energy More Slowly than Air Does

Water can absorb a lot of heat energy without much change in temperature. If air and water absorb the same amounts of energy, the air heats up much more quickly than the water. This can be demonstrated by adding heat energy to two balloons, one filled with air and one with water. As soon as heat energy makes contact with the latex, the air balloon will pop immediately, but the water balloon will not.

One reason for this is that water contains far more molecules than air does in the same amount of space (volume). In water, the heat energy is dispersed among many molecules; each water molecule absorbs some of the energy and moves a little faster. The water molecules in the balloon continue to absorb energy without getting much warmer. The latex stays cool and doesn't pop. In the same volume of air, which contains fewer molecules, each air molecule gets a bigger share of the energy to absorb and moves much faster. As the air molecules heat up, they move farther apart from one another, zip off in different directions, and collide with others, creating a rise in temperature. The faster the molecules move, the higher the temperature. When the air molecules in the balloon can't absorb any more energy—which happens almost immediately—the heat energy is absorbed by the latex matter of the balloon, which quickly bursts.

(This text is revised from an ocean science curriculum (*1*).)

(continues)

The Ocean: A Giant Heat Reservoir *(2 of 2)*

Water Molecules Absorb Heat Energy Differently from Other Molecules

In other kinds of matter, such as rock, molecules heat up quickly compared to water even if that matter has approximately the same number of molecules in the same amount of space as water.

A water molecule is made up of three atoms: two hydrogen and one oxygen. The bonds <u>between the oxygen and hydrogen atoms in a water molecule</u> (1), unlike the bonds between atoms in most other kinds of molecules, are very flexible and can absorb a lot of energy before increasing in temperature. This allows them to wobble in place as they absorb energy, rather than zipping off in different directions, colliding with other molecules and causing the temperature to increase.

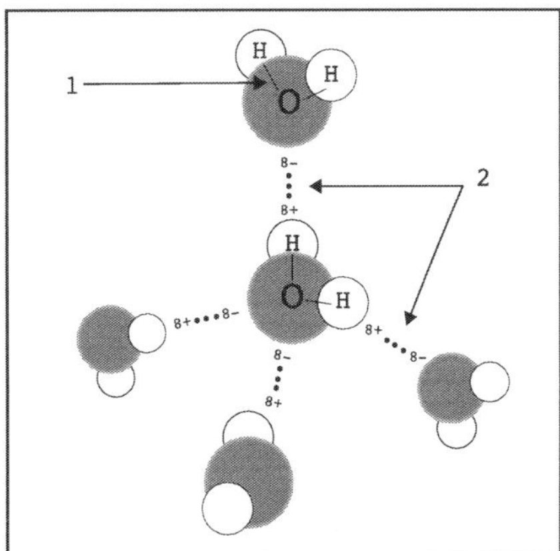

Hydrogen bonds <u>between adjacent water molecules</u> (2), although not as strong as the bonds that hold the atoms within a molecule together, are still strong and flexible enough to keep water molecules from separating from one another easily. Water can bond to as many as four other water molecules, which helps strengthen and stabilize the bonds between them even more.

Because it takes more energy to break the bonds between adjacent liquid water molecules than between molecules of most other substances, water can absorb more energy before the bonds break and the water molecules start moving faster and bumping into one another, causing the temperature to increase.

The Ocean Acts as a Heat Reservoir

About 70% of the surface of Earth is covered by ocean, and the ocean can be up to seven miles deep. That's a lot of water on the planet, and all that water can absorb a lot of heat energy. This makes our ocean a vast heat reservoir.

Because water can absorb an enormous amount of energy before heating up, the ocean helps keep our planet from getting too hot. And when the ocean releases (transfers) heat energy back into the cooling air, such as at night or in winter, it also helps keep our planet from being too cold. The ocean keeps temperatures on our planet moderate enough to sustain life.

---

1. Lawrence Hall of Science, Ocean Sciences *Curriculum Sequence for Grades 6-8: The ocean-atmosphere connection and climate change.* (The Regents of the University of California, Berkeley, 2014).

This page may be duplicated for educational use.

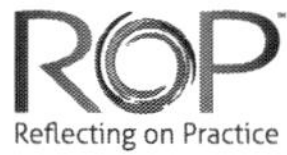

Key Ideas from the Literature: Tension, Power, and Family

(A) TENSION between Everyday and Scientific Views

Every specialized kind of human activity, every subject area and field—whether it's serving in Congress, refereeing baseball, designing handbags, or doing science—has its own language and ways of talking (*1*). The language of science is an example of specialized language, not only because of its use of specific science definitions, but also because of the way science is spoken and written using idiosyncratic terms, meanings, and sentence structure.

For instance, whereas in everyday language the word "theory" is often used to refer to a guess or a hunch (*2*), in science, the word "theory" is understood to mean "a well-elaborated body of scientific knowledge that explains a large group of phenomena." Additionally, scientific language has a preference for passive voice (e.g., "the earth was uplifted" rather than "pressure lifted the earth up"; or "what element is being represented?" in place of "what element are we representing?"). One effect of this grammatical construction is that people disappear as actors or agents from science (*1*). Learning science, therefore, is more than making sense of one's experiences and observations; it includes recognizing and employing the language of the scientific community (*3*).

Learning science requires understanding the scientific views and <u>speaking the language of science</u>, which educators need to model for learners. However, there is a tension between *directing the conversation to communicate the views of science* and *allowing learners to voice their views and be equal contributors to the conversation* (*4*). The opportunity for learners to talk and share their thinking is necessary for learning. But learners' understanding, their everyday views, and the language they use to articulate their thinking often don't align with scientific views and the language of science. No surprise, since learners are grappling with ideas and concepts while also learning how to talk about those ideas using scientific language.

The fundamental point is that <u>meaningful learning requires learners to make connections between their own ideas and the scientific views</u>. Learners need the opportunity both to articulate their everyday ideas and to apply and explore newly learned scientific ideas for themselves, through talk and other actions (*4*). The educator is needed to help learners make those connections.

- Where in your practice is the tension between everyday and scientific views most prevalent?
- What Educator Moves can you make to ease that tension?

© 2021 Regents of the University of California

Key Ideas from the Literature: Tension, Power, and Family *(continued)*

(B) POWER & SCAFFOLDING in Conversations

Power. Opportunities to talk are important for learners, allowing them to share, clarify, and distribute knowledge among peers. Peer talk unfolds as learners, unhindered by the inherent asymmetry of educator-learner interaction, ask questions, explain, and formulate ideas with each other (5).

The egalitarian structure of peer groups may be more conducive to idea generation, and elaboration, than educator-learner set-ups; topics are more far-ranging and learners freely add to one another's comments (6). In the presence of an adult educator, learners tend to generate more explanations and higher levels of reasoning, speak one at a time, and limit themselves to responding to the educator's prompts (6, 7). The educator is prompting learners to explicate their comments and articulate relationships between ideas. This contrast between the way learners talk with and without the presence of an educator reflects the power relations that exist between educator and learners.

There are power relations in all social structures of human activity: school, government, family, museums, etc. (8, 9). Power exists only when a person(s) can affect the outcomes of another (8). For instance, the teacher in a classroom may determine the topics of study, pace of instruction, and grades. The teacher has power over students each time they act in response to what the teacher, does—e.g., answer a question posed by the teacher. There can be power relations between people of different race and gender, adults and children, and more- and less-knowledgeable individuals. The existence of a power relation is not inherently good or bad; its merit is determined by how that power is used to affect the behavior of the others.

- In what situations can power relations be a factor in informal settings?
- What Educator Moves can you make to address those?

Scaffolding. There is a distinct difference between the quality/quantity of learning that occurs on one's own and that which can be achieved with guidance from more capable others. This difference is called the *zone of proximal development* (the zpd, or "zone"). It represents the distance between what can be achieved on one's own and what more can be accomplished with the assistance of someone more experienced (10). *Scaffolding* is a strategy for working within a learner's zpd. It describes how an educator "[controls] elements of the task that are initially beyond the learner's capacity"—giving information in smaller chunks, synthesizing the most important developments, etc.—to permit the learner to "concentrate on and complete only those elements that are within his range of competence" (11). As the learners' capacity increases, the educator gradually withdraws assistance, so that learners have the chance to complete the task themselves.

- With a group of learners in an ephemeral interaction such as in an informal learning environment, how can the educator work within the learners' zpd?

(continues)

     This page may be duplicated for educational use.

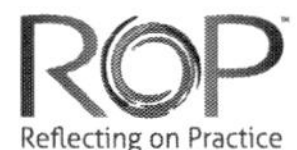

Key Ideas from the Literature: Tension, Power, and Family *(continued)*

(C) FAMILY CONVERSATIONS in Informal Learning Environments

Families are intergenerational social groups whose members share close relational ties. They have motivations and agendas for why, when, and how they visit informal learning environments, which in turn may affect their behavior during their visits (*12, 13*). Their agendas and goals may or may not align with the institution's intentions for their visitors. Children are more persistent, sustain engagement, and spend more time at exhibits when they are with adults in their group than when alone or with peers only (*14, 15*).

Adult contributions to family conversations. More talk goes on in adult-child groups than in children-only groups; however, adults do most of the talking in the first configuration (*14*). Adults read and interpret labels, serving as mediator and translator for the children (*16*). In their talking, adults direct children to manipulate the exhibit in order to generate evidence for observations and comparisons; draw children's attention to relevant evidence; guide children's talk to make connections between ideas and observations; and provide explanations, analogies, and descriptions of unobservable principles (*14, 16*).

Adults also explain science ideas by casting the children's experiences in causal terms, make connections with prior knowledge and personal histories, and introduce abstract principles (*14, 17*). Adults tailor their level of interaction at the exhibit to the children's characteristics, including the child's gender, interests, and knowledge of the content (*14, 18*). And in some cases, adults, despite talking equally to boys and girls about how to use the exhibits and the evidence gathered at the exhibit, tend to "explain science" more often to boys than to girls during interactions in a science center (*15, 19*).

In sum, adults provide explanations and actively scaffold learning experiences at informal science learning environments, and the children benefit because it calls attention to details they might not notice, connects concepts children might not know or realize are relevant, and provides language for expressing their thinking. However, in this dynamic the children typically don't actively contribute to the conversation. If they were actively participating in the conversation, the experience would be more memorable and likely more meaningful for long-term retention and retrieval of the ideas.

Joint talk in family conversations. Children are more likely to recall objects and concepts they encountered from a visit to a museum when *both they and their parent(s) talked about it during the visit* (*20*). Children's recall is more complete and descriptive when they've participated in conversations that allow them to elaborate on details and express their emotions and opinions (*21, 22*). Parents can model and/or encourage this elaboration with questions that prompt the child to reflect and make connections, rather than asking yes/no or factual questions to which parents already know the answer.

(continues)

Key Ideas from the Literature: Tension, Power, and Family *(continued)*

Prior science learning experiences provide valuable conversational content for adults and children to discuss jointly during exploration at informal science learning environments (*23*). It's been found that memories generated from out-of-school settings are almost exclusively the sources of prior experiences that come up in informal science learning environments. These sources include (a) everyday experiences (e.g., sap being drawn from a maple tree reminds a child of maple syrup at breakfast); (b) designed spaces (e.g., relating similarities between plants on the trail now with a memorable experience at a museum during a family vacation); and (c) programs for science learning (e.g., knowing how to spot poison oak on the trail from summer camp). Families use these prior experiences in their conversations for a variety of reasons: to remind each other about a shared experience, to prompt a child to display knowledge gained from the prior experience, or to explain or defend an idea by referencing the prior experience.

- What Educator Moves can you make to support learning for both adults and children in visiting families?

References

1. J. L. Lemke, *Talking Science: Language, Learning, and Values.* (Ablex Publishing, Westport, CT, 1990).

2. S. Michaels, A. W. Shouse, H. A. Schweingruber, *Ready, Set, SCIENCE!: Putting Research to Work in K-8 Science Classrooms.* (National Academy Press, Washington, DC, 2007).

3. R. Driver, Students' concepts and the learning of science. *Intl. J. Sci. Educ.* **11**, 481–490 (1989).

4. P. H. Scott, E. F. Mortimer, O. G. Aguiar, The tension between authoritative and dialogic discourse: A fundamental characteristic of meaning making interactions in high school science lessons. *Science Education* **90**, 605–631 (2006).

5. L. P. Rivard, S. B. Straw, The effect of talk and writing on learning science: An exploratory study. *Science Education* **84**, 566–593 (2000).

6. K. Hogan, B. K. Nastasi, M. Pressley, Discourse patterns and collaborative scientific reasoning in peer and teacher-guided discussions. *Cognition and Instruction* **17**, 379–432 (1999).

7. T.-J. Lin *et al.*, Less is more: Teachers' influence during peer collaboration. *Journal of Educational Psychology* **107**, 609 (2015).

8. M. Foucault, The subject and power. *Critical Inquiry* **8**, 777–795 (1982).

9. S. Hall, in *Discourse Theory and Practice: A Reader*, M. Wetherell, S. Taylor, S. Yates Eds. (Sage, London, 2001), chap. 7, pp. 72–81.

10. L. Vygotsky, *Mind in Society: The Development of Higher Psychological Processes.* (Harvard University Press, Cambridge, MA, 1978).

11. D. Wood, J. S. Bruner, G. Ross, The role of tutoring in problem solving. *Journal of Child Psychology* **17**, 89–100 (1976).

12. T. Moussouri, Negotiated agendas: families in science and technology museums. *International Journal of Technology Management* **25**, 477–489 (2003).

13. A. Briseno-Garzon, D. Anderson, A. Anderson, Entry and emergent agendas of adults visiting an aquarium in family groups. *Visitor Studies* **10**, 73–89 (2007).

(continues)

This page may be duplicated for educational use.

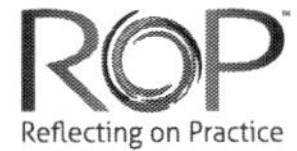

Key Ideas from the Literature: Tension, Power, and Family *(continued)*

14. K. Crowley *et al.*, Shared scientific thinking in everyday parent-child activity. *Science Education* **85**, 712-732 (2001).

15. H. T. Zimmerman, S. Reeve, P. Bell, Family sense-making practices in science center conversations. *Science Education* **94**, 478-505 (2010).

16. M. Tare, J. French, B. N. Frazier, J. Diamond, E. M. Evans, Explanatory parent-child conversation predominates at an evolution exhibit. *Science Education* **95**, 720-744 (2011).

17. D. B. Ash, Dialogic inquiry in life science conversations of family groups in a museum. *J. Res. Sci. Teach.* **40**, 138-162 (2003).

18. S. Palmquist, K. Crowley, From teachers to testers: How parents talk to novice and expert children in a natural history museum. *Science Education* **91**, 783-804 (2007).

19. K. Crowley, M. A. Callanan, H. R. Tenenbaum, E. Allen, Parents explain more often to boys than girls during shared scientific thinking. *Psychological Science* **12**, 258-261 (2001).

20. M. Tessler, K. Nelson, Making memories: The influence of joint encoding on later recall by young children. *Consciousness and Cognition* **3**, 307-326 (1994).

21. M. D. Leichtman *et al.*, Talking after school: Parents' conversational styles and children's memory for a science lesson. *Journal of Experimental Child Psychology* **156**, 1-15 (2017).

22. C. A. Haden, Talking about science in museums. *Child Development Perspectives* **4**, 62-67 (2010).

23. L. McClain, H. T. Zimmerman, Prior experiences shaping family science conversations at a nature center. *Science Education* **98**, 1009-1032 (2014).

Module 3.
Session 3: Whole-Community Video Reflection on Learning Conversations

Session Overview

This reflection session examines **facilitating conversations to support learning** in actual practice.

Participants revisit two Research Discussions on learning and talk to **deepen their understanding of the importance of talk for learning.** This rereading of text also primes participants for the upcoming Video Reflections. They're introduced to **Educator Moves for learning conversations**, and practice using the moves during a whole-group observation of video from a volunteer presenter.

Finally, discussions of observation data and the Video Reflection experience as a whole reinforce how educators can **facilitate learning conversations in their activities and programs**.

Session Objectives

- Revisit key ideas on learning and talk.
- Recalibrate Video Reflection tools and experience.
- Introduce Educator Moves—Learning Conversations.
- Practice using Reflective Practice tools in a Video Reflection for a member of the community.

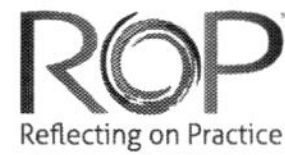

SESSION AGENDA		
Task **Routine**	**Description**	🕐 **Estimated Time** **(in minutes)**
Introduction *Session Objectives*	The goals and objectives of the session are introduced.	2
Research Discussion Sentence-Phrase-Word: *Revisiting Learning & Talking*	Participants revisit big ideas from two previous Research Discussions. They engage with and make meaning from those ideas with a particular focus on capturing the essence of the text.	35
Retrieve & Connect Table Talk: *Reflecting on Video Reflections*	Participants reflect on their experiences with Video Reflections and articulate emerging changes in their practice. A new Educator Moves tool focused on Learning Conversations is introduced and discussed.	25
Break option		10
Hands-on *Video Reflection in Whole Community*	Participants engage in another whole-group video reflection as they observe a colleague's practice.	55
From Learners' Perspective Turn & Talk: *The Video Reflection Experience*	Participants discuss their experiences with Video Reflection.	10
Retrieve & Connect Minute Paper: *Personal Commitment to Reflective Practice*	Participants expand on their commitment to reflective practice in their Learning Journals.	5
Continue the Learning	Review preparation steps for video reflections in critical-colleague groups.	8
	Total Estimated Time	150 mins. (2.5 hrs.)

MATERIALS

Recurring for all Sessions

- See Module 1 Session 1

For this Session

- Video clip (from an educator-presenter, see **Getting Ready** #4)
- External speakers
- 2 Posters for **Roles** and **Agreements** (reuse from Module 1 Session 2, **Getting Ready** #4)

For each small group

- Chart paper and markers

For each participant

- Handouts (1 copy each) from Appendix A, *Tools for Reflective Practice*
  - **Observation Instrument**
  - **Educator Moves—Learning Conversations**
  - **Checklist for Reflection Session**
- Handouts (replacement copies as needed)
  - **Key Ideas from the Literature: How People Learn** (from Module 2 Session 1)
  - **Key Ideas from the Literature: Talking to Learn** (from Module 3 Session 1)
  - **Protocol for Video Reflection** (Appendix A)
  - **Feedback Chart** (Appendix A)
  - **Reflection Exercise Worksheet to Prepare for Feedback** (Appendix A)

Participants can reuse the **Checklist** handout if the process for determining roles and rotation schedules is established and easily accessible publicly.

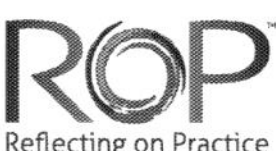

Getting Ready

1. Consider informing participants before the session that they will be revisiting **Key Ideas from the Literature: How People Learn** and **Key Ideas from the Literature: Talking to Learn**, in case some individuals want to review the reading material ahead of time.

2. Review the handouts and readings from Session 1 for Modules 2 and 3. Review the **Hands-on** steps for *Practice a Video Reflection*, if needed, to recall how to model the steps explicitly.

3. Duplicate handouts.

4. Assign a presenter. Ask or assign someone from the community to do their Video Reflection with the whole group.

    - Ask the presenter to complete the **Reflection Exercise Worksheet to Prepare for Feedback** worksheet.

    - Offer assistance as needed.

    - Ask the presenter to provide you the video clip at least one day before the video session.

5. Review the details gathered in the community documents, **Educator Moves** and **Institutional Practice** from Video Reflections on prior knowledge (if that hasn't been done already).

    - Determine how collected details are to be addressed (e.g., discussing a change to an institutional practice that was brought up, sharing how to apply an Educator Move).

    - Consider how the process for collecting these details across the Video Reflections need to be refined, if at all.

6. The **Sentence-Phrase-Word Research Discussion** collectively offers six ideas for discussion. If the community is smaller than 12, it won't be possible to reread all the ideas. Either predetermine which ideas you want the community to reread, or allow the community to decide for themselves as they make their choices and form groups.

7. Be ready to give updates during **Continue the Learning** regarding processes for Video Reflections in small, critical-colleague groups. Questions to consider for updates:

    - Are participants staying in the same small groups, or are they forming new groups?

    - Are there issues with video recording? (e.g., logistics of collecting and storing videos, responses from learners to being recorded, etc.)

Consider inviting a member of the community to be the guide during the Video Reflection. Doing so offers members of the community leadership opportunities. Review the process to prepare the guide for leading the whole community and remind them to follow the protocol closely. Be sure the presenter is aware of who will be the guide for their video reflection.

- Have there been concerns with preparing for the Video Reflections? (e.g., naming and framing problems, selecting a clip, timeliness of getting materials to the guide)

- Have there been obstacles with doing the Video Reflections? (e.g., making time, finding the space, having the equipment to watch together)

- Does the process for collecting the roles and rotation schedules for each group make sense?

8. Prepare the video clip. Be sure it's cued up and ready to play from the point where the presenter wants to begin sharing. Know where in the video to stop sharing. Test the sound before everyone arrives.

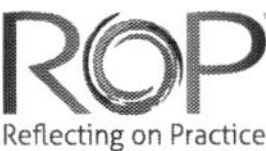

Session 3 Step by Step

Introduction
Session Objectives

1. **Introduce the session.**
   Share that this Video Reflection will be done in whole community and focuses on Learning Conversations. It continues the discussion on how people learn and the importance of talk by looking more closely at actual practice. Doing another video reflection for the module all together offers the community a chance to recalibrate the reflective practice process and practice using the new set of Educator Moves.

2. **Display and introduce the session objectives:**

   - Revisit Key Ideas on learning and talk.

   - Recalibrate Video Reflection tools and experience.

   - Introduce Educator Moves—Learning Conversations.

   - Practice using Reflective Practice tools in a Video Reflection for a member of the community.

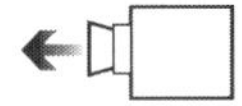

Sentence-Phrase-Word: *Revisiting Learning and Talking*

1. **Retrieve Key Ideas from the Literature handouts from Modules 2 and 3.**
   Ask participants to retrieve their copies of the **Key Ideas from the Literature: How People Learn** and **Talking to Learn** for use during the research discussion. If needed, distribute replacement copies.

2. **Recall Research Discussion and routine.**
   Remind participants they used this **Sentence-Phrase-Word** Research Discussion in a previous session to engage with and make meaning from the text again—with particular focus on capturing the *essence* of the text: <u>what speaks to them</u>.

3. **Display and call out the initial steps of the routine:**

   a. **Choose and reread one of the big ideas.** Tell participants to choose one of the ideas (or assign them one) from the **Key Ideas from the Literature: How People Learn** or **Key Ideas from the Literature: Talking to Learn** handouts. Ask them to read their

2 minutes

35 minutes — Research Discussion

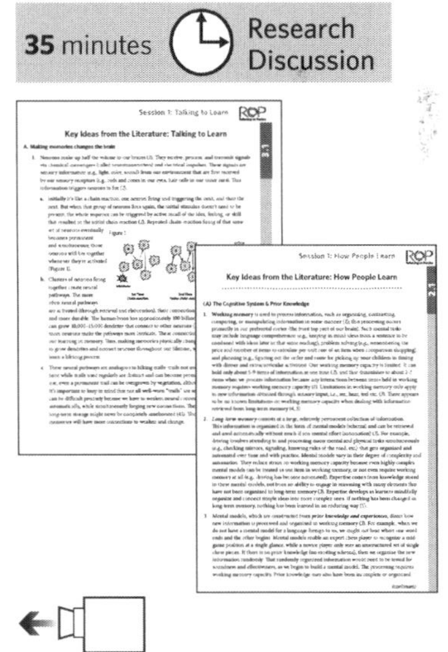

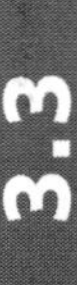

If participants are choosing, make sure there are at least two people for any one topic.

chosen (or assigned) idea using the "active reading" steps they've used before.

b. After rereading their idea, make three selections.

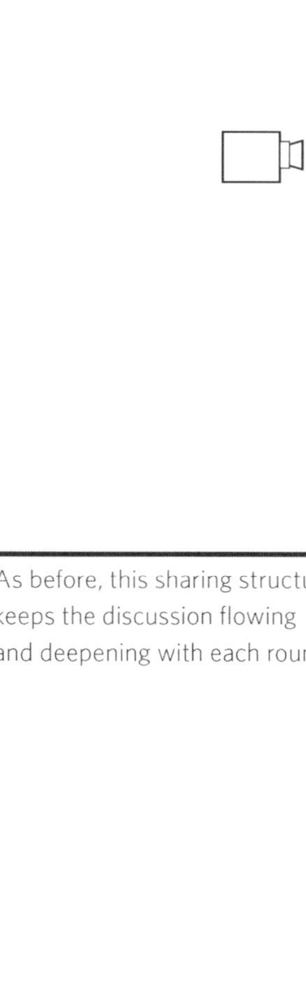

- ◻ Select a **sentence** <u>that captures the idea</u>. Everyone's selection will be different and a reflection of their experiences.

- ◻ Select a **phrase** <u>that helped</u> <u>them gain a deeper understanding of the idea</u>. The phrase should come from a different sentence.

- ◻ Select a **word** <u>that has either caught their attention or struck them as powerful or important</u>.

c. Allow time and encourage silence for active reading and selections.

4. Form small groups and distribute chart paper.
If groups haven't already formed to reread a specific idea together, form them now based on participants' selections—all those who chose idea (A) form a group, (B) another, and so on. Keep the groups small. If many participants chose the same idea, make multiple small groups. Distribute chart paper for each group.

5. Begin rotation to share, record, and discuss.
Tell participants to share and record their ideas, explaining why they selected their sentences, phrases, and words. Remind them the discussion should emphasize their <u>reasoning</u> for selecting what's of interest, so it's okay if people chose the same text.

Remind participants that the purpose of this Sentence-Phrase-Word reread discussion is to think more deeply about these ideas on learning and talk as a prelude to their Video Reflection on learning conversations. Display the steps to have them share in a rotation as follows:

a. First participant shares a **sentence** and explains why it was chosen.

b. Record the first selection on the chart paper.

c. The group comments on and discusses the selection.

d. Then the next person shares, records, and discusses, and so on until everyone has shared a sentence.

e. This process repeats about selected **phrases**, and finally, about selected **words**.

f. Group prepares to share the essence of its idea with the whole community, drawing on the recorded selections.

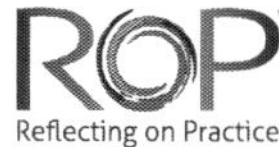

6. **Circulate and monitor.**
   Listen in on small-group conversations, paying attention to what participants are saying and not saying. Monitor to ensure that all groups are working at about the same pace.

7. **Call for a small-group review.**
   As groups finish their rounds, remind them to look at their recorded selections and think about their discussion. As a group, they should determine the essence of their idea to share with the whole community.

8. **Open a whole-group share.**
   Call everyone back together and invite each small group to share the gist of its chosen or assigned idea. Suggested prompts:

   - What's the essence of your idea?

   - What's the powerful takeaway?

9. **Provide quiet time to write down ideas to remember for transition into Video Reflection.**
   Ask participants to write down a few things on learning and talk that they want to keep at the front of their minds during the upcoming Whole-Community Video Reflection.

Table Talk: *Reflecting on Video Reflections*

25 minutes — Retrieve & Connect

1. **Sum up the Video Reflection experience to date.**

   a. Each person has set and presented one problem from their practice that they are investigating, and received focused feedback on that problem from the community (either in whole or small group).

   b. Together, everyone has offered the community 5-6 learning opportunities to deepen our collective practice: two in whole community and 3-4 in their critical-colleague groups.

   c. They will have another 4-5 learning opportunities from video reflections: one in whole community, now, and 3-4 in their critical-colleague groups, afterwards.

   d. In this round of video reflections, they will use a new set of **Educator Moves** focused on the topic of this module: Learning Conversations.

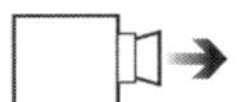

Participants might bring up obstacles due to logistics and equipment. Some of those concerns might be better addressed in **Continue the Learning** as they prepare for the next round of Video Reflections in small groups.

If available, offer an update on the changes in **Institutional Practices** that have come from these reflections. Remind participants of the process for gathering and addressing these items.

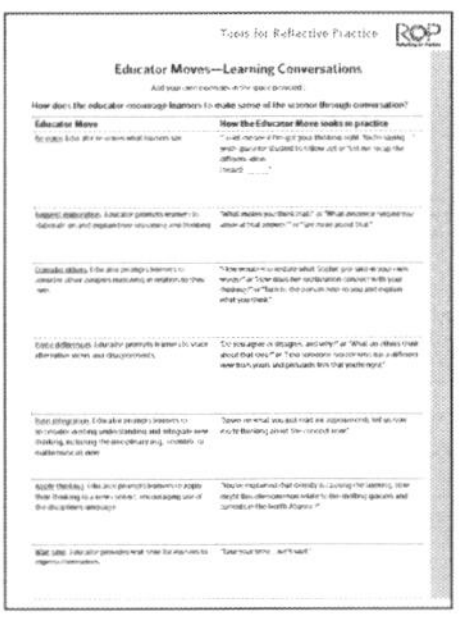

2. **Form small groups to discuss Video Reflection experiences.**
   Ask participants to talk in small groups about their experience with Video Reflections in Module 2. Display the following prompts:

   a. *What did you notice in yourselves, peers, and learners?*

   b. *How might (or did) you **do** things differently, or **think** differently about how you do things?*

   c. *What tweaks, if any, to the Tools for Reflective Practice tools should the community consider? (includes the Community Agreements and Roles & Responsibilities)*

3. **Begin whole-group share.**
   After 5-7 minutes of small-group discussion, call time and invite participants to share with the community. Use the Discussion Map. As they share, document the following on the board, in separate columns:

   - Comments about what they notice and how they may be thinking and doing things differently, so the whole community can see the collective changes that may be emerging.

   - Suggestions, if any, for tweaks to the Tools for Reflective Practice—both revisions that meet consensus and those still up for discussion. If the agreed-upon revisions can be quickly incorporated into the Video Reflection, consider using them in this very session, to test them out.

4. **Distribute "Educator Moves—Learning Conversations" handout.**

   a. Remind participants that the Educator Moves they used last time were focused on <u>prior knowledge</u>. For this module, the Moves will focus on **Learning Conversations**. The handout lists actions educators may take to guide learning conversations, with possible reasons for each move.

   b. Remind participants that they refer to **Educator Moves— Learning Conversations** when they <u>think</u> about what they noticed in the video clip, and consider whether the observed actions could be interpreted as any of these moves. During their discussion, they add examples and elaborate or revise the reasons to reflect the group's understanding of each Educator Move.

   c. Remind participants of the process for collecting their additions and revisions to the **Educator Moves—Learning Conversations** throughout each Video Reflection in this module.

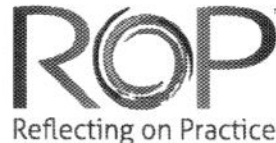

5. **Address clarification questions.**
   Give participants a few minutes to review the handout and invite
   them to ask for clarification, if needed. Remind them that they'll
   have a chance to practice with the tool in just a moment, so they
   shouldn't get too caught up in the details just yet.

Video Reflection in Whole Community

55 minutes 🕐 Hands-on

1. **Introduce the whole-group Video Reflection.**
   Tell participants it's time again to use the tools and ideas they just
   reviewed and discussed.

2. **Have tools and posters ready.**
   Affix the two posters (**Roles** and **Agreements**) in a prominent
   location. Remind participants to retrieve their copy of tools used
   previously that will be used now. Distribute replacement copies as
   needed.

   - **Protocol for Video Reflection.**

   - **Educator Moves—Learning Conversations**

   - **Feedback Chart**

3. **Distribute new copy of Observation Instrument.**

4. **Call attention to the Reflection Session posters.**
   Direct everyone's attention to the two posters, Roles &
   Responsibilities, and Community Agreements for Video
   Reflection. Give them a moment to remind themselves of the
   details, including any revisions the community just suggested.

5. **Use protocol to direct the process.**
   Remind participants that while the guide uses the **Protocol for
   Video Reflection** to lead the group through the Video Reflection,
   everyone should also use it to follow along with the process. If a
   member of the community will be the guide, then introduce
   them now.

6. **Guide introduces the educator-presenter.**
   Turn over the responsibility to the guide. Remind the guide to
   introduce the presenter and thank the presenter for doing their
   Video Reflection with the whole community.

7. **Start the Whole-Community Video Reflection.**
   Follow the steps in the protocol to guide the community through the
   Video Reflection. Be sure to use the Tools for Reflective Practice as
   indicated in the Protocol.

Don't forget to record details for
Educator Moves.

3.3

8. **Add action items for changes in institutional practices in Step 11.**
   If it's possible for some practices to be addressed immediately, then consider it. Otherwise, remember to share the process and invite participants to join the effort for making institution-level change, as appropriate.

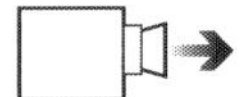

From Learners' Perspective · **10** minutes

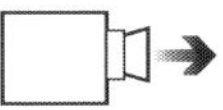

This conversation gives participants the chance to voice any additional refinements to the process. If suggestions in the **Retrieve & Connect** were used during this Video Reflection, now is a good time to consider whether those suggestions should be kept, further refined, or discarded.

Turn & Talk: *The Video Reflection Experience*

1. After the Whole-Community Video Reflection, ask participants to talk with someone next to them about their experience. Display the suggested prompts:

   a. *What did you notice about yourself and peers in this experience?*

   b. *What modifications would you suggest to make the experience better?*

2. **Facilitate whole-group share.**
   As participants share their thinking, listen to what they share and encourage the community to listen closely. Invite clarifications and elaborations, as well as different perspectives. Finally, summarize the ideas that were shared.

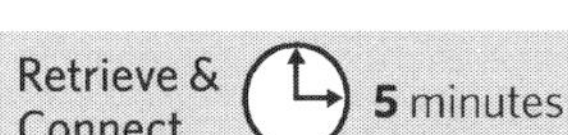

Retrieve & Connect · **5** minutes

Minute Paper:
Personal Commitment to Reflective Practice

Ask participants to find and review their entries in their Learning Journals for Module 2 Session 3's Minute Paper, "**Personal Commitment to Reflective Practice**." Ask them to date a new entry.

Display the following prompts. Ask them to pick a couple of questions to consider and write a response.

a. *What are you learning about your own practice? What interesting, perplexing, and curious revelations are you encountering in your practice?*

b. *What continue to be your concerns about Video Reflections? What new concerns are emerging?*

c. *What are you enjoying most about sharing your practice with your peers?*

d. *What outcome(s) seems to be emerging?*

e. *What can you do differently, or continue to do, to reflect on your practice?*

Continue the Learning

8 minutes

1. **Updates for the process on Video Reflections in small groups.**
   Share any changes to the processes for Video Reflections in small groups that haven't already been discussed (see **Getting Ready** #7).

2. **Display and review reminders for Video Reflections.**
   Distribute handouts and offer reminders, as needed.

   a. Small groups arrange amongst themselves when and where to meet for their Video Reflections.

   b. **Checklist for Reflection Sessions.**

      ☐ Step 1 should be completed now. The roles and rotation schedule should be recorded on the community document before participants depart.

      ☐ Steps 2 and 3 are completed prior to the small-group meetings.

   c. **Reflection Exercise Worksheet to Prepare for Video Reflections.**

      ☐ This worksheet guides them through Steps 2 and 3 in the **Checklist**.

      ☐ The situation that participants investigate in this round of Video Reflections pertains to "talk" and "learning conversations."

      ☐ Each presenter completes the worksheet prior to their Video Reflection.

      ☐ Acknowledge the reflection exercise is not an easy task. Encourage participants to reach out to colleagues or the facilitators for assistance, even if it's just to listen.

3. **All Video Reflections in small groups need to be completed before the next whole-community session.**
   Share the final date.

❖

MODULE 4.
Objects and Design

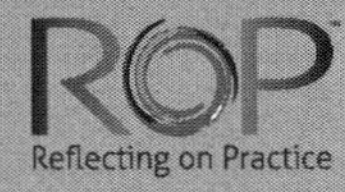

Overview

In informal science learning environments, in which a team of educators may teach the same set of programs and activities, consistency across educators' teaching practice may be inextricably tied to **how these experiences are designed**. In Module 4, participants integrate insights and ideas from previous modules to consider how their understanding of how people learn is applied in their existing designs. Participants apply these insights and information to revise the learning experiences they teach and design.

Session 1 assesses the use of **objects** in participants' programs and activities. It begins with recognizing that objects are used in many learning experiences in informal learning environments, and offers a set of categories for organizing and talking about these objects. Empirical studies are used to provide participants with deeper insights into **how learners interact with objects** at informal learning environments, and to consider the **limitations and possibilities** for learning from the different types of objects. The hands-on activity encourages and challenges participants to use insights gained from the Research Discussion to rough design a learning experience using one of four types of objects to embody what they know about how people learn and to spark new thinking about possibilities and limitations of different objects to support learning.

Session 2 brings together the research and tools for supporting learning discussed throughout the program, including **Thinking Moves** and **Thinking Routines**, with the specific goal of incorporating these ideas into the design of learning experiences. Participants reflect on the learning research that underlie each of the tools and pedagogy. Calling on their experiences in *Reflecting on Practice* as concrete examples, participants work in small groups to reflect on the design of a learning experience at their institution, **share ideas for revisions**, and **receive input** from colleagues using the design tool, **Reflection Exercise Worksheet for Rapid Design Feedback**.

Session 3 offers two ways for participants to further reflect on their practice. The opening **whole-community Video Reflection** focuses on using objects to support learning. Following that, participants use the **Reflection Exercise Worksheet for Critiquing Objects in Learning Experiences** to critique how they currently foster conversation and engagement with objects in learning experiences they offer. They are then challenged to think of ways to leverage these objects to more effectively support learning.

Once again, participants meet in their **critical-colleague groups to do their own Video Reflections** on the module's focus, the use of objects to support learning.

Module 4.
Session 1: Teaching with Objects

Session Overview

This first session in Module 4 invites participants to think critically about the **objects** they use in learning experiences. To appreciate how informal learning environments offer access to "special" objects, participants read case studies that examine **how learners interact with and learn from different types of objects**. Participants are then challenged to apply these ideas to design an experience using one of four types of objects to capitalize on these objects' potential to support learning, engagement, motivation, and understanding.

Session Objectives

- Consider what objects are and how to use them to support learning.
- Discuss the limitations and possibilities of different types of objects to support learning.
- Examine how interactions with learners are greatly influenced by the design of the object-based experience.

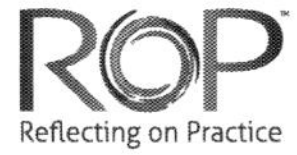

SESSION AGENDA		
Task **Routine**	**Description**	🕐 **Estimated Time** **(in minutes)**
Introduction *Session Objectives*	The goals and objectives of the session are introduced.	2
Retrieve & Connect Think-Pair-Share: *Teaching with "Things and Stuff"*	A discussion about teaching with objects is initiated with the prompt, *How do you use objects in your practice?*	8
Research Discussion Case Studies: *How Learners Interact with Objects*	Small groups discuss five case studies focused on visitor interactions with different types of objects.	50
Break option		8
Hands-on *Playing with Objects*	Participants use ideas from the research discussion to rough design a learning experience, focusing on four types of objects.	70
Let's Talk Practice Connect, Extend, Challenge, & Apply: *Pushing the Possibilities & Limitations*	Participants record their thinking on using objects to support learning and specific actions they'd like to try in their practice.	10
Continue the Learning	Participants are given preparation work for the subsequent session.	2
	Total Estimated Time	150 mins. (2.5 hrs.)

MATERIALS

Recurring for all Sessions

- See Module 1 Session 1

For each small group

- Handout (1 copy)
  - **Key Ideas from the Literature: How Learners Interact with Objects**
- Materials for one type of object (See **Getting Ready** #6)
  - Group One: 1-2 real whole squid—frozen and thawed or fresh
  - Group Two: 3-4 videos of squid—especially capturing prey
  - Group Three: 1 model of squid
  - Group Four: 5-6 photographs of squid and different species of cephalopods, including some capturing prey
- Materials for arts and crafts, e.g., scissors, colored paper, glue, tape, pipe cleaners, etc.

For each participant

- Handouts (1 copy each)
  - Background information about squid feeding behavior (See **Getting Ready** #1)
  - **Object Types & Features**
  - **Reflection Exercise Worksheet for Rapid Design Feedback** and the **Reflection Questions** (if leading Session 2 next, Appendix B)
  - **Reflection Exercise Worksheet for Critiquing Objects in Learning Experiences** (if leading Session 3 next, Appendix B)

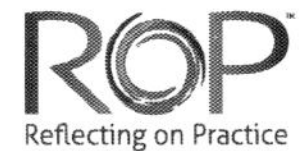

Getting Ready

1. Distribute background reading on squid feeding behavior a few days prior to the session so participants have some knowledge about squids for the Hands-on task. There is limited time during the session for this task, so being introduced to the information beforehand will be helpful.

   - Background information about feeding behavior of squids can be found from various sources on the Internet (e.g., https://teara.govt.nz/en/octopus-and-squid).

   - Either print out the information, or direct participants to the link.

2. Review the handouts and slide deck for this session, and additional suggested readings.

   For the facilitator:

   - *Learning Science in Informal Environments* (Chapter 5: "Science Learning in Designed Settings")

   Continue the Learning suggested for participants:

   - *Surrounded by Science* (Chapter 5: "Interest and Motivation: Steps Toward Building a Science Identity")

3. Duplicate handouts.
   For the Research Discussion,

   - Decide on 3-4 studies that are most relevant to your community's work; there are five studies from which to choose.

   - Make one set of studies for each small group and paper-clip them together for distribution (4 studies = small groups of 4). The studies will be distributed among members in each group.

4. To emphasize the specialness of objects, the slide deck features real examples of objects in a variety of informal learning environments. Feel free to substitute with your own examples.

5. Determine small-group formations for the **Research Discussion** and **Hands-on** tasks. Conversations in the former will lead into the latter; the goal is to foster a diversity of viewpoints in the conversations while allowing participants to deepen their discussions from one task to the next. It's important to strategize how to form the groups: you may want participants to stay in the same groups for both tasks, or have them form new groups for each task. Things to consider for your planning:

> The squid was chosen for two reasons. It is an intriguing organism, and many species are sustainably caught for people to eat. For more information, look up "squid" in Seafood Watch (www.seafoodwatch.org). Second, the four types of objects for the Hands-on task are easily attainable. If another organism is preferred or more easily accessible, feel free to substitute it. Be sure to gather materials for all four types of objects.

4.1

a. *Group size*:

- Less than six people per group is ideal.

- There are five investigations for the Research Discussion. Not all of the studies need to be reviewed by the community.

- There are four types of objects for the Hands-on; assign one group for each type of object. If needed, a couple of groups can work with a duplicate set of objects that are easily acquired; such as having two groups use the video.

b. *Same group/new group*:

- <u>Same groups for both the Research Discussion and Hands-on</u>:
  Advantage: participants may continue their conversations from one routine to the next, and thus **go deeper** into their thinking.

- <u>Changing groups between the two tasks</u>:
  Advantage: may invite different perspectives for participants to consider and **disseminate ideas**—especially useful if you choose not to do a whole-group share at the end of the Research Discussion.

c. *Educators' differences*: For all sessions in Module 4, if your community consists of educators from different departments, they may interact with different audiences, use different objects, and/or teach different types of programs, activities, classes, etc. Consider what might be most interesting and helpful for pushing on the practice of each department's members, particularly for tasks that involve participants revising activities, critiquing current use of objects, and extending beyond current practice. Options for grouping:

- <u>Keeping department members together in small groups</u>:
  Advantage: no time lost in understanding the context of other departments' objects and use. *Disadvantage*: minimizes the opportunity for "fresh eyes" and different perspectives.

  This arrangement may be preferred if the *Reflecting on Practice* sessions are the only or rare occasions in which department members can discuss and review their programs and activities as a unit.

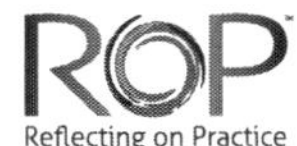

- Mixing department members across small groups:
  Advantage: participants can get a fresh perspective from
  colleagues outside of their department. *Disadvantage*: time
  is needed to contextualize the objects' use in each
  department's program.

  This arrangement may be preferred if the individual
  departments have other occasions to share and plan
  together beyond the *Reflecting on Practice* sessions.

6. Gather materials for four types of objects to explore squid (or
   other organism, if preferred) for the Hands-on task.

   - *Real.* Whole squids can be purchased fresh or frozen in many
     grocery stores, seafood stores, or bait shops. Preserved squid
     can be purchased from biological supply companies.

   - *Model.* Squid models, replicas, and toys can be purchased or
     made from simple, inexpensive materials. Make sure they are
     anatomically correct.

   - *Video.* There are many free videos of squids (and other
     cephalopods) swimming and feeding available on the Internet.
     Here are some examples.

     - http://www.youtube.com/watch?v=vT9fJLlFeKU

     - http://www.youtube.com/
       watch?v=OBg0k9GbHiw&feature=related

     - http://www.youtube.com/
       watch?v=URrXDJy1SGk&feature=related

     - http://www.youtube.com/watch?v=yTaEzlnw-
       LM&feature=related

   - *Photographs.* There are many images of squids and other
     cephalopods available on the Internet. We recommend images
     at least 500 dpi in resolution so the image prints fully on a piece
     of paper. Printing in color is ideal.

7. Make three reference posters on chart paper for the Hands-on task:

   - **Five Foundational Ideas on Learning** (Use same brief titles as
     Walkabout in Module 1 Session 1)

   - **Learning Cycle Design Framework** (List the phases and a brief
     description as shown on the **Reflection Questions** handout for
     the Learning Cycle (Appendix B)

Participants make this chart to organize their observation notes from the skits for each type of object. Title the chart for the type of object. If there are multiple skits for one type of object, they should add notes to the same chart.

For the whole-group discussion that follows all the skits, make four Observation Charts, one for each type of object, to record the community's observations and ideas. You may prefer to make these Observation Charts on a whiteboard, rather than chart paper.

For both tasks, decide whether to let participants choose their own partner/groups, or be assigned. Offering choice is helpful for getting commitment to engage in the task, though there may be work the community is doing or needs to get done that would benefit from assigning groups.

Partners and small groups should be somewhat familiar with the learning experiences and objects being reviewed. This arranging can help participants feel the tasks are relevant to their work and they have meaningful experiences to contribute.

- "Observation Chart"

	Type of Object	
	Possibilities	**Limitations**
Talk	What are learners and educator(s) able to talk about?	What are learners unable to talk about that would've been helpful?
Doing	• What are learners doing with the objects? • What information are learners able to get from their interaction with the object?	• What are learners unable to do with the objects that would've been helpful? • What information is the learner unable to get from their interaction with the object?

8. Preparation for the next session in **Continue the Learning**. Decide which session the community will do after this one, and choose the appropriate slide in the slide deck.

   a. Session 2 in this Module focuses on designing learning experiences. How objects are used in informal learning environments is very much defined by the design, which in turn, can influence the educator's teaching practice. This session would be a sensible next step for staff whose work involves developing programs, activities, exhibits, etc. If Session 2 will be next, distribute copies of the **Reflection Exercise Worksheet for Rapid Design Feedback** and the **Reflection Questions** (Appendix B) for participants to complete in pairs in advance of the session. During the session, pairs partner up to form small groups of four to share their work-in-progress and receive quick feedback.

   b. Session 3 in this Module is a deeper reflection on how objects are used and the Educator Moves involved. Prior to the Video Reflection in whole community, participants critique how objects are used in learning experiences. They will work in small groups for this task, which they should form prior to the session. If Session 3 will be next, distribute copies of the **Reflection Exercise Worksheet for Critiquing Objects in Learning Experiences** (Appendix B) for participants to complete individually. During Session 3, they will work in small groups to compare their individual reviews.

Session 1 Step by Step

Introduction
Session Objectives

1. **Introduce the session.**
   Share that in this session participants will investigate how *objects* can be used to encourage learners to ask and answer questions and make sense of the educational content.

2. **Display and review the session objectives**:

   - Consider what objects are and how to use them to support learning.

   - Discuss the limitations and possibilities of different types of objects to support learning.

   - Examine how interactions with learners are greatly influenced by the design of the object-based experience.

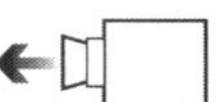

2 minutes

Think-Pair-Share: ***Teaching with "Things and Stuff"***

1. **Introduce the idea of *objects*.**
   Share that we all likely teach using "things and stuff"—otter pelt, live snake, images of Mars, ball on a stick to model the Moon, etc. Some of those are accessible and commonplace (e.g., ball on a stick), and others are "special" (e.g., otter pelt). Collectively, we're going to refer to these things and stuff as "objects." Here are two definitions to consider:

   a. The *Oxford English Dictionary* defines *object* as "a material thing that can be seen and touched."

   b. *Objects* are the "special" things and stuff—artifacts, exhibits, animals, specimens, etc. that informal learning environments have and offer access to. Objects are the raison d'être for the institution.

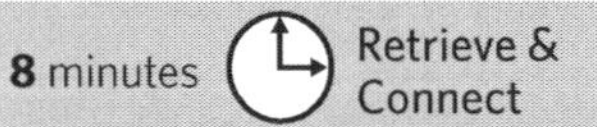

8 minutes — Retrieve & Connect

The term *objects* is deliberately vague. This keeps our discussion broadly inclusive as educators consider the objects they use and their role in learning and teaching in informal learning environments. There are philosophical and ethical issues pertaining to objects in the museology field—such as what constitutes an object and whose object is being represented by whom—that we don't address here. These issues and their implications to practice need to be addressed. In this session, however, the focus is to think critically about the things we use and how we leverage them to support learning.

2. **Start Think-Pair-Share.**

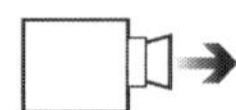

a. **Display the question for <u>Think</u>.** Ask participants to think about the following prompt:

How do you use objects *in your teaching to support learning?*

b. **Form Pairs.** After two minutes, ask participants to pair up with someone and discuss their thoughts.

3. **Facilitate a whole-group share.**
After about five minutes, invite participants to share their partner conversations. Facilitate a brief discussion to gather diverse viewpoints.

Research Discussion  **50** minutes

Case Studies:
How Learners Interact with Objects

1. **Transition to Research Discussion.**
Segue into the Research Discussion by making the following points about objects to set the context.

a. There's been a shift beyond exclusive focus on the objects and the knowledge that experts deem to be important, towards the meaning and connections visitors make from their interactions with the objects. Nonetheless, access to such objects remains an important part of why people visit these settings.

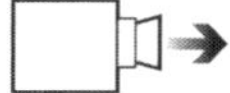

b. Use the PowerPoint slides as a visual complement as you share some collective thinking about such objects:

- They are often described as unique, original, genuine, or authentic.

- They may be collected for their natural, cultural, and/or historical significance, or be ascribed significance when they are collected and preserved.

- They may be built specifically for interaction with phenomena (e.g., an exhibit built to display colored shadows to explore how humans perceive color, such as at San Francisco's Exploratorium).

- They may be one of a kind (e.g., the most complete remains of a Dodo bird at the Oxford University Museum of Natural History), or an example of their kind (e.g., giant sequoias at Yosemite National Park in California).

- They may be living or nonliving.

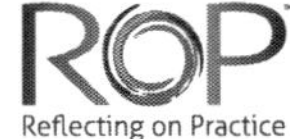

4.1

- They may…

  - evoke memories and sentiments. (For instance, photographs of the Vietnam/American war as captured by photojournalists traveling with the soldiers, on exhibit at the War Remnants Museum in Ho Chi Minh City, Vietnam.)

  - capture the intangible. (Such as a video recording of Tupac Shakur's last live performance at the House of Blues, exhibited at the Rock & Roll Hall of Fame in Cleveland.)

  - tell a story. (For example, the Alexander Fleming Laboratory Museum at St. Mary's Hospital in London uses reconstructions, displays, and videos to tell the story of Fleming's work and the discovery and development of penicillin.)

  - be digitized and made available through virtual exhibits and archives to extend access to the objects any time, anywhere for people with Internet connection.

- To aid our reflections and discussions about objects, it will be convenient to organize them by *type* (or common characteristics), and be reminded that their *features* may be starting points for conversations and elaborations with our learners.

c. Objects have cultural, historical, scientific, artistic, etc. information. Informal learning environments offer learners access to these objects for learning that may or may not be readily accessible to the general public. The question to ponder throughout this session is: *in our learning experiences, to what extent do we make space for learners to get this information themselves and talk about what they think about that information?*

2. **Set up the Case Studies Research Discussion.**

   a. Tell participants they'll be divided into small groups and given the same set of studies to read and discuss. These are empirical studies that investigate learners' mostly un-facilitated interactions with objects in informal learning environments. Participants will read a synopsis of each study and consider the limitations and possibilities for learning from different types of objects.

   b. After the research discussion, participants will work in small groups to design an experience that reflects some of the ideas discussed in the reading. Keep the upcoming design task in mind during this research discussion task.

The text for this research discussion is different from previous written materials. In all other readings, the text synthesizes and organizes many studies and ideas into a few big topics. This reading features details from a small sample of investigations.

Case study research is not sampling research (Stake, 1995). This method aims to understand one case in-depth, or may include several cases for a slightly broader representation. The sample size remains small and the researcher doesn't make generalizations from the investigation. The reader considers the details to generalize the findings with their own experiences and knowledge.

While studies in this Research Discussion are not case study investigations, we use them here as "cases" from a small sample of research in a particular context.

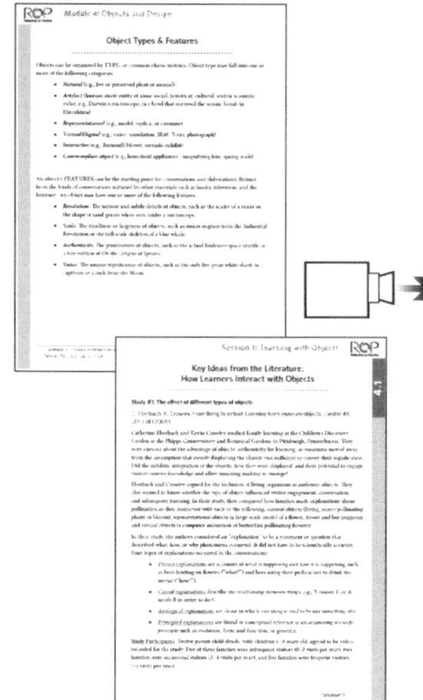

3. **Form groups and distribute handouts.**

   Direct formation of groups based on your plan for fruitful discussions. (See **Getting Ready** #5.) To each participant, distribute one copy of **Object Types & Features** handout. To each small group, distribute one set of **Key Ideas from the Literature: How Learners Interact with Objects**.

4. **Display and review the instructions on the slide:**

   a. Divide up the studies within the group so that everyone has at least one to read.

   b. On a blank page in their Learning Journal, draw a line down the middle to make two columns. Title the page **Objects in Learning Experiences**. Label one column **Limitations** and the other **Possibilities**.

   c. After reading, share a brief overview with your group and discuss the following questions for each study.

      ▫ What are the types of objects featured in this study? What is your experience with using them to support learning (personally and professionally)?

      ▫ From your experiences, what do you recall about how learners engage with and talk about these types of objects? How do the study findings add to your experiences regarding the limitations and learning potentials from these objects?

      ▫ What are new possibilities for learning from these types of objects that you hadn't considered before?

   d. Consider the limitations and possibilities for learning for the different types of objects featured in the case studies. For each type of object, add details in the respective column in your Learning Journal; this information will be a reference guide in the following hands-on task.

   e. Discuss any questions that emerge.

   f. (Optional) Be ready to share big ideas from the small-group discussion with the community.

5. **Circulate and monitor the pace of discussions.**

   It will be quiet at first, as everyone reads. Let participants know they should start sharing whenever the group is ready. Monitor across the groups to ensure they're working at about the same pace.

6. **Share discussion points.**

   As you circulate, bring the following ideas into the small-group conversations where appropriate.

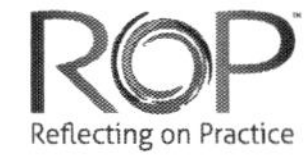

- *Observing is a foundational practice in science* (Study 3). Observing is a learned skill—knowing what to look for and using what you notice to further your inquiry. This is true of what participants are experiencing as they observe one another's practice.

- *Educators can spark visitor conversations* (Study 3). Visitors have their own agendas when they visit informal learning environments. Prompting visitors to connect objects to their prior knowledge and experiences may spark more conversations.

- *Prior knowledge informs science thinking and science reasoning* (Studies 3 & 5). Families engage in science talk and science reasoning in their conversations at informal learning environments. It may not all be accurate, or academically rigorous, but they do engage in it.

7. **(Optional) Begin whole-group share.**
   Call everyone back together and invite each group to share a couple of big ideas that emerged from their discussion. Model the Discussion Map as you facilitate. Suggested prompts:

   - What new connections to your knowledge and experience emerged for you that broadened your thinking in new directions?

   - From these connections, what generalizations can you make about learning with objects?

Playing with Objects

(A) Rough design experiences featuring objects (25 minutes)

1. **Post the reference posters.**
   Post the two posters in a prominent place that can be viewed by everyone in the room: **Learning Cycle Design Framework** and **Five Foundational Ideas on Learning**.

2. **Connect Research Discussion to Hands-on task.**
   Explain to participants that this hands-on task offers a chance to begin messing around with ideas from the studies in the Research Discussion, as they consider further, the limitations and possibilities of different types of objects.

3. **Introduce the task.**
   Tell participants they will work in four groups to *rough design* a learning experience to teach visitors about the *feeding behavior of squid*. Display and explain the following.

Rather than looking at it from a deficit perspective (i.e., what they aren't doing or are doing wrong), recognize what visitors <u>are</u> doing— talking and reasoning—and make their abilities explicit to them. This position takes a strengths perspective: people have knowledge and skills from their everyday lives that are useful, meaningful, and relevant (Gonzalez, Moll, & Amanti, 2005; Vélez-Ibáñez & Greenberg, 1992).

If you've decided to reconfigure the groups for the upcoming Hands-on routine, the new small groups will have a mix of people who can carry ideas from the research discussion into the next routine. In that case, it may not be necessary to do a whole-group share at this point.

70 minutes   Hands-on

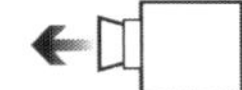

Direct people's attention to the reference posters (Five Foundational Ideas on Learning and Learning Cycle Design Framework) for their use in the task. Remind them they have the details for these posters in handouts from previous sessions.

Using the learning cycle as a common design structure for the experience, the performers can act out one or multiple phases, e.g., two minutes in Exploration Phase and then three minutes in Concept Invention.

Participants can continue in the same groups from the Research Discussion, or reconfigure the groups to mix experience, perspectives, and personalities.

a. *Rough design* is a rough draft, and as such, it's expected that there will be gaps and inconsistencies in the design. However, the general outline of the design should follow the Learning Cycle Design Framework, with opportunities for learners to participate in ways that embody what we know about how people learn (Five Foundational Ideas on Learning).

b. Each group decides the type of experience to design, e.g., cart activity, school program, livestream online, auditorium demo.

c. Each group will be assigned one type of object to feature in their experience: model, real, video, or photo.

d. Everyone has 15 minutes to rough design an experience with a science content focus on exploring the feeding behavior of squid.

e. Everyone will share their rough design in a 5-minute skit that acts out what the educator and learners would do with the object to explore the content.

4. **Articulate purpose of the activity.**
Acknowledge that 15 minutes is not enough time to create a great experience and when they design learning experiences, they aren't constrained to using only one type of object. Remind participants that this is just a rough design and everyone will be in the same situation. Display and explain two purposes for this activity.

- *Application.* Play with our new connections on learning with objects in work we do creating learning experiences. Use the "Objects in Learning: Limitations & Possibilities" details in your Learning Journals and the **Objects: Types & Features** handout to aid your engagement in this task. Two key questions to keep in mind: *What information can learners get from the object* (i.e., what can they talk about)? *How can they get that information from the object* (i.e., what can they do)?

- *Imagination.* Spark new thinking that pushes on the possibilities and limitations of four different types of objects for supporting learning. Draw from your experience about how learners talk and engage with these types of objects to broaden your generalizations.

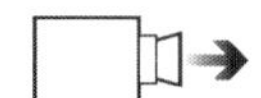

5. **Display instructions.**
Explain that their rough design will be a general outline of the whole experience, e.g., a 50-minute school program, and their 5-minute skit performs only part(s) of the whole experience. They will need to decide:

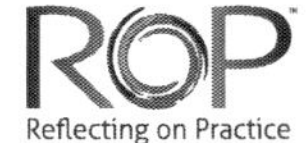

- *What type of experience will the group design?* Feature the object in a classroom-based program, cart activity on gallery floor, at an exhibit, exploration in the discovery lab, show in the auditorium, etc.

- *Who is the audience?* School group, family, general public, adults, etc.

- *What will learners do with the object to explore the concept? What will educator(s) do to support that engagement and discourse?* Everyone in the group should have an active role in the skit, there can be more than one "educator," and group members can be the "learners." Everyone in the community can also play the role of learners, e.g., they can be the general public in an auditorium, learners at home for a livestream, or students in a school group.

- *For the skit, what phase(s) of the learning cycle will be performed?*

6. **Form small groups and distribute materials.**
   Designate or let participants choose their groups. Assign each group a type of object and distribute materials accordingly. Serve as timekeeper or display and start a countdown clock once everyone has their materials.

7. **Circulate as small groups design.**
   While participants are designing their experiences, circulate to offer additional supplies and answer questions pertaining to the task. Call their attention to the timer to keep on track.

Groups don't have to use the arts and crafts materials. These materials are made available to be helpful, but caution participants to not let those materials distract them from the task to feature their assigned object.

(B) Perform skits featuring objects (25 minutes)

1. **Give a reminder about skit presentations.**
   Remind participants that each group will perform a 5-minute glimpse of the whole experience. Performers will need to provide the context of the experience.

   - What type of experience did the group design?
   - Who is the audience?
   - Who are the educator(s) and learners?
   - What phase(s) of the learning cycle does the skit occur in?

2. **Post "Observation Chart" and give instructions for Learning Journal.**

   a. Post one of the "Observation Charts." Let participants know that while they watch the performances, they will gather information about the types of objects, similar to the details they noted from the Research Discussion. If possible, they should write down evidence from the skits.

b. Instruct participants to make four copies of this chart in their Learning Journals, <u>titling each with the name of one of the object types</u>. Tell them they will use the charts to organize their observation notes from the skits for each type of object. If there are multiple skits for one type of object, they should add notes to the same chart.

3. **Restate the purposes of the task.**
Remind participants that the task is not to determine whether one type of object is better than the other, or compare effectiveness of the designs presented by one group or another. The purposes are to *apply ideas* from the Research Discussion and *spark imagination* about how each type of object allows for different ways of engaging and interacting. Hopefully, they will notice new possibilities not obvious before and find solutions for limitations.

4. **Orchestrate skit performances.**
Keeping to time can be challenging, as participants are excited to share their designs. Do the following:

   - Groups perform their skits, one skit at a time.

   - Participants should take notes on the "Observation Charts" in their Learning Journals during the skit.

   - Be strict about ending each presentation on time (or early) so that every group can share their design.

   - Give participants a minute to clean up their notes and transition between groups.

5. **Groups clean up their materials.**
After all the groups have finished, ask them to move their materials to the side of the room out of the way.

(C) Compare the objects (20 minutes)

1. **Participants compare notes, with partner or in small groups.**
Ask participants to discuss and compare their observation notes with a partner, or in a small group, in preparation for the whole-group discussion. They have about 10 minutes.

2. **Considerations for the partner/small-group discussion.**
Offer the following reminders and recommendations for participants to consider in their discussion.

   - These skits performed rough designs that they only had 15 minutes to prepare. Focus on the objects, not the skits or designs.

Either display a 5-minute count down or hold up a 1-minute warning sign for performers to see as their time approaches the end.

- Don't let the skits or rough design constrain or limit your ideas, but rather use them as a springboard to discuss limitations and possibilities for the objects.

- Draw on their perspective as the designers to offer insights on what they felt they could and couldn't do with the object, regardless of whether the sentiment showed up in their design or skit.

- Connect with the study findings from the research discussion.

3. **Begin whole-group share.**
   Explain that this whole-group conversation will consolidate the community's thinking about the objects. Focus on one object at a time. Record their comments in the "Observation Chart" for each object.

4. **Push for elaboration and listening.**
   Encourage the participants to elaborate on and offer evidence for their thinking. Invite them to listen closely, consider what peers share, and ask questions of each other.

Connect, Extend, Challenge, & Apply: *Pushing the Possibilities and Limitations*

10 minutes — Let's Talk Practice

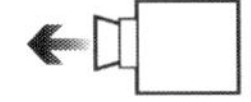

1. **Display the Let's Talk Practice questions.**
   Invite participants to capture their thinking from this session in individual entries in their Learning Journals. Ask them to look across the information about objects in the handout, worksheet, and chart.

   - **Connect.** *How are the ideas and information discussed in this session connected to what you already know?*

   - **Extend.** *What new possibilities for the types of objects you use did you notice? What potential solutions for limitations might you want to explore further?*

   - **Challenge.** *What still seems challenging or confusing? What questions or wonderings do you have?*

   - **Apply.** *What one or two changes will you make in your practice around the use of objects to apply your new ideas or understandings?*

2. **(Optional) Conduct a whole-group share.**
   If time permits, invite individuals to share their ideas with the community.

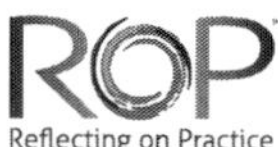

2 minutes

Continue the Learning

1. **Suggested reading.**

   If you've decided to assign the reading, share it now.

2. **Advance preparation for Module 4 Session 2.**

   If you've decided to do Module 4 Session 2 next, distribute the **Reflection Exercise Worksheet for Rapid Design Feedback** and the **Reflection Questions** and provide the following instructions.

   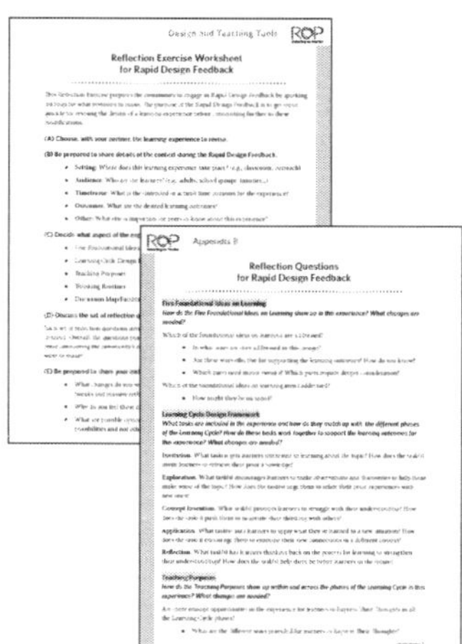

   a. Explain that the next session is focused on designing learning experiences, specifically examining and revising the community's own work. The **Let's Talk Practice** task in Session 2 involves getting input quickly for revisions before committing further to modifications.

   b. Everyone needs to do some pre-thinking with a partner to prepare for this task. During the next session, two pairs will partner up to form a group of four to do the *Micro Lab for Rapid Design Feedback*.

   c. Now, pair up with someone to select the learning experience the two of you would like to revise, are currently revising, or are developing new.

   d. Before the next session, work with your partner to complete the **Reflection Exercise Worksheet for Rapid Design Feedback**.

   e. Tell participants that they can skip the Thinking Moves and Thinking Routines questions on the **Reflection Questions** handout for now.

3. **Advance preparation for Module 4 Session 3.**

   If you've decided to skip Session 2 and do Session 3 next, then distribute the **Reflection Exercise Worksheet for Critiquing Objects in Learning Experiences** worksheet and provide the following instructions.

   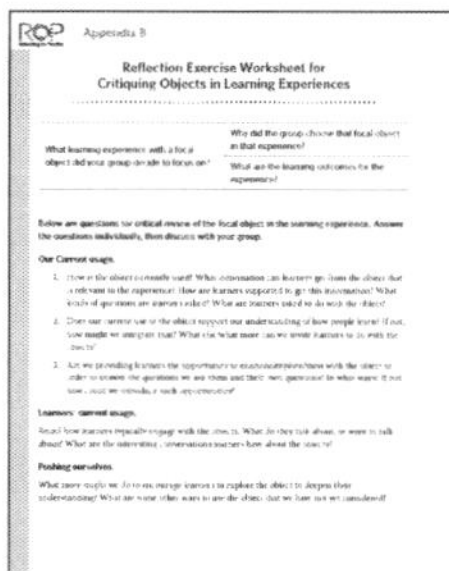

   a. Session 3 includes the familiar Video Reflection in whole community. A volunteer will be needed.

   b. There will also be a hands-on task to critique objects. This task focuses on how objects are used in the learning experiences your community offers.

   c. Everyone needs to do some preparation for that task. The task will be done in small groups, which they will form now.

d. Within their groups, discuss and decide which learning experience they would like to review in light of how the objects are used.

e. Individually, complete the **Reflection Exercise Worksheet for Critiquing Objects in Learning Experiences**. Remind them to use the materials from Session 1. They will share and discuss their independent reviews during the next meeting (Session 3).

Materials from this session include handouts and self-generated charts.

- **Key Ideas from the Literature: How Learners Interact with Objects**

- **Objects: Types & Features**

- "Objects in Learning Experiences: Limitations & Possibilities" notes from Research Discussion

- "Observation Charts" for the skits

Object Types & Features

Objects can be organized by **TYPE**, or common characteristics. Object type may fall into one or more of the following categories:

- *Natural* (e.g., live or preserved plant or animal)

- *Artifact* (human-made entity of some social, historical, cultural, and/or scientific value, e.g., Darwin's microscope, rice bowl that survived the atomic bomb in Hiroshima)

- *Representational* (e.g., model, replica, or costume)

- *Virtual/Digital* (e.g., video, simulation, SEM, X-ray, photograph)

- *Interactive* (e.g., Bernoulli blower, tornado exhibit)

- *Commonplace object* (e.g., household appliances, magnifying lens, spring scale)

An object's **FEATURES** can be the starting point for conversations and elaborations distinct from the kinds of conversations initiated by other materials such as books, television, and the Internet (*1*). An object may have one or more of the following features:

- *Resolution.* The minute and subtle details of objects, such as the scales of a snake or the shape of sand grains when seen under a microscope.

- *Scale.* The smallness or largeness of objects, such as steam engines from the Industrial Revolution or the full-scale skeleton of a blue whale.

- *Authenticity.* The genuineness of objects, such as the actual *Endeavor* space shuttle or a first edition of *On the Origin of Species*.

- *Value.* The unique significance of objects, such as the only live great white shark in captivity or a rock from the Moon.

---

1. G. Leinhardt, K. Crowley, in *Perspectives on Object-Centered Learning in Museums*, S. G. Paris, Ed. (Lawrence Erlbaum Associates, Mahwah, NJ, 2002), pp. 301–324.

    This page may be duplicated for educational use.

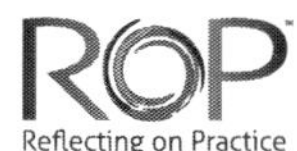

Key Ideas from the Literature:
How Learners Interact with Objects

Study #1: The effect of different types of objects

Catherine Eberbach and Kevin Crowley (*1*) studied family learning at the Children's Discovery Garden at the Phipps Conservatory and Botanical Gardens in Pittsburgh, Pennsylvania. They were curious about the advantage of objects' authenticity for learning, as museums moved away from the assumption that merely displaying the objects was sufficient to convey their significance. Did the exhibits' integration of the objects, how they were displayed, and their potential to engage visitors convey knowledge and allow meaning-making to emerge?

Eberbach and Crowley argued for the inclusion of living organisms as authentic objects. They also wanted to know whether the *type* of object influenced visitor engagement, conversation, and subsequent learning. In their study, they compared how families made explanations about pollination as they interacted with each of the following: *natural* objects (living, insect pollinating plants in bloom); *representational* objects (a large-scale model of a flower, flower and bee puppets); and *virtual* objects (a computer animation of butterflies pollinating flowers).

In their study, the authors considered an "explanation" to be a statement or question that described what, how, or why phenomena occurred. It did not have to be scientifically accurate. Four types of explanations occurred in the conversations:

- <u>*Process explanations*</u> are accounts of *what is happening and how it is happening*, such as bees landing on flowers ("what?") and bees using their proboscises to drink the nectar ("how?").

- <u>*Causal explanations*</u> describe *the relationship between things*, e.g., X causes Y, or A needs B in order to do C.

- <u>*Analogical explanations*</u> are those in which *one thing is said to be like something else*.

- <u>*Principled explanations*</u> are literal or conceptual *reference to an organizing scientific principle* such as evolution, form and function, or genetics.

<u>Study Participants</u>. Twelve parent-child dyads, with children 6–9 years old, agreed to be video-recorded for the study. Five of these families were infrequent visitors (0–2 visits per year), two families were occasional visitors (3–4 visits per year), and five families were frequent visitors (>5 visits per year).

(continues)

This page may be duplicated for educational use.

Key Ideas from the Literature:
How Learners Interact with Objects *(continued)*

<u>Study Design</u>. Parent-child dyads at the garden were instructed to participate with all three types of objects (natural, representational, and virtual), taking as little or as much time as desired. The families were told that the objects would give them an opportunity to learn about pollination. Their experiences were video-recorded and their conversations were transcribed.

<u>Key Findings</u>.

- Learners mostly talked about biological processes: 78% of the explanations related to pollination.

- 71% of the explanations were accurate.

- Families made references to shared experiences, e.g., watching television programs, pets, gardening, and school.

- Explanations were predominantly *process* (62%), with some *analogical* (20%) and *causal* (12%), and the fewest were *principled* (6%).

- Learners made more and longer *process explanations* when exploring **representational** and **virtual** objects than when exploring **natural** objects.

- Learners made more connections to school when exploring **representational** objects.

- Learners made more connections to everyday experiences when exploring **natural** objects than when exploring **virtual** objects.

---

1. C. Eberbach, K. Crowley. From living to virtual: Learning from museum objects. *Curator* **48**, 317-338 (2005).

(continues)

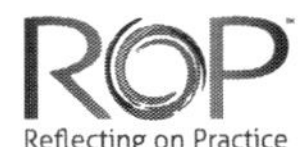

4.1

Key Ideas from the Literature:
How Learners Interact with Objects *(continued)*

Study #2: Making the objects

Kimberly Sheridan, Erica Rosenfeld Halverson, Breanne Littis, Lisa Brahms, Lynette Jacobs-Priebe, and Trevor Owens (*1*) conducted comparative case studies at three makerspaces: *Sector67* in Madison, Wisconsin; *Mt. Elliott Makerspace* in Detroit, Michigan; and *Makeshop* at the Children's Museum of Pittsburgh in Pennsylvania. Empirical studies had primarily focused on what young people learn through specific making activities within a makerspace, e.g., building circuits into textiles or using Scratch (programming language) for interactive media design. Sheridan and her colleagues argued that makerspaces and the maker movement offer greater learning potential than these activities suggest. The team wanted to understand how different makerspaces functioned as learning environments.

<u>Study Sites</u>. Both adults and children use these three makerspaces. *Sector67* has expensive technical equipment for participants to use, including 3-D printers, kilns, oscilloscopes, and woodworking tools, and new equipment and tools are purchased on the basis of input from the community. Participants are primarily entrepreneurs (making products for sale), hobbyists (making objects in their free time), and novices (learning to be makers); they either pay a one-time fee for use or have a membership. *Mt. Elliott Makerspace* is located in a church basement; its mission is to be a thriving model for under-resourced neighborhoods. It has areas for bike repair, woodworking, electronics, and silk-screening, as well as a kitchen and a computer lab. Areas and tools are added as interest in (and resources for) activities change. Participants range from toddlers to seniors from the local neighborhood, although most are 8–19 and a few core adults also participate. Participation is open to all when the space is open. *Makeshop* has floor space at a children's museum, and is a collaboration between the Carnegie Mellon's Entertainment Technology Center and the University of Pittsburgh Center for Learning in Out-of-School Environments (UPCLOSE). Areas are carefully designed to introduce young children to making, e.g., woodwork, digital media and programming, and sewing. The space is facilitated by teaching artists who have expertise in an area of making; these facilitators offer individualized support for making projects and lead group workshops. Participants are visitors to the museum: toddlers to teens and their accompanying adults.

<u>Study Design</u>. Researchers visited the three makerspaces over a 12-month period. They collected more than 150 hours of field observations and interviews, and conducted extensive review of the makerspace resources—websites, online forums, and video and photo documentation of products.

(continues)

This page may be duplicated for educational use.

Key Ideas from the Literature:
How Learners Interact with Objects *(continued)*

<u>Key Findings</u>.

- *Time spent on projects* was the most striking difference across the sites: participants spent from weeks to years in *Sector67*, from hours to weeks in *Mt. Elliott*, and from minutes to hours at *Makeshop*. This range reflected the relationship between the makers and the space—members of the community either sharing and collaborating or dropping in for one-off experiences.

- *Learners worked iteratively* with their ideas, materials, tools, and processes to make their objects.

- *Making the objects cut across disciplines.* Sewing occurred alongside electronics, computer programming took place next to woodworking and welding. The very proximity of these diverse disciplines to one another cultivated innovation; at *Makeshop,* a young girl making a blanket for her American doll bed fluidly shifted from sewing to circuitry to reusing recycled materials to make a nightlight to go with the blanket.

- *Learning was purposeful and authentic.* All three sites offered a variety of ways in which to learn: structured workshops; experts training novices to use specific equipment; solicited and unsolicited feedback from peers and others; guidance from a facilitator; watching others; and free exploration. For example, the young girl at *Makeshop* was watching a boy make a flashlight with a light circuit while she was sewing. When she finished her blanket, she consulted with her mom about how to make a light, and then figured out how to make a working light through trial and error. In most of these cases, the learning was self-directed; the learners learned what they needed in order to make the objects they wanted.

---

1. K. Sheridan et al., Learning in the making: A comparative case study of three makerspaces. *Harvard Educational Review* **84**, 505-531 (2014).

(continues)

 This page may be duplicated for educational use.

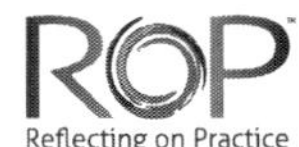

Key Ideas from the Literature:
How Learners Interact with Objects *(continued)*

Study #3: Using tools for engaging in science talk

Heather Toomey Zimmerman, Lucy Richardson McClain, and Michele Crowl (*1*) studied how families use magnifiers during trail walks at Penn State's Shaver's Creek Environmental Center in central Pennsylvania. They identified a need to understand how families engage in scientific thinking in outdoor spaces where labels are limited (or cannot be provided) and the setting naturally changes (seasonally, annually). The researchers focused on magnifiers as a tool to support scientific thinking, because of the magnifiers' popularity in outdoor education programming and their prominence in life-science practices of observation.

Observing is a foundational practice in science, but it is often mistakenly thought to be a simple undertaking (*2*). Novices use observation to collect information, but may not be able to distinguish between relevant and irrelevant features, or to connect features to disciplinary concepts, and may conflate observable evidence with their own beliefs. Experts use observation throughout the scientific inquiry. They hone in on relevant features; use disciplinary concepts to focus, organize, and connect their observations; reason through use of observation data; and coordinate observation data with hypothesis testing. Novices need guidance to move along the continuum towards more expert observation skills.

<u>Study Participants</u>. Seven families (24 people total) allowed researchers to video-record their walks during naturalist-led public programs.

<u>Study Design</u>. The families were recorded at the nature center as they participated in naturalist-led walks, although they were encouraged to explore the trail at their own pace. They were offered a selection of magnifiers to take with them on the walk: binoculars, hand lenses, and/or bug boxes. They were not given instructions on how to use the tools, but would have been provided instructions on request (none of the families in this study requested help). Their walks were video-recorded for 45–120 minutes each, after which researchers selected instances when the magnifiers were used for in-depth analysis. These instances (2 hours total) were then transcribed verbatim, retaining slang terms (e.g., "wanna") and marked pauses (e.g., "uhh"), and relevant gestures and actions were described.

Key Ideas from the Literature:
How Learners Interact with Objects *(continued)*

<u>Key Findings</u>.

- Families used the magnifiers in their own ways, not just for meaningful conversations about plants and animals in their local environment as intended by the nature center.

- Almost half (47%) of the use of magnifiers was not connected to a science topic; this occurred mostly at the beginning of the walk, or the first time a family member held the tool. In these instances, usage was free-form exploration (e.g., using the binoculars to look around without dialogue or discussion), for dramatic and imaginary play, and/ or to signify authority (i.e., the person holding the magnifier felt or was perceived to be "in charge").

- More than half (53%) of the magnifiers' use was connected to a science topic, and this happened most often midway through the walk. In each case, the magnifier was used to *identify* a natural object. In three quarters of those instances, families went on to *describe* what they saw—color, location, quantity, shapes. In about 38% of <u>those</u> instances, families were *interpreting* the object they identified and *applying* ideas. This talk involved a family member (adult or child) explaining the significance of the phenomenon or object they'd found, and making connections to prior experiences that were primarily out-of-school.

- Talk initiated by the naturalist (e.g., the naturalist stopped in front of a skunk cabbage in the wetland area and encouraged families to look carefully for "signs of spring") often did not spark family talk beyond identifying. The majority of science talk that involved *describing* and *interpreting* and *applying ideas* was family-initiated.

1. H. T. Zimmerman, L. R. McClain, M. Crowl. Understanding how families use magnifiers during nature center walks. *Research in Science Education* **43**, 1917-1938 (2013).

2. C. Eberbach, K. Crowley, From everyday to scientific observation: How children learn to observe the biologist's world. *Review of Educational Research* **79**, 39-68 (2009).

(continues)

This page may be duplicated for educational use.

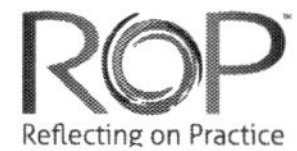

Key Ideas from the Literature:
How Learners Interact with Objects *(continued)*

● ●

Study #4: Nature of the objects for promoting conversations

Jill Hohenstein and Lynn Uyen Tran (*1*) studied the use of questions in exhibit labels to generate explanatory conversations at an object-based gallery in an urban science museum in the United Kingdom. They pointed to the significance of making explanations for conceptual learning, and the use of open-ended *Wh* questions for promoting further dialogue and reflection. Hohenstein and Tran recognized that visitors read and used exhibit labels, and wondered if and how a provocative question—*"Why is this here?"*—might catalyze conversation among visitors. They examined conversations at the locations of three artifactual objects that differed in their *resolution*—more specifically, in their **physical complexity** and **self-explanatory** natures.

For the purposes of this study, Hohenstein and Tran defined explanation as a statement that involves reasoning related to causal relations, processes, scientific principles, and analogies (*2*). Explanations are different from other categories of informal talk, such as *identification*, in which a speaker calls out or names objects or parts of objects, or *description*, in which a speaker elaborates upon details of the object. Explanations that arise in everyday conversation present excellent opportunities for children to articulate and revise their theories of scientific phenomena, with guidance from parents and other adults (*3*). **Physical complexity** refers to the intricate details of an object. **Self-explanatory** describes the efficacy of an object in conveying its significance (idea, concept) solely by being observed or manipulated, without interpretive guidance.

<u>Study Participants</u>. Visitors (children and adults) who entered the camera's view near the three exhibits of interest were video recorded for the duration of their time at each exhibit.

<u>Study Design</u>. Video data were collected at three exhibits and conversations were transcribed. The three exhibits were selected because they varied in historical era, popularity, and type: one object was a model of a Victoria-era workshop; another was the cross-section of an Austin Mini Cooper from the 1959 Auto Show; and the third was a common rice bowl that had survived the atomic bombing of Hiroshima in World War II. Three labeling conditions were applied: Current Label, using the existing, unaltered museum label text; Added Questions, using the museum's label plus a provocative guiding question (Why is this here?); and Simplified Text, using a reader-friendly version of the original text plus the provocative question.

Key Ideas from the Literature:
How Learners Interact with Objects *(continued)*

• •

<u>Key Findings</u>.

- Visitors remained at the Victorian workshop significantly longer than at either of the other two exhibits; they stayed the shortest and talked the least at the rice bowl.

- Adding the provocative question increased explanations and open-ended questions at two of the three exhibits. There was no change in visitor response at the rice-bowl exhibit, but the question generated positive learning conversations at the other two objects: visitors at the Victorian workshop offered more *causal* explanations (relationship to or dependence on things to one another), whereas at the Mini Cooper, they asked more *open-ended questions.*

- The *resolution* (e.g., the physical complexity and self-explanatory nature of objects) was different across the three objects.

  - The Victorian workshop is the most physically complex. The machines are visible; visitors push a button to see how the real workshop would've operated. This complexity made the workshop the most self-explanatory of the three.

  - The cross-section of the Mini Cooper is almost equally complex, but less self-explanatory than the workshop. Visitors have to infer—by looking at the cross-section—how the transverse placement of the engine revolutionized car making.

  - The rice bowl is the least complex and self-explanatory. The bowl is remarkably intact after surviving the atomic bomb, and sand has fused onto its surface. It is placed next to photographs of an atomic mushroom cloud.

- At the Victorian workshop, most visitor explanations were directed at visible details about the mechanisms driving the machinery. Questions and explanations at the Mini Cooper focused on why certain sections had been removed and what was made visible as a result; few visitors talked about the engine or car design. For the rice bowl, visitors primarily talked about content from the exhibit label about the significance of the fused sand.

---

1. J. M. Hohenstein, L. U. Tran. The use of questions in exhibit labels to generate explanatory conversation among science museum visitors. *Intl. J. Sci. Educ.* **29**, 1557–1580 (2007).

2. F. C. Keil, Explanation and understanding. *Annual Review of Psychology* **57**, 227–254 (2006).

3. K. Crowley et al., Shared scientific thinking in everyday parent-child activity. *Science Education* **85**, 712–732 (2001).

(continues)

     This page may be duplicated for educational use.

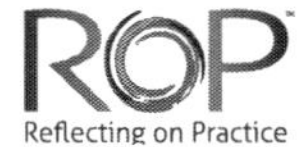

Key Ideas from the Literature:
How Learners Interact with Objects *(continued)*

Study #5: Family engagement in scientific reasoning

James Kisiel, Shawn Rowe, Melanie Ani Vartabedian, and Charles Kopczak (*1*) studied family interactions at touch-tank exhibits in four aquariums on the west coast of the United States. They pointed out that families engage in scientific reasoning during visits to informal science learning environments, but most studies have occurred at physical science–based interactive exhibits. Kisiel and his colleagues wondered about the opportunities for encouraging scientific reasoning where the experience includes both live organisms *and* physical interaction. They argued that touch-tanks offer children and adults the chance to interact with (particularly, touch) the organisms and talk with one another and aquarium staff about the experience. The multi-sided, multi-user characteristics of a touch-tank might be a valuable feature at which families can engage in **scientific reasoning** during visits to aquariums.

Scientific reasoning refers to generating and assessing evidence in order to make explanations and arguments. Evidence is not mere recall of facts and figures; producing evidence involves asking questions, formulating hypotheses, and designing and implementing experiments. Katherine McNeill and Dean Martin (*2*) offer a useful set of definitions for these components of scientific reasoning:

- A *claim* is a statement that answers a question or problem.

- *Evidence* consists of data that support the claim (either from investigations that learners conduct firsthand or from others' research and publications).

- *Reasoning* provides a justification for why or how the evidence supports the claim, and often includes scientific principles or ideas that learners apply to make sense of the data.

<u>Study Participants</u>. Forty-one families participated in the study. A family was defined as a multigenerational group of at least one adult and one child 3–17 years old. Group size ranged from two to eight, with mean of four. About one third of families were first-time visitors at the particular study site, though 95% indicated that they had visited a touch-tank exhibit before.

(continues)

Key Ideas from the Literature:
How Learners Interact with Objects *(continued)*

• •

<u>Study Design</u>. Families were invited to participate and their interactions at the touch-tank exhibit were video-recorded for the length of their stay at the exhibit. Their conversations were transcribed for analysis. Families stayed an average of 15 minutes (ranging from 5 to 26 minutes). After their experience at the touch-tank, families were interviewed for information regarding their demographics and their aquarium-visiting behavior. Aquarium staff members were also interviewed, to gather information on the institutions' perspectives of these spaces.

<u>Key Findings</u>.

- Families engaged in a wide variety of activities and discourse at touch-tank exhibits (e.g., identification, gathering information), including scientific reasoning activities: making claims, challenging claims, and confirmation activities.

- Families made claims based on their observations (e.g., animal behavior or appearance) or prior knowledge, followed by an assertion or question.

- Families mostly acknowledged claims without dispute, though sometimes the validity of a claim was challenged. The challenge led either to stronger support for the original claim or adjustment of the initial claim.

- Families took on evidence-seeking activities to confirm the challenging counterclaim. These included consulting an authority, seeking evidence, and conducting tests.

- None of the touch-tank exhibits featured explicit interpretive goals or staff guidance about how scientists make sense of the organisms on display, i.e., how science works.

- Aquarium staff viewed the purpose of the touch-tank to be (a) promoting stewardship, (b) promoting learning, (c) providing unique experiences, and/or (d) providing an institutional attraction. Although promoting learning could include outcomes related to scientific reasoning or participating in the process of science, statements from staff focused on understanding and awareness of animal features and behavior, as well as their roles within the ecosystem.

<hr>

1. J. Kisiel, S. Rowe, M. A. Vartabedian, C. Kopczak. Evidence for family engagement in scientific reasoning at interactive animal exhibits. *Science Education* **96**, 1047–1070 (2012).

2. K. L. McNeill, D. M. Martin, Claims, evidence, and reasoning. *Science and Children* **48**, 52 (2011).

 This page may be duplicated for educational use.

Module 4.
Session 2: Designing Experiences for Learning

Session Overview

This session brings together the research and tools for supporting learning discussed throughout the *Reflecting on Practice* program, with the specific goal of incorporating these ideas into the design of learning experiences. Participants work with colleagues to review selected activities used at their institution, and then initiate revisions to better align with and incorporate the Learning Cycle framework, Thinking Routines, Teaching Purposes, and Five Foundational Ideas.

Session Objectives

- Review *Reflecting on Practice* Tools focusing on how each one supports learning and engagement for important takeaways to bring into educators' practice.

- Introduction to Thinking Moves and Thinking Routines that have been used to support learning throughout the program.

- Redesign a chosen learning experience to reflect ideas from research emphasized in *Reflecting on Practice*.

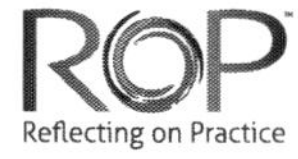

SESSION AGENDA		
Task Routine	Description	**Estimated Time** (in minutes)
Introduction *Session Objectives*	The goals and objectives of the session are introduced.	3
Retrieve & Connect Quick Write: *Designing Learning Experiences*	Participants respond to a prompt focused on incorporating what they've learned into the design of experiences to support all learners.	7
Let's Talk Practice *What Thinking Looks Like*	The community is introduced to Thinking Moves and Thinking Routines that have been used to support learning throughout the program.	20
Hands-on *Design & Teaching Tools Match Up*	The community reviews *Reflecting on Practice* tools for designing experiences to support learning and engagement. Participants reflect on the learning research that underlie each of the tools and pedagogy.	45
Break option		10
Let's Talk Practice Micro Lab: *Rapid Design Feedback*	Participants work in small groups to share inklings for revising their learning experiences and receive input before they commit further to modifications.	60
Continue the Learning	Participants are tasked with continuing to revise one of their existing lessons.	5
	Total Estimated Time	150 mins. (2.5 hrs.)

4.2

MATERIALS

Recurring for all Sessions

- See Module 1 Session 1

For this Session

- **Read-Aloud for the Facilitator: What Does Thinking Look Like?**

For each small group

- **Card Sets for *Design & Teaching Tools*** (See **Getting Ready** #4)
  - 2 paper-clips
  - Envelope
- Handouts (1 copy each from Appendix A)
  - **Educator Moves—Prior Knowledge**
  - **Educator Moves—Conversation**
  - **Educator Moves—Objects**
- Handouts (1 copy each from Appendix B)
  - **Learning Cycle Design Framework**
  - **Discussion Map**
  - **Facilitation Approaches**
  - **Teaching Purposes**

For each participant

- Handouts (1 copy each from Appendix B)
  - **Thinking Moves & Thinking Routines**
  - **Reflection Exercise Worksheet for Critiquing Objects in Learning Experiences**

A detailed handout for the Thinking Routines is available on the Companion Website **(www. routledge.com/CW/Tran)**. Some participants may find the additional information helpful for deciding how to use particular Thinking Routines in their learning designs. Review and decide whether this handout would be useful for your community. It may work well to have a few copies available for participants to use as needed.

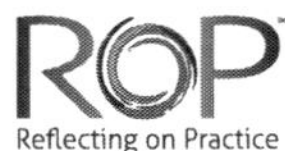

Getting Ready

1. Follow up to ensure that all participants complete the **Reflection Exercise Worksheet for Rapid Design Feedback** with their partner prior to the session. They don't need to have ideas for revision fully developed prior to the session, but they should do some pre-thinking so the time spent during the session is productive for everyone.

2. Duplicate handouts.

3. Decide on the make-up of small groups for the Micro Lab. Cluster two or three pairs together to make small groups of four to six.

   a. Ideally, all groups should be the same size so that the whole community progresses through the Micro Lab together.

   b. Keep in mind that groups with three pairs would have less time for open discussion than those with only two pairs. In such a case, clustering pairs that are revising similar types of experiences (e.g., cart activity) can aid them in building on one another's experiences and ideas.

   c. It may be desirable to cluster particular pairs together due to a variety of reasons: experiences they're revising, personalities, types of work, etc.

4. Prepare sets of cards for the hands-on activity, *Design & Teaching Tools Match Up*

   a. Decide on the number of small groups of 3-4 people to determine how many sets of Design & Teaching Tools Card Sets to make (one set per group).

   b. Duplicate the Card Sets and Instruction Cards for Design & Teaching Tools for the number of small groups (e.g., 4 groups = 4 copies).

   c. For each copy, cut along the dotted line for Card Set #1, and paper-clip the 9 cards together; cut along the dotted line for Card Set #2, and paper-clip the 9 cards together. Cut along the dotted line for the Instruction Cards.

   d. For each set, place a paper-clipped set of Card Set #1 and Card Set #2, along with the corresponding Instruction Card, into an envelope.

You may want to have some replacement copies of the **Reflection Exercise Worksheet for Rapid Design Feedback** in case they are needed.

4.2

Session 2 Step by Step

3 minutes

Introduction
Session Objectives

1. **Introduce the session.**
   Share that in this session, participants will review the research on learning and teaching they've read and discussed in previous sessions, and purposefully incorporate those ideas into the design of one of their activities or programs.

2. **Display and introduce session objectives.**

   - Review *Reflecting on Practice* Tools focusing on how each one supports learning and engagement for important takeaways to bring into your practice.

   - Introduction to Thinking Moves and Thinking Routines that have been used to support learning throughout the program.

   - Redesign a chosen learning experience to reflect ideas from research emphasized in *Reflecting on Practice*.

Retrieve &
Connect 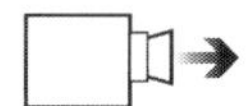 **7** minutes

Have everyone respond to the prompt, even if they can't find (or don't have) the initial entry.

Quick Write: *Designing Learning Experiences*

Take time for Learning Journal entries.
Have participants locate and review their Quick Write journal entry from Module 2 Session 1: *how do we design experiences to support all our learners?* Display and explain the reflection task:

- Review their original entry.

- Reflect on the conversations and experiences on learning they've been engaging in.

- Date a new entry.

- Expand on their initial thinking in any of the following ways:

  □ Elaborate with more details on what they wrote initially

  □ Add new ideas they hadn't considered before

  □ Revise what they wrote; perhaps their views have changed

  □ Pose questions they might be wondering; for instance, they might still struggle with how to make space for conversations in their designs

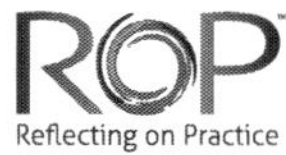

What Thinking Looks Like

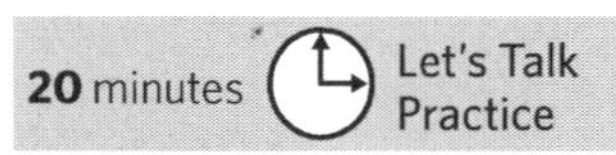

20 minutes — Let's Talk Practice

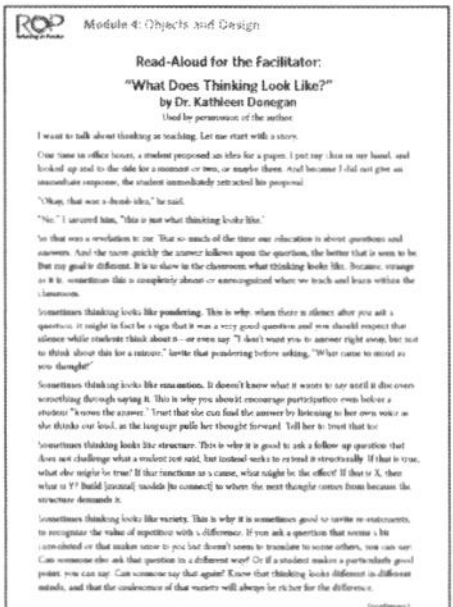

1. **Introduce "What Does Thinking Look Like?"**

   a. Tell participants that, to introduce another Teaching Tool, you'll read aloud an essay written by an English professor at the University of California, Berkeley and delivered to her peers during a professional learning experience for university faculty in April 2017. Ask participants to sit back and listen as you read.

   b. After the reading, ask participants to jot down a few ideas that came to them as they listened. Possible prompts include:

      ◻ What connections can they make?

      ◻ What do they think about the thesis of the essay: when we teach, it may not be obvious to our learners *what thinking looks like.*

      ◻ What do they think about Professor Donegan's description of what thinking looks like?

   c. Ask for volunteers to share some of their thoughts.

2. **Emphasize key point about thinking.**
   If this point doesn't come up from participants' reactions to the reading, share how this essay illuminates the need to make thinking recognizable to our learners, and that we can do this through our interactions with them.

   ■ As reflected in Professor Donegan's story, it may not be obvious to our learners what thinking looks like when we teach: pausing to consider an idea before responding, or asking a question to acquire additional information to consider, etc.

   ■ In teaching, we tend to keep thinking invisible to learners. For instance, all the thinking and rethinking involved in designing an activity takes time and careful thought—but learners only experience the finished product, as if it only takes 50 minutes to understand a concept. Another example: When we select the first or quickest answers to a question we've posed, it implies that these are the right and only answers, as if thinking is about fast recall of information.

3. **Introduce Thinking Moves.**
   Ask (rhetorically), *When we ask our learners to "think," what are we asking them to do? When we consider what kinds of thinking we want to promote, and what is useful when learners are trying to understand concepts, what comes to mind?*

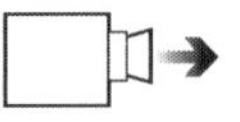

This list and the term "thinking moves" are borrowed from Project Zero's Visible Thinking (Ritchhart, Cheuch, & Morrison, 2011).

Display a short list of "thinking moves" shown to be integral to understanding. Explain that thinking moves are **actions** that learners take when they think and work towards understanding. Acknowledge that the list is not exhaustive.

- Observe closely and describe what's there

- Build explanations and interpretations

- Reason using evidence

- Make connections

- Consider different viewpoints and perspectives

- Capture the heart (core) of a concept and form conclusions

- Wonder, spark curiosity, and ask questions

- Uncover complexity and going below the surface of things

4. **Explain how Thinking Moves complement Educator Moves.**
   Point out that the Educator Moves we've been using in the Video Reflections focus our attention on what the educator does to support learning. The Thinking Moves listed here complement the educator's actions and focus on what we want our learners to be doing.

5. **Highlight the importance of making thinking recognizable to learners.**

   - Thinking Moves make the act of thinking explicit for both learners and educators. It specifies the actions that learners can take when they think and work towards understanding, and it reminds us of what we need to do (or not do) to support thinking.

   - This process can help uncover misconceptions learners may hold, and affords us and them a window into what they're understanding, how they're understanding it, and what they don't yet understand. As educators, we can use this information to determine what additional experiences we may need to provide to help them understand or reach a deeper understanding. Learners can use this information to monitor and regulate their own learning.

6. **Segue to Thinking Routines.**

   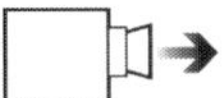

   a. Display and explain that "Thinking Routines" are different sequences of steps that learners take to make their thinking actions visible to themselves and others. The sequences emphasize learners' own ideas as the starting point for learning. Educators integrate these sequences into the experiences they design and facilitate. These routines have been used throughout the *Reflecting on Practice* program.

b. Distribute the **Thinking Moves & Thinking Routines** handout. Notice that each Thinking Move can be enacted through multiple Thinking Routines. Educators should begin by considering what Thinking Moves their learners should engage in and then select the Thinking Routines.

c. Mention that some of the routines (Think-Pair-Share, Turn & Talk, Walkabout, etc.) may be familiar to many educators, sometimes under different names (e.g., Walkabout = Gallery Walk = Chalk Talk). In *Reflecting on Practice*, these routines may vary slightly from what participants have previously experienced. Other routines may be less familiar—for example, 3-2-1 Bridge or Pictures & Words; the former from Ritchhart et al.'s work (2011), and the latter devised from the authors' work in the *Reflecting on Practice* program. Many other excellent routines exist but aren't listed here simply because they weren't used in the program.

d. If the detailed version is made available, then point out that version offers general instructions on how to lead many different Thinking Routines based on how the routine was carried out in the *Reflecting on Practice* program. Participants are of course welcome to customize the routines for their own contexts.

7. Allow quiet time to review.
Give participants a few minutes to look over the handout and ask any questions they may have.

8. Offer details to clarify and specify.
Share these details where appropriate in the discussion:

- What qualifies these sequences as *thinking routines* is their deliberate design to make thinking "visible" for our learners, so that they (and we) can specify what they're doing when they're "thinking."

- The Thinking Routines are *teaching strategies* that educators use to help learners take ideas out of their heads and manipulate them aloud, with others, and recognize how thinking is happening.

- The Thinking Routines are *learning strategies* that learners can embody as part of how they think and learn, because the routines emphasize uncovering one's *own* ideas as the starting point and then continuing to make connections between ideas.

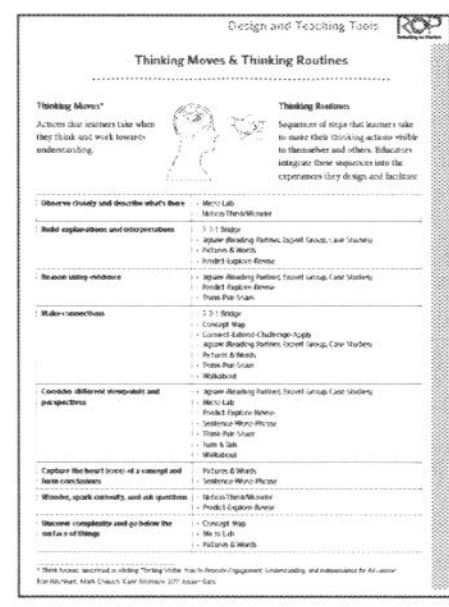

The active process of learning in the Five Foundational Ideas on Learning involves thinking.

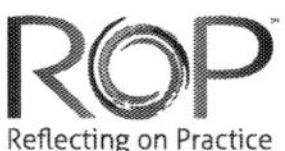

 Hands-on · **45** minutes

Design & Teaching Tools Match Up

1. **Introduce the activity.**

   a. Explain that participants will work in table groups to discuss the Design and Teaching Tools from *Reflecting on Practice* that are used throughout the program.

   b. The purpose of this activity is for the community to be reminded of the tools, just as they transition into reviewing and revising the learning experiences they design.

   c. Let participants know that there is one copy of each tool on their tables for reference, if they don't have their own copy accessible.

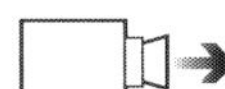

2. **Display and review the instructions for the activity.**

   a. Every group will receive one envelope with two sets of cards and instructions.

   b. They'll work together to follow the instructions and discuss the cards, one set at a time.

   c. Each of the sets includes discussion prompts on the instruction card to address before moving on to the next card set.

   d. They have 30 minutes total.

3. **Distribute Card Sets and monitor the table groups.**

   a. Remind participants to complete the first two steps for Card Set #1 before going on to Card Set #2.

   b. Let table groups work at their own pace but monitor their progress to keep them moving along. Groups that finish early can consider the prompts for the whole-group share that will come next.

4. **Hold a brief whole-group share.**
   Ask participants to share some of what emerged from their table discussions. Possible prompts:

   - Which tools are used most commonly by members of the community? Provide examples.

   - Which tools remain difficult to use? Is there a common theme for the reason?

- Which tools engendered the most discussion within the table group: disagreements, questions, ideas?

- What was most surprising or interesting about the conversations during the activity?

5. **Potential concerns and experiences to keep in mind.**
   Members of the community will likely have a range of experiences applying the suite of tools for design and teaching used in *Reflecting on Practice* into their practice. Some thoughts to consider and connect as you listen to their comments.

   - The tools are integrated into every aspect of the program; thus, the facilitation team has been modeling them each time they lead something from the program.

   - For the participants, the tools were usually called out explicitly in the **From Learners' Perspective** task, and then they were invited to consider how to use the tools in the **Let's Talk Practice** and **Continue the Learning** tasks. Sometimes, they were asked to share their experiences with a tool in the **Retrieve & Connect** task at the start of a session. In general, the tools are practical applications of the ideas on learning they read and discussed in the **Research Discussions** and used in the design and instruction of the **Hands-on** tasks they experienced as learners.

   - Acknowledge that some tools may have been easier to integrate into their practice than others, and that ease may be due to how the experiences were designed in the first place. For instance, it may be difficult to use the Discussion Map when the design didn't include prompts, allot time for learners to talk and listen, or give learners something to talk about.

   - Concede that these tools are made available for the community to use, but participants aren't required to adopt the tools into their practice. The tasks and experiences from *Reflecting on Practice* sessions model the tools, analyze their designs, and relate them to the research on learning. Participants are invited to use the tools in their own work to apply what they're learning. Participants should feel free to use other preferred tools that similarly support learning. It's important, however, that they don't dismiss these tools because they feel they're too difficult to use. Instead, push on participants to consider if the tools need to be tweaked to be more useful to them.

<table>
<tr><td>Let's Talk Practice   🕐 **60** minutes</td><td></td></tr>
</table>

Micro Lab: *Rapid Design Feedback*

1. **Begin with individual reflection in Learning Journals.**
   Ask participants to retrieve their notes from the **Reflection Exercise Worksheet for Rapid Design Feedback** (assigned in Continue the Learning in Session 1 of this module). Explain the following:

   a. That task offered participants a chance to begin thinking about how to revise the design of learning experiences to use the tools and apply the ideas on learning the community has been discussing.

   b. Now, have partners recall their conversation and review what they wrote for the questions in Section E.

      □ What changes do you want to make? It's possible there may be a combination of minor tweaks and massive reconsiderations.

      □ Why do you feel those changes are needed?

      □ What are possible options for the changes, including why you're suggesting these possibilities and not others?

2. **Emphasize purpose of the Micro Lab and form small groups.**
   Remind participants that the purpose of this Micro Lab is to get some input quickly for the revisions they would like to make to the learning experience before they commit further to those modifications. Cluster pairs into groups of four or six; have pairs count off 1-2(-3).

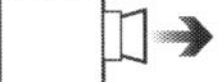

3. **Begin Micro Lab.**
   Display and review the instructions briefly. Tell participants who will be the timekeeper, and then begin the routine.

   a. **Share:** Ask *Pair 1s* to share the context for the learning experience and proposed revisions to the group for 7 minutes. No one speaks except the speaker. Everyone else listens and take notes.

   b. **Silence:** Allow 60 seconds of silence for everyone to take in and mentally review what was heard.

   c. **Share for round 2:** Repeat a. and b. for *Pair 2s* (then *3s*).

   d. **Begin small-group discussions.** Groups may now have an open discussion for 20-25 minutes to offer feedback on what they heard.

4. Partners consider the feedback.

Give pairs about 10 minutes to confer about the small-group discussion for their revisions. Remind them they aren't obligated to take any of the suggestions provided. The group conversation should have furthered their thinking on the revisions they want to make.

5. (Optional) Conduct a brief whole-group share.

Invite groups to share key points from their discussions and reflections for the community to consider.

Continue the Learning

2 minutes

1. Recommended task:

Invite participants to continue to work with their partner to make revisions to the learning experience.

2. Reminder: Whole-Community Video Reflection coming up.

Remind participants that the next session will include a Whole-Community Reflection on this module's focus, **Objects and Designing Experiences for Learning**. Between now and then, you'll ask or assign one volunteer participant to share a video-clip with the whole group.

3. Advance preparation for Critiquing Objects in Module 4 Session 3.

a. Explain that the next session also includes a hands-on task to critique objects. This task focuses on how objects are used in the learning experiences your community offers.

b. Everyone needs to do some preparation for that task. The task will be done in small groups, which they will form now.

c. Within their groups, discuss and decide which learning experience they would like to review in light of how the objects are used.

d. Individually, complete the **Reflection Exercise Worksheet for Critiquing Objects in Learning Experiences** worksheet. Remind them to use the materials from the first session in this module. They will share and discuss their independent reviews during the next meeting.

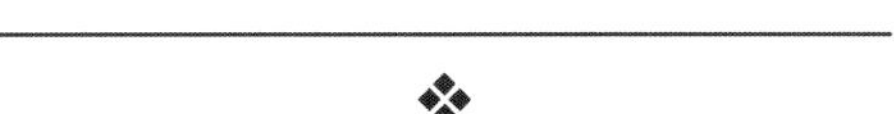

Materials from Module 4 Session 1 include handouts and self-generated charts.

- Key Ideas from the Literature: How Learners Interact with Objects

- Objects: Types & Features

- "Objects in Learning Experiences: Limitations & Possibilities"

- "Observation Chart" for the skits

Read-Aloud for the Facilitator:

"What Does Thinking Look Like?"
by Dr. Kathleen Donegan
Used by permission of the author.

I want to talk about thinking as teaching. Let me start with a story.

One time in office hours, a student proposed an idea for a paper. I put my chin in my hand, and looked up and to the side for a moment or two, or maybe three. And because I did not give an immediate response, the student immediately retracted his proposal.

"Okay, that was a dumb idea," he said.

"No," I assured him, "this is just what thinking looks like."

So that was a revelation to me. That so much of the time our education is about questions and answers. And the more quickly the answer follows upon the question, the better that is seen to be. But my goal is different. It is to show in the classroom what thinking looks like. Because, strange as it is, sometimes this is completely absent or unrecognized when we teach and learn within the classroom.

Sometimes thinking looks like **pondering**. This is why, when there is silence after you ask a question, it might in fact be a sign that it was a very good question and you should respect that silence while students think about it—or even say, "I don't want you to answer right away, but just to think about this for a minute." Invite that pondering before asking, "What came to mind as you thought?"

Sometimes thinking looks like **emanation**. It doesn't know what it wants to say until it discovers something through saying it. This is why you should encourage participation even before a student "knows the answer." Trust that she can find the answer by listening to her own voice as she thinks out loud, as the language pulls her thought forward. Tell her to trust that too.

Sometimes thinking looks like **structure**. This is why it is good to ask a follow-up question that does not challenge what a student just said, but instead seeks to extend it structurally. If that is true, what else might be true? If that functions as a cause, what might be the effect? If that is X, then what is Y? Build [mental] models [to connect] to where the next thought comes from because the structure demands it.

Sometimes thinking looks like **variety**. This is why it is sometimes good to invite re-statements, to recognize the value of repetition with a difference. If you ask a question that seems a bit convoluted or that makes sense to you but doesn't seem to translate to some others, you can say: Can someone else ask that question in a different way? Or if a student makes a particularly good point, you can say: Can someone say that again? Know that thinking looks different in different minds, and that the coalescence of that variety will always be richer for the difference.

(continues)

     This page may be duplicated for educational use.

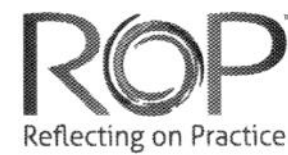

When we prepare for class, we are thinking alone and silently; this is true of both students and professors. But in the class, we can model that thinking together. Share it. Prefer process over performance. We can use active thinking to break down the cones of silence around private preparations. Professors, of course, are professional thinkers, and students are novices. This is why professors must invite students into the practice of thinking together. We are so much more knowledgeable than them that if the only criterion is knowledge, students will forever be on the bottom rung. But if another criterion is thinking in real-time, thinking in public, then we have doubled what we can teach.

This is what thinking looks like, I told my student. And then his idea grew, and our relationship grew, and he grew, and I grew too. Because I realized that if students see and value the real-time working of the mind, if they see and recognize and aspire to that, they will all the more want the knowledge to think with. And all the more, want to be in the classroom and in the company where thinking together happens.

Kathleen Donegan is a 2015 Distinguished Teaching Award recipient from University of California, Berkeley. This address was given in April 2017.

Card Sets for Design & Teaching Tools Match Up

Card Set #1 ✂------	Card Set #2 ✂------
Tool	**Function**
Discussion Map	A guide for educators to use to facilitate conversations that emphasizes opportunities for learners to express their thoughts and listen deeply to what is shared by the community.
Teaching Purposes	A set of reasons for facilitating different segments within a learning experience that together support the learning outcomes.
Facilitation Approaches	A framework that educators use to orchestrate the exchange of ideas to ensure learners have opportunity to voice their thinking and consider ideas.
Learning Cycle Design Framework	A progression for designing learning experiences that models science inquiry.
Thinking Moves	Actions that learners take when they think and work towards understanding.
Thinking Routines	Sequences of steps that learners take to make their thinking actions visible to themselves and others. Educators integrate these sequences into the experiences they design and facilitate.
Educator Moves— Prior Knowledge	A set of actions that educators take to encourage learners to retrieve their prior knowledge. These actions are observable to aid reflective practice.
Educator Moves— Learning Conversation	A set of actions that educators take to encourage learners to articulate their thinking for themselves and for others to consider. These actions are observable to aid reflective practice.
Educator Moves— Objects	A set of actions that educators take to encourage learners to engage with objects to gather information and make connections. These actions are observable to aid reflective practice.

     This page may be duplicated for educational use.

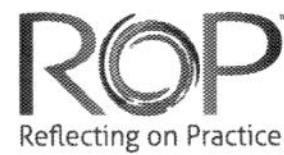

Instruction Cards for Design & Teaching Tools Match Up

Card Set #1 Instructions

1. **Lay out the 9 cards** so everyone in your group can access them. Each card bears the _name_ of a Design & Teaching Tool used in the _Reflecting on Practice_ program. (Three tools are newer to the community.)

2. For the six tools you know, discuss the following prompts:

    a. _What is the tool? When have I used it? How have I used it?_

    b. _Which tools are more typically used in Video Reflections, but can double as a Design and Teaching Tool?_

    c. _For each tool, how does using that tool support learning?_

3. **Record your responses** in your Learning Journal. If there is a tool you haven't used, consider when you might want to try it out (or what you can do to make it possible for you to use the tool).

Card Set #2 Instructions

1. **Lay out the 9 cards** so everyone in your group can access them. Each card is a _description of the function_ of one tool.

2. **Match each description to the tools** in Card Set #1. _Can you match the descriptions for the new tools?_

3. As you do the matching, discuss the following prompts:

    a. _Which tools have become familiar to you? Why is that?_

    b. _Which tools remain difficult for you to use? Why is that?_

    c. _Which tools do you have questions about? What are you wondering?_

Module 4.
Session 3: Whole-Community Video Reflection on Objects

Session Overview

This session has two reflection and application focuses. For the first, participants continue their study of teaching with objects through Whole-Community Video Reflection on how an educator at their institution uses **objects to support learning**. They're introduced to a new tool for this purpose: **Educator Moves for objects**. The Let's Talk Practice provides the opportunity to apply these ideas to their current and future use of objects.

For the second, participants revisit their initial thoughts on how people learn, and articulate in writing how their thinking has changed and why. They then reflect on what they want to incorporate into their practice about designing learning experiences to support learning.

Session Objectives

- Introduce Educator Moves—Objects.
- Practice using this tool in a Video Reflection for a member of the community.
- Critique how objects are currently used in our learning experiences and reflect on future use incorporating Educator Moves.

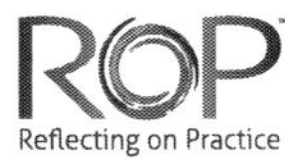

SESSION AGENDA		
Task **Routine**	**Description**	**Estimated Time** **(in minutes)**
Introduction *Session Objectives*	The goals and objectives of the session are introduced.	2
Let's Talk Practice Table Talk: *Educator Moves—Objects*	Participants reflect on and discuss their experiences with Video Reflections and articulate emerging changes in their practice. A new Educator Moves tool focused on using objects to support learning is introduced and discussed.	10
Hands-on *Whole-Community Video Reflection*	Participants engage in their fourth whole-group video reflection as they observe a colleague's practice. This provides another opportunity to use all five Video Reflection tools.	55
Break option		10
Let's Talk Practice Table Talk: *Critiquing Objects in Learning Experiences*	Small groups reflect on their current usage of objects in a learning experience they chose to investigate. They explore additional potential applications of ideas presented in the research discussions and Educator Moves as they consider their use of the objects in the future.	50
Retrieve & Connect 3-2-1 Bridge: *Ideas on Learning Revisited*	Participants revisit their initial ideas from Module 1 Session 1 on how they think people learn and add what they want to remember about designing learning experiences to support learning.	20
Continue the Learning	Review preparation steps for video reflections in critical-colleague groups.	3
	Total Estimated Time	150 mins. (2.5 hrs.)

4.3

MATERIALS

Recurring for all Sessions

- See Module 1 Session 1

For this Session

- Video clip (from an educator-presenter, see **Getting Ready** #2)
- External speakers
- 2 Posters for **Roles** and **Agreements** (reuse from Module 1 Session 2, **Getting Ready** #4)

For each small group

- Chart papers
- Markers, multiple colors

For each participant

- Handouts (1 copy each) from Appendix A, *Tools for Reflective Practice*
    - **Observation Instrument**
    - **Educator Moves—Objects**
    - **Checklist for Reflection Sessions**
- Handouts (replacement copies as needed)
    - **Reflection Exercise Worksheet for Critiquing Objects in Learning Experiences** (Appendix B)
    - **Protocol for Video Reflection** (Appendix A)
    - **Feedback Chart** (Appendix A)
    - **Reflection Exercise Worksheet to Prepare for Feedback** (Appendix A)

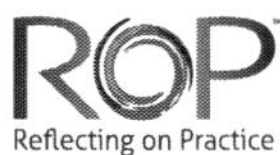

Getting Ready

1. Follow up to ensure that all participants complete the **Reflection Exercise Worksheet for Critiquing Objects in Learning Experiences** prior to the session. Their responses don't have to be detailed, but they should do some pre-thinking so the time spent during the session is productive for everyone.

2. Determine a presenter. Ask or assign someone from the community to do their Video Reflection with the whole group. If there is a third facilitator on the team, then that person should be the presenter.

   - Ask the presenter to complete the **Reflection Exercise Worksheet to Prepare for Feedback.**

   - Offer assistance as needed.

   - Ask the presenter to provide you the video clip at least one day before the video session.

3. Duplicate handouts.

4. Review the details collected for **Educator Moves** and **Institutional Practice** from Video Reflections on learning conversations (if that hasn't been done already). Determine how those details will be addressed in the community's work. Consider if the process for collecting details across the Video Reflections needs to be refined.

5. Consider whether there are additional updates to the processes for Video Reflections in small groups to share during **Continue the Learning**. Refer to consideration questions in **Getting Ready** #7 in Module 3 Session 3. Additional questions:

   - Are there members in the community left out of the experience? Why are they excluded? What can be done to support them?

   - Are there new staff who need to be familiarized with the process? Perhaps a separate conversation with these individuals may be needed.

   - Are there shifts in individual and/or community practices that can be highlighted?

6. Prepare the video clip. Be sure it's cued up and ready to play from the point where the volunteer presenter wants to begin sharing. Know where in the video to stop sharing. Test the sound before everyone arrives.

Call for a volunteer or a couple of volunteers to be the guide(s) for this Whole-Community Video Reflection. It will be instructive for the facilitators and the community to see one of their colleagues facilitate the Video Reflection Protocol effectively.

4.3

Session 3 Step by Step

 **2** minute

Introduction
Session Objectives

1. **Introduce the session.**

   Share that the first half of this session is a Whole-Community Video Reflection focused on objects, and that it continues the discussion on how the community uses objects to support learning. As with the last two whole-group video-reflections, participants will practice using a new tool and consider refinements of the tools and process, while they give feedback to a colleague.

   The latter half of the session will be an opportunity for participants to work with their small group to critique how objects are currently used in their programs and activities. They will also explore additional potential applications of the research discussions and Educator Moves in their use of the objects in the future.

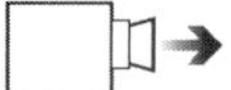

2. **Display and introduce the session objectives:**

   - Introduce Educator Moves—Objects.

   - Practice using this tool in a Video Reflection for a member of the community.

   - Critique how objects are currently used in our learning experiences and reflect on future use incorporating Educator Moves.

Let's Talk Practice · **10** minutes

Table Talk: *Educator Moves—Objects*

1. **Sum up the Video Reflection experience to date.**

   a. Each person has set and presented **two** problems from their practice, and received focused feedback on those problems from the community (either in whole or small group).

   b. Together, everyone has offered the community from 9 to 11 learning opportunities to deepen our collective teaching practice.

   c. They will now have another 4-5 learning opportunities from video reflections: one in whole community, now, and 3-4 in their critical-colleague groups, afterwards.

   d. In this round of video reflections, they will use a new set of Educator Moves.

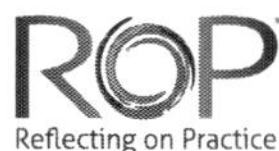

2. **Distribute "Educator Moves—Objects" handout.**

   a. Remind participants that the Educator Moves they used previously were focused on <u>prior knowledge</u> and <u>learning conversations</u>. For this module, the Moves will focus on **objects**. The handout lists actions educators may take to use objects to support learning, with possible reasons for each move.

   b. Tell participants to refer to **Educator Moves—Objects** when they <u>think</u> about what they noticed in the video clip, and consider whether the observed actions could be interpreted as any of these moves. During their discussion, they can add examples and elaborate on or revise the reasons to reflect the group's understanding of each Educator Move.

   c. The community will continue to add to and revise **Educator Moves—Objects** throughout each Video Reflection in this module.

3. **Address clarification questions.**
   Give participants a few minutes to review the handout and invite them to ask for clarification, if needed. As with the previous Educator Moves, they'll have a chance to practice with the tool in just a moment, so they shouldn't get too caught up in the details just yet.

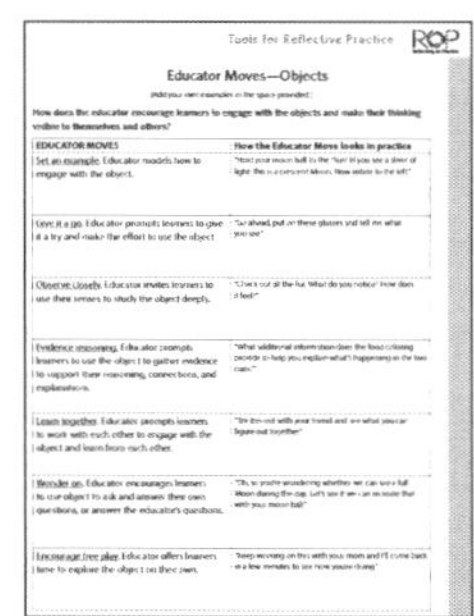

Share with participants that they can review the same video clip later to look for the other moves (prior knowledge and learning conversations) as well.

4.3

Whole-Community Video Reflection

55 minutes   Hands-on

1. **Review Video Reflection tools and process, as needed.**
   Tell participants that, as in previous modules, they'll do the first Video Reflection as a whole community before going on to use the tools with their critical colleagues. By now, video reflections should be familiar to most everyone.

2. **Have tools and posters ready.**
   Affix the two posters (**Roles** and **Agreements**) in a prominent location. Remind participants to retrieve their tools for Video Reflection. Distribute replacement copies as needed.

   ■ **Protocol for Video Reflection**

   ■ **Educator Moves—Objects**

   ■ **Feedback Chart**

3. **Distribute a new copy of the "Observation Instrument."**

4. **Call attention to the Reflection Session posters.**
   Direct everyone's attention to the Roles & Responsibilities, and Community Agreements for Video Reflection posters. Give them a moment to remind themselves of the details. Invite them to offer revisions, if needed.

5. **Use protocol to direct the process.**
   Remind participants that while the guide uses the **Protocol for Video Reflection** to lead the group through the Video Reflection, everyone should also use it to stay in step with the process. If a member of the community will be the guide, then introduce them now.

6. **Guide introduces the educator-presenter.**
   Turn over the responsibility to the guide. Remind the guide to introduce the presenter and thank them for doing their Video Reflection with the whole community.

7. **Start the Whole-Community Video Reflection.**
   Follow the steps in the protocol to guide the community through the Video Reflection. Be sure to use the Tools for Reflective Practice as indicated in the Protocol.

8. **Add action items for changes in institutional practices in Step 11.**
   If it's possible for some practices to be addressed immediately, then consider them now. Otherwise, remember to share the process and invite participants to join the effort for making institution-level change, as appropriate.

Don't forget to record details for Educator Moves.

Let's Talk Practice 50 minutes

1. **Introduce the task.**
   Ask participants to get into the small groups they formed in preparation for this Critiquing Objects task. Remind them of the connections they made and steps they took in preparation.

   a. Session 1 of this module prompted the community to consider different types of objects, how the objects spark conversation and engagement, and the limitations and possibilities of different types of objects.

   b. Prior to this meeting, the community sorted into small groups to reflect on current practices and imagine future possibilities for using the objects in their work in light of these ideas they discussed.

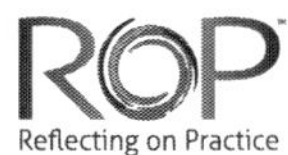

   c. Each group chose a learning experience to interrogate how their focus objects are used currently.

   d. The groups selected different types of objects, but collectively these objects represent the variety of "things and stuff" used in their work at the organization.

   e. Members of the group started the interrogation of the object individually using the **Reflection Exercise Worksheet for Critiquing Objects in Learning Experiences**.

2. **Display and explain the task.**
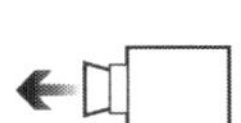
This Let's Talk Practice finishes the interrogation of those objects. The purpose is twofold: a) to examine how the objects are currently used and what the experiences are like for learners, and b) to consider whether current usage reflects the community's new thinking about objects and whether more can be done with the objects to support learning.

3. **Display and review the two-stage instructions.**
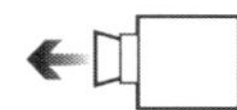
Distribute chart paper and markers. Explain the following.

   a. Stage 1. Collective recall about the object.

       ☐ Share out individual reviews in the first two sections of the **Critiquing Objects in Learning Experiences** worksheet to generate a collective sense of how the focal object is currently used by educators and learners.

       ☐ Ensure everyone's reviews are voiced and considered.

       ☐ Discuss and record on chart paper the extent to which your current usage and learners' usage reflect the community's new thinking about objects.

       ☐ Review the **Educator Moves—Objects** to determine whether and how any Educator Moves currently show up in your practice. Add these notes to the chart paper.

   b. Stage 2. Imagining new possibilities.

       ☐ Share out individual reviews for the third section (Pushing Ourselves) of the **Critiquing Objects in Learning Experiences** worksheet to begin sparking ideas for change.

       ☐ Ensure that everyone's individual reviews are voiced and considered.

       ☐ Discuss and record on chart paper the types of changes participants want to make.

4.3

> ❑ For each **Educator Move**, discuss what change is needed to include that Move or address it more effectively in the future. Add these notes to the chart paper.

4. **Circulate and monitor the pace of discussions.**
   Monitor across the groups to ensure they're working at about the same pace, so that all will be ready for the whole-group share.

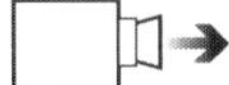

5. **Conduct a whole-group share.**
   When there's about 15 minutes left for this task, call time and invite participants to the whole-community discussion. Display the following two questions and give participants a moment to think about them before taking responses:

   - *What are the focal objects in the learning experience your group reviewed? What changes to how the objects are used does your group suggest ?*

   - *What revelations emerged from discussion within your groups?*

Retrieve & Connect | 20 minutes

3-2-1 Bridge: *Ideas on Learning Revisited*

1. **Start with a review of 3-2-1 Bridge.**
   Have participants take out their Learning Journals and find their previous **3-2-1 Bridge** entry for "Ideas on Learning" from Module 1 Session 1. Remind them that they've thought, read, and talked *a lot* about learning since starting the *Reflecting on Practice* program; this routine makes their journey explicit. Have them reread their earlier entries and consider the changes they've made in their thinking since then, and why.

2. **Update the routine.**
   Tell participants that, as before, in this routine they'll be given a prompt and asked to record their thoughts about it in three phases. This time, their focus will be on comparing prior understanding to current thinking. This is an opportunity to record their thoughts freely, and there's truly no right or wrong response. Ask participants to draw a line under their previous entries.

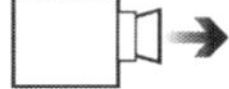

3. **Display the prompt:**

   How do people learn?

4. **Ask participants to <u>compare 3 words</u>.**
   In their Learning Journals, have participants review the words they wrote for the first 3-2-1 Bridge, then write down three words that come to mind now when they think about how people learn. They

 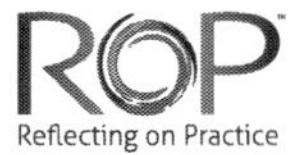

needn't generate new words if none comes to mind, but they do need to consider and write down *why* the words have or haven't changed.

5. **Have them <u>compare 2 questions</u>.**
   Now have them review their two original questions. Tell them to cross off questions that have been answered. If one or both questions have *not* been answered, ask them to consider whether they're still wondering about them. If they have new questions, have them write those down.

6. **Ask them to <u>compare 1 analogy or metaphor</u>.**
   Tell participants to review their previous analogy or metaphor and consider whether to revise it to reflect what they now think about how people learn. Again, the analogy/metaphor doesn't need to change, but they do need to consider and write down why it has or hasn't changed.

   Display the definitions as a reminder: 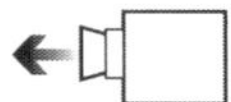

An **analogy** is a <u>comparison</u> in which an idea or a thing is compared to another thing that is quite different from it. It "explains" that idea or thing by comparing it to something that is familiar.

Example: How a doctor diagnoses diseases is like how a detective investigates crimes.

Metaphor is a <u>figure of speech</u> that makes an implicit, implied, or hidden <u>comparison</u> between two things that are unrelated but share some common characteristics. In other words, a resemblance between two contradictory or different objects is made based on one or more common characteristics.

Example: It's going to be clear skies from now on. (This implies that clear skies aren't a threat and life is going to be without hardships.)

---

7. **Transition to Turn & Talk.** 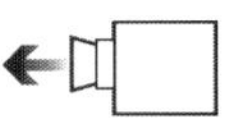
   Have participants turn to someone next to them and discuss their 3-2-1 comparisons. Keep these discussions to partners or trios to ensure that everyone has ample time to talk. Display questions for discussion:

   a. *What changes did you make or not make?*

   b. *What prompted the changes? What made changes unnecessary?*

    c. *How has your thinking about how people learn evolved and/or deepened?*

    d. *What change(s) will you make in your practice to apply ideas and revelations from discussions in this module?*

8. **(Optional) Conduct a whole-group share.**
   If time permits, invite partners to share key points from their discussion.

 **3** minutes

Continue the Learning

1. **Review the preparation and processes for Video Reflections in critical-colleague groups, as needed.**

   a. Share any changes to the processes for Video Reflections in small groups (see **Getting Ready** #5)

   b. Distribute the handouts (**Checklist for Reflection Sessions** and **Reflection Exercise Worksheet to Prepare for Feedback**), as needed.

   c. Be sure the roles and rotation schedules for each group are recorded on the community document before participants depart.

   d. Presenters must complete the **Reflection Exercise Worksheet to Prepare for Feedback** prior to their Video Reflections.

   e. The situation that participants investigate in this round of Video Reflections pertains to "objects to support learning."

   f. Reiterate the challenge of the reflection task, and draw attention to the community's accomplishments. Encourage participants to reach out to colleagues or the facilitators for assistance, even if it's just to listen.

2. **Post completion date.**
   Small groups make their own schedules, but all Video Reflections should be done by the date posted.

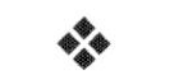

References

Ash, D. (2004). Reflective scientific sense-making dialogue in two languages: The science in the dialogue and the dialogue in the science. *Science Education, 88*(6), 855–884.

Baron, D. (2007). Critical friendship: Leading from the inside out. *Principal Leadership, 7*(9), 56–58.

Billett, S. (2002). Critiquing workplace learning discourses: Participation and continuity at work. *Studies in the Education of Adults, 34*(1), 56–67.

Billett, S. (2004). Workplace participatory practices: Conceptualising workplaces as learning environments. *Journal of Workplace Learning, 16*(6), 312–324.

Bransford, J. D., Brown, A. L., & Cocking, R. R. (Eds.). (2000). *How people learn: Brain, mind, experience, and school.* Washington, DC: National Academy Press.

Bransford, J. D., Vye, N., Stevens, R., Kuhl, P., Schwartz, D., Bell, P., . . . Sabelli, N. (2006). Learning theories and education: Towards a decade of synergy. In P. A. Alexander & P. Winne (Eds.), *Handbook of educational psychology* (2nd ed.). Mahwah, NJ: Erlbaum.

Briggs, M. (2019). A more challenging mystery tube for teaching the nature of science. *The Physics Teacher, 57*(5), 300–303.

Brown, J. S., Collins, A., & Duguid, P. (1989). Situated cognition and the culture of learning. *Educational Researcher, 18*(1), 32–42.

Bybee, R. W., Taylor, J. A., Gardner, A., Van Scotter, P., Powell, J. C., Westbrook, A., & Landes, N. (2006). The BSCS 5E instructional model: Origins and effectiveness. *Colorado Springs, Co: BSCS, 5*, 88–98.

Chadwick, I. C., & Raver, J. L. (2015). Motivating organizations to learn: Goal orientation and its influence on organizational learning. *Journal of Management, 41*(3), 957–986.

Costa, A. L., & Kallick, B. (1993). Through the lens of a critical friend. *Educational Leadership, 51*, 49–51.

Crossan, M. M., Lane, H. W., & White, R. E. (1999). An organizational learning framework: From intuition to institution. *Academy of Management Review, 24*(3), 522–537.

Dawson, E. (2014a). Equity in informal science education: Developing an access and equity framework for science museums and science centres. *Studies in Science Education, 50*(2), 209–247.

Dawson, E. (2014b). "Not designed for us": How science museums and science centers socially exclude low-income, minority ethnic groups. *Science Education, 98*(6), 981–1008.

Elkjaer, B., & Wahlgren, B. (2005). Organizational learning and workplace learning—similarities and differences. In E. Antonacopoulou, P. Jarvis, V. Andersen, B. Elkjaer, & S. Høyrup (Eds.), *Learning, working and living: Mapping the terrain of working life learning* (pp. 15–32). Hampshire: Palgrave Macmillan.

Ellenbogen, K. M., Luke, J. J., & Dierking, L. D. (2004). Family learning research in museums: An emerging disciplinary matrix? *Science Education, 88*(S1), S48–S58.

Falk, J. H., & Dierking, L. D. (2010). The 95 percent solution. *American Scientist, 98*(6), 486–493.

Fenichel, M., & Schweingruber, H. A. (2009). *Surrounded by science: Learning science in informal environments.* Washington, DC: National Academies Press.

Fuller, A., & Unwin, L. (2004). Expansive learning environments: Integrating organizational and personal development. In H. Rainbird, A. Fuller, & A. Munro (Eds.), *Workplace learning in context* (pp. 126–144). London: Routledge.

Fuller, R. G. (Ed.) (2002). *A love of discovery: Science education—The second career of Robert Karplus.* New York: Kluwer Academic/Plenum Publishers.

Garibay, C. (2009). Latinos, leisure values, and decisions: Implications for informal science learning and engagement. *The Informal Learning Review, 94,* 10–13.

Gonzalez, N., Moll, L., & Amanti, C. (Eds.). (2005). *Funds of knowledge: Theorizing practices in households, communities, and classrooms.* Mahwah, NJ: Lawrence Erlbaum Associates, Inc.

HHMI BioInteractive (Producer). (2019, October). How Science Works. Retrieved from https://www.biointeractive.org/classroom-resources/how-science-works

Iaccarino, M. (2003). Science and culture. *EMBO Reports, 4*(3), 220–223. Retrieved from http://onlinelibrary.wiley.com/doi/10.1038/sj.embor.embor781/full

Lave, J., & Wenger, E. (1991). *Situated learning: Legitimate peripheral participation.* Cambridge, UK: Cambridge University Press.

Lawrence Hall of Science. (2014). *Ocean Sciences Curriculum Sequence for Grades 6–8: The ocean-atmosphere connection and climate change.* Berkeley: The Regents of the University of California.

Lehr, J. L., McCallie, E., Davis, S., Caron, B. R., Gammon, B., & Duensing, S. (2007). The value of "dialogue events" as sites of learning: A research framework. *International Journal of Science Education, 29*(12), 1467–1487.

Leinhardt, G., & Crowley, K. (2002). Objects of learning, objects of talk: Changing minds in museums. In S. G. Paris (Ed.), *Perspectives on object-centered learning in museums* (pp. 301–324). Mahwah, NJ: Lawrence Erlbaum Associates.

Marsick, V. J., & Watkins, K. E. (2001). Informal and incidental learning. *New Directions for Adult and Continuing Education, 2001*(89), 25–34.

Martin, L. W., Tran, L. U., & Ash, D. B. (2019). *The reflective museum practitioner: expanding practice in science museums.* New York, NY: Routledge.

Mazzocchi, F. (2006). Western science and traditional knowledge. *EMBO Reports, 7*(5), 463–466.

Meltzoff, A. N., & Decety, J. (2003). What imitation tells us about social cognition: A rapprochement between developmental psychology and cognitive neuroscience. *Philosophical Transactions of the Royal Society, 358,* 491–500.

NASA Science. (N.D.). About the Moon: Misconceptions. Retrieved from https://moon.nasa.gov/about/misconceptions/

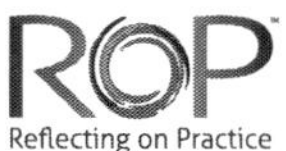

National Academies of Sciences, E., & Medicine. (2018). *How people learn II: Learners, contexts, and cultures*. Washington, DC: The National Academies Press.

National Research Council. (2007). *Taking science to school: Learning & teaching science in grades K-8*. Washington, DC: National Academies Press.

National Research Council (Ed.) (2009). *Learning science in informal environments: People, places, and pursuits*. Washington, DC: National Academies Press.

New Oxford American. (n.d.) Critical. In *New Oxford American Dictionary*. Retrieved December 21, 2020, from https://www.lexico.com/definition/critical

NGSS Lead States. (2013). *Next Generation Science Standards: For states, by states*. Washington, DC: The National Academies Press.

Ritchhart, R., Church, M., & Morrison, K. (2011). *Making thinking visible: How to promote engagement, understanding, and independence for all learners*. San Francisco, CA: John Wiley & Sons.

Rock, D. (2008). SCARF: A brain-based model for collaborating with and influencing others. *NeuroLeadership Journal, 1*(1), 44–52.

Sandell, R., Dodd, J., & Garland-Thomson, R. (2010). *Re-presenting disability: Activism and agency in the museum*. Abingdon, UK: Routledge.

School Reform Initiative. (2017a). Feedback provided during protocols. Retrieved from http://schoolreforminitiative.org/doc/warm_cool_hard.pdf

School Reform Initiative. (2017b, March 30). Tuning protocol. Retrieved from https://www.schoolreforminitiative.org/download/tuning-protocol/

Scott, P. H., Mortimer, E. F., & Aguiar, O. G. (2006). The tension between authoritative and dialogic discourse: A fundamental characteristic of meaning making interactions in high school science lessons. *Science Education, 90*(40), 605–631.

Sleeter, C. E. (2012). Confronting the marginalization of culturally responsive pedagogy. *Urban Education, 47*(3), 562–584.

Stake, R. E. (1995). *The art of case study research*. Thousand Oaks, CA: Sage Publications.

Stoll, L., Bolam, R., McMahon, A., Wallace, M., & Thomas, S. (2006). Professional learning communities: A review of the literature. *Journal of Educational Change, 7*(4), 221–258.

TEDx Talks (2015, April 2). Learning styles & the importance of critical self-reflection | Tesia Marshik | TEDxUWLaCrosse. Retrieved from https://www.youtube.com/watch?v=855Now8h5Rs

Tran, L. U. (2019). The how and why of reflective practice for science museum professionals. In L. M. W. Martin, L. U. Tran, & D. B. Ash (Eds.), *The reflective museum practitioner: Expanding practice in science museums* (pp. 7–22). New York: Routledge.

Tran, L. U., Gupta, P., & Bader, D. (2019). Redefining professional learning for museum education. *Journal of Museum Education, 44*(2), 135–146.

Tran, L. U., Werner-Avidon, M., & Newton, L. R. (2013). Successful professional learning for informal educators: What is it and how do we get there? *Journal of Museum Education, 38*(3), 333–348.

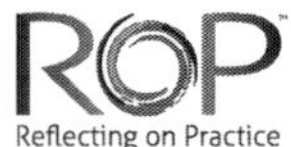

Tversky, A., & Kahneman, D. (1981). The framing of decisions and the psychology of choice. *Science, 211*(4481), 453–458.

Understanding Science. (2010). Mystery tubes. Retrieved from https://undsci.berkeley.edu/lessons/mystery_tubes.html

Unwin, L. (2004). *Taking an expansive approach to workplace learning: Implications for guidance.* Derby: Centre for Guidance Studies, University of Derby.

Vélez-Ibáñez, C. G., & Greenberg, J. B. (1992). Formation and transformation of funds of knowledge among US-Mexican households. *Anthropology & Education Quarterly, 23*(4), 313–335.

Vygotsky, L. (1978). *Mind in society: The development of higher psychological processes.* Cambridge, MA: Harvard University Press.

Webster-Wright, A. (2009). Reframing professional development through understanding authentic professional learning. *Review of Educational Research, 79*(2), 702–739.

Wood, D., Bruner, J. S., & Ross, G. (1976). The role of tutoring in problem solving. *Journal of Child Psychology, 17*(2), 89–100

Reflecting on Practice
Appendix A
Tools for Reflective Practice

Reflection Sessions in *Reflecting on Practice* 316

Checklist for Reflection Sessions 318

Reflection Exercise Worksheet to Prepare for Feedback 320

Protocol for Video Reflection 322

Observation Instrument 324

Feedback Chart 326

Educator Moves—Prior Knowledge 328

Educator Moves—Learning Conversations 330

Educator Moves—Objects 331

Reflection Sessions in *Reflecting on Practice*

Reflective practice is a journey that is both individual and collaborative. It's our own practice that we develop and deepen, but we don't do it alone.

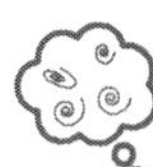 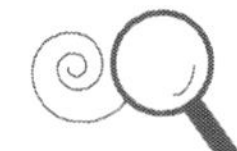 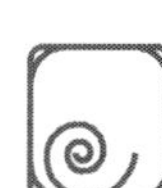 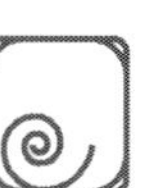 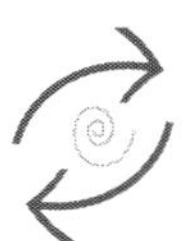

Reflect on a situation in my work	**Set a problem to investigate**	**Prepare to seek help for my investigation**	**Present my problem & work sample**	**Reexamine the problem under investigation**
Confront my beliefs & assumptions about learning & my learners	• Name what's problematic • Frame my perspective	Select representative work sample that gives pertinent information about the situation & problem	• Peers offer help • Listen to & consider their feedback	• Make changes • Rename what's problematic • Reframe my perspective • Reconsider my beliefs & assumptions

Reflection Sessions with peers

1. Recall ideas on learning
2. Use Tools for Reflective Practice
3. Provide productive feedback

• Be kind to the presenter
• Be curious about the work
• Be appreciative of the experience

The purpose of **reflection sessions** is to provide productive feedback to one another as we work to investigate and solve problems in our practice. We achieve this by observing and discussing work samples from actual practice. In these sessions, we develop shared meanings and language, deepen our practice together, and strengthen our collective efforts towards supporting our learners.

Reflection sessions stretch the social relationships of our professional learning community. The sessions require us to make our practice public, pose meaningful questions, offer substantive feedback, and hold one another accountable for meeting the needs of our learners. These sessions rely on us believing that **we can solve our own problems**, if given the appropriate support. That support includes trusted critical friendships—friends who believe that **together we are more capable** of knowing what we need to know and learning what we need to learn, than when we journey alone.

(continues)

     This page may be duplicated for educational use.

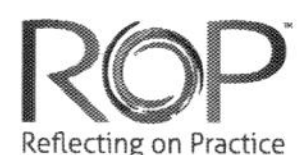

Reflection Sessions in *Reflecting on Practice* (continued)

Roles & Responsibilities

Everyone has a designated role. Each role has responsibilities.

EDUCATOR-PRESENTER: The professional who seeks help from colleagues and presents a work sample during a reflection session.	**PEERS:** Colleagues who discuss & provide feedback on the presenter's work sample.	**GUIDE:** A peer who leads a reflection session. Rotate this role between presenters or rounds of reflection sessions.
• Provides work sample & focus question for peers to consider. • Listens with open mind & whole heart to feedback; reflects on peers' observations. • Analyzes one's own learning. • Considers what changes to make & why.	• Contributes substantive feedback to the discussion. • Is respectful and appreciative of colleagues. • Recasts or recalls inappropriate comments or questions. • Listens with care and compassion. • Participates, but doesn't dominate.	• Facilitates all aspects of the protocol & ensures all steps are accomplished. • Keeps peers on track so that conversation is productive. • Is protective of the presenter, & recasts or recalls inappropriate comments or questions. • May need to probe for cool feedback comments. Without them, the presenter may not benefit from the experience.

Community Agreements for Reflection Sessions

- **Be kind to the presenter.** By making their work public, colleagues expose themselves to comments and critiques that may not be familiar to them. This vulnerability is to be cared for by everyone in the group.

- **Be curious of the work.** Without thoughtful and probing questions, the presenter won't benefit from the feedback experience. Listen closely to understand one another. Different ideas and perspectives expand our thinking.

- **Be appreciative of the experience.** Every person has a role in the process that should be valued and supported. Disagree with ideas, not people. Together, members in our community learn from one another, solve our problems, and become stronger.

Checklist for Reflection Sessions

The purpose of **reflection sessions** is to provide productive feedback to the presenters as they work to investigate and solve problems in their practice. The journey is incremental. The **presenters** set the problem and provide work samples for peers to examine. **Peers** use a protocol to guide them through a conversation to ensure feedback is focused on evidence and questions provided by the presenters.

- Video Reflections are reflection sessions that use video clips of practice as work samples and the **Protocol for Video Reflection** to guide the conversation.
- One Video Reflection can take up to 60 minutes.
- A session can accommodate a maximum of two Video Reflections (2 hours max.).

Step 1. Convene with your small group to determine meeting times and roles.

Meeting Time	Roles*	
1 Date Time	**Presenter 1** **Guide 1**	**Presenter 2** **Guide 2**
2 Date Time	**Presenter 3** **Guide 3**	**Presenter 4** **Guide 4**
3 (if needed) Date Time	**Presenter 5** **Guide 5**	**Presenter 6** **Guide 6**

*Everyone needs to have at least one turn in all three roles (Presenter, Guide, and Peer)

Step 2. Prepare for your turn as presenter.

- Complete the **Reflection Exercise Worksheet to Prepare for Feedback**.
- Consult with facilitators or your guide if assistance is needed.

Step 3. Provide representative work sample, problem statement, and focus question to your guide.

This information should be provided before your group meets so your guide can be prepared to lead the group to give you productive feedback. The work sample for videos is a 5- to 7-minute clip.

Step 4. Participate in small-group Video Reflection.

- Make new copies of the **Observation Instrument** for each presenter's reflection for each group member (two presenters = two copies of Observation Instrument per peer).
- Use all **Tools for Reflective Practice** during the video reflection.
- Post the *Community Agreements* and *Roles & Responsibilities.*

(continues)

 This page may be duplicated for educational use.

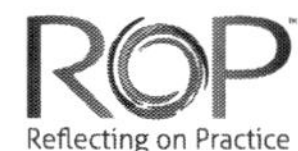

Checklist for Reflection Sessions *(continued)*
List of Tools for Reflective Practice

- **Community Agreements** to document how we will support and interact with each other in the community. *Your group may want to duplicate the poster displayed during the whole-group interactive sessions, or each person can refer to their own copy from the handout,* **Reflection Sessions** in *Reflecting on Practice.*

- **Protocol for Video Reflection** to guide peers through the problem-solving conversations to ensure feedback is focused on evidence and questions provided by the presenters. *The guide will need to have a copy of the Protocol; everyone else will follow along with their own copy and keep up with the steps as the guide sets the pace.*

- **Observation Instrument** is used during the second viewing of the clip, when you simultaneously watch the video and write down what you **NOTICE** (Step 5 in Protocol). After the viewing, there's quiet time to process your observations and record your notes in the **THINK** and **WONDER** columns (Step 7 in Protocol). *You'll need a clean copy of the Observation Instrument for each presenter's video reflection.*

- **Educator Moves** are used when thinking and wondering about the observations you make of your own and your peers' practice. It's a resource to remind you of the kinds of actions you might want to take and common language to describe those actions. Educator Moves are specific to the topic of each module—Module 2: Prior Knowledge, Module 3: Conversations, and Module 4: Objects & Design. *Used in Steps 7 and 8 of Protocol.*

- **Feedback Chart (Giving and Receiving Feedback)** to ensure feedback given is critical and productive, and feedback is received with open mind and heart. Productive feedback is a gift of information given to colleagues because you care for them and their ongoing learning and self-improvement. *Used in Step 8 of Protocol.*

Reflection Exercise Worksheet to Prepare for Feedback

This reflection exercise guides the presenter to set the problem and prepare for soliciting feedback from peers. Reflective practice is an ongoing, iterative journey of investigating one's own practice, individually and with colleagues. We constantly tweak and fine-tune our ways of thinking and doing.

Answer the following questions to (A) **set a problem** to investigate and (B) **seek help** from colleagues.

(A) Reflect on a problem in your practice to investigate.

Problem setting is the iterative process of naming and framing. **Naming** describes the situation and specifies how it's troublesome. **Framing** is how you "look" at the situation and involves articulating your position, attitude, or perspective. This process is neither linear nor easy to do, though it is tremendously important. Focusing on solutions without truly understanding what the problem is in the first place leaves us frustrated, exhausted, and no closer to solving the problem.

1. **What aspects of the learning/teaching situation will I direct my attention to for the moment?**
   In Module 2, those aspects pertain to "learners' prior knowledge." (In Module 3, they concern "learning conversations," and in Module 4, "using objects to support learning.")

   - For instance (in Module 2), *I want to examine how I address learners' prior knowledge. I feel there never seems to be enough time to consider all their diverse experiences.*

2. **How am I *framing the situation*?**
   Your frame affects how you approach, perceive, or understand the situation, and then, make decisions. Reflective practice requires deliberately framing and reframing one's practice in light of the consequences of one's actions, principles, beliefs, values, biases, and assumptions.

   - For instance, *I've noticed learners have a range of prior knowledge that's not always entirely scientifically accurate.* I frame the situation on prior knowledge in learning and teaching as …

… *assessing learners prior knowledge is central to being a learner-centered educator.*	… *learners retrieve memories to strengthen connections.*
I activate their prior knowledge so … *I know how much they know. Knowing my audience allows me to adjust the content I teach them.*	I activate their prior knowledge so … *they know what they already know and are more aware of what they're unsure of. My audience can then determine how this experience is relevant to them.*

3. **What is troublesome about the situation? What is my sense of the problem?**
   Specifying how the situation is troublesome helps to articulate what you believe is the problem. Your frame of reference affects what you see to be troublesome. Further investigation, individually and with peers, may reveal a different frame for consideration, or that what you sense is the problem now is not really the problem.

     This page may be duplicated for educational use.

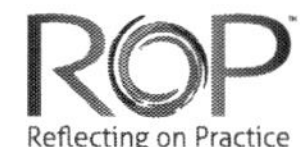

Reflection Exercise Worksheet to Prepare for Feedback *(continued)*

It's **troublesome that**, sometimes,	It's **troublesome that**, sometimes,
1. *there isn't enough time for me to hear all their prior knowledge.* 2. *the prior experiences learners share don't relate to the topic; their tangents are a distraction.* 3. *their prior knowledge is wrong. In speaking it out loud, learners become more convinced they're right and confuse other people.*	1. *there isn't enough time for learners to recall their prior knowledge.* 2. *learners recall prior knowledge that isn't relevant to the topic; this tangent makes them realize they're wrong.* 3. *their wrong answers are held so strongly that they have a hard time considering a different viewpoint. Their frustration may limit how others engage.*
The **problem is** *I can't be learner-centered in my teaching because it's not possible for me to hear or attend to all learners' prior knowledge.*	The **problem is** *I can't make space for all learners to struggle and make new connections when the experiences are ephemeral.*

(B) Prepare for feedback to help investigate the problem you've articulated.

1. **Recall an occasion, example or opportunity (an instance) that can provide more information for understanding and investigating the problem.**

   - Is there an upcoming event that can be video recorded?

   - Is there a pilot, prototype, or script that can be reviewed?

2. **Individually review your instance (e.g. video, prototype).**

   - What information about your practice and problem does this instance provide?

   - What about it makes you feel it's troublesome (or not going "right")?

   - What do you think needs fixing? Why do you feel this needs fixing?

3. **Choose a representative sample from your instance for peers to provide feedback.**

   - For videos, the sample would be a 5- to 7-minute clip from the video.

4. **Be prepared to describe the context to your peers.**

   - Refer to Step 2 in the **Protocol for Video Reflection** for the details to provide.

   - Use complete sentences for frame of reference and problem statement: *My **frame** of reference for accessing learners' prior knowledge in learning is that it's important for learners to retrieve memories to strengthen connections. The **problem** is I can't make space for all learners to struggle and make connections when the experiences are ephemeral.*

Protocol for Video Reflection

1. **Review protocol (2 minutes).** Guide reviews this protocol, including the agreements and everyone's roles and responsibilities.

A. Presenter sets the problem

2. **Describe the context (5 minutes).** Educator-presenter (presenter) describes the context for the video and what aspects of their teaching practice they want the group to focus on. Peers may ask questions to clarify the focus question.

   a. <u>Setting</u>: In what learning experience is the interaction occurring?

   b. <u>Audience</u>: Who are the learners? Adults, school group/grade, family…?

   c. <u>Timing</u>: When in the interaction is the clip taking place?

   d. <u>Teaching Purpose</u>: What was the teaching purpose in this interaction?

   e. <u>Learning Outcome</u>: What was your desired learning outcome?

   f. <u>Problem</u>: What's the problem in your practice you're investigating? What's your frame of reference for examining this problem?

   g. <u>Focus Question</u>: What help are you seeking?

3. **First watch (7 minutes).** Just watch the clip. No writing, no feedback.

4. **Recall the feedback focus (1 minute).** Brief pause to consider what aspect of the practice is the focus of the feedback. What's the problem the presenter is seeking help to solve?

5. **Second watch (7 minutes).** Peers use the **Observation Instrument** to collect evidence in the <u>Notice</u> column. Be specific. Optional, note the time on the video.

6. **Clarifying questions (3 minutes max).** Chance for peers to ask the presenter clarifying questions that have brief, factual answers, e.g., have these learners visited before?

B. Peers offer feedback

7. **Consider possible inferences and implications (5 minutes).** Individually, peers review data in the <u>Notice</u> column and enter thoughts in the <u>Think</u> and <u>Wonder</u> columns. Refer to **Educator Moves** to determine if actions from the Notice column could be inferred as an Educator Move. Refer to the **Feedback Chart** to determine which items in the Think column are warm and cool feedback.

(This protocol is inspired by the Tuning Protocol (1).)

(continues)

     This page may be duplicated for educational use.

Protocol for Video Reflection *(continued)*

8. **Discuss video observations, questions, reflections, and feedback (20 minutes).** Peers discuss what they noticed, thought, and wondered, and which Educator Moves were present. Start with warm feedback. Always support your feedback with observation data.

   a. Presenter physically moves outside of circle, remains silent, and takes notes. Refer to **Feedback Chart**.

   b. Guide ensures discussion remains focused on area of emphasis requested by presenter.

   c. Consider the **Educator Moves**. *What moves <u>did</u> the presenter take? What seemed to be the effect of the moves?*

C. Presenter responds

9. **Respond and conclude (5 minutes).** Presenter responds to any comments or questions without being defensive. Presenter reflects out loud on ideas that came to mind while listening to the feedback. Peers are silent. Guide may clarify or lend focus.

D. Group reflection

10. **Two takeaways (3 minutes).** Everyone thinks holistically about the group conversation, and writes down two action items. *What is one thing I want to take into my own practice? What is one thing I want us to discuss further regarding our institutional practices?*

11. **Community actions (2 minutes).** The group generates a list of the community action steps they want to have further conversations about in the future.

Reminders for Guide

<u>**Be assertive about keeping time.**</u> A protocol that doesn't allow for all the components will do a disservice to the presenter, the work presented, and the peers. Don't let any one person monopolize.

<u>**Be protective of educator-presenters.**</u> By making their practice public, presenters are exposing themselves to input that may be unfamiliar. Inappropriate comments or questions should be recast or withdrawn. Keep discussion centered on the work presented and the focus question posed by the presenter.

<u>**Be vigilant about substantive discourse.**</u> Many presenters may be accustomed to blanket praise. Without thoughtful but probing questions and comments, they won't benefit from the video reflection. Remember, presenters should be able to revise their work productively based on this discussion—building on its strengths (warm feedback) and bringing it more closely in line with the changes they seek (cool feedback).

---

1. School Reform Initiative, Tuning protocol. (2017).

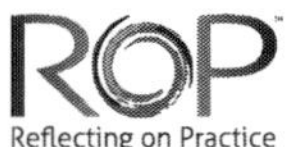

Observation Instrument

← Step 5 → ← Step 7, use Educator Moves →

NOTICE.	THINK.	WONDER.
What do you see and hear happening?	What do you think is going on? What can be inferred, assumed, speculated about, or reasoned from the observed actions?	What does that make you wonder? What suggestions and implications for teaching can be derived based on observations and inferences?

This page may be duplicated for educational use.

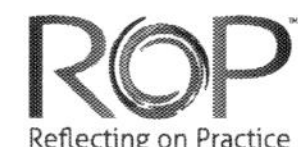

Observation Instrument—Examples

NOTICE What do I see or hear happening?	THINK What could the reason be?	WONDER How could this be addressed?
Few or no learners respond to questions the educator asks.	1. Learners don't feel comfortable answering the question. 2. Learners don't have enough information to answer confidently. 3. Educator asked a question with only one right answer.	1. Ask learners to turn and talk to their neighbors before asking them to share their answer with the whole class. 2. Provide learners an experience (e.g., reading, activity) from which they can acquire information to answer the question. 3a. Rephrase the question to have more than one possible answer. 3b. Pose the question rhetorically, in a context that doesn't involve a group discussion.
7:30 Educator writes one-word answers on board as learners volunteer them, then educator orally elaborates on their answers. 8:00 A little bit of calling on specific learners, a lot more of getting learners to "shout out" answers without educator acknowledging who said them—educator simply repeats what he hears "from the crowd."	Reinforces that educator wants short phrases, since he doesn't ask anyone to explain further, and doesn't put any of his own elaboration into writing.	If you're hoping for more in-depth answers... 1. Acknowledge whoever volunteers a specific response and ask for elaboration, and/or 2. Ask "can anyone follow up with more details about what _____ just said?" 3. Write key words from the reasoning of the first few questions to elicit more details about the next one(s).
Educator asks a question and then answers it herself almost immediately.	1. The educator doesn't think learners know the answer. 2. Educator asked a rhetorical question. 3. Educator is nervous about the silence from non-responsive learners.	1. Consider whether learners have had enough experiences/acquired enough information to answer the question. 2. If it's a rhetorical question, let learners know that. 3. Be comfortable with silence. Give learners time to process the question and come up with an answer. Practice waiting silently for 3–5 seconds after asking a question before talking again.

This page may be duplicated for educational use.

Feedback Chart

· ·

Giving Feedback

Feedback is a gift of information given to colleagues because you care for them and their ongoing learning and self-improvement. Show them you care through your actions and thoughtful feedback during the reflection sessions. **Be curious about the work. Be kind to the presenter. Be appreciative of the experience.** Feedback is a learning opportunity for you, as well. You learn to notice details and dig deeper into reflecting on your own practice.

All feedback must be based on observation data presented, and informed by the ideas on learning that the community explores together. Both types of feedback must be offered.

	Warm	**Cool**
Actions	• Recognizing • Highlighting	• Discerning • Probing
Examples	• "I heard her tell learners to talk with each other before asking for volunteers. I think that offered every learner the chance to participate, since she only had time for 3 volunteers to share with the whole group." • "At minute Y, I heard him say X, which acknowledged learners' prior experiences that he solicited earlier. I think he's helping the learners make the connections to entice them to stay at the exhibit longer."	• "I noticed 8 out of the 10 learners that the presenter called on to share their ideas with the whole group were girls. I think there might have been a gender imbalance in that group conversation. I wonder if it's just this group, or if it's something in the task or topic that is more appealing to girls? What's the gender ratio in that group? What are the boys doing, or not doing? Is the presenter deliberately calling on girls? Do we need to have gender parity in every whole-group conversation? I assume the selected learners are all self-identifying as girls, but perhaps not."
Purpose	Warm feedback helps the presenter build on specific areas of strength in their work.	Cool feedback takes a closer look at what occurred and offers different vantage points for consideration.
Nature	Warm feedback is *reassuring*. It lets the presenters know they have abilities upon which to leverage.	Cool feedback is *stretching*. It lets the presenters know where and how they can change their practice.

 This page may be duplicated for educational use.

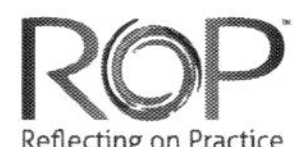

Feedback Chart *(continued)*

Receiving Feedback

Feedback is a gift of information about ourselves from another person's vantage point, through their lens of experiences and knowledge. It's a perspective that we cannot see for ourselves. It's difficult to regulate our own learning and change our practice without this information.

Feedback can be hard to take in. It might make us aware of the biases, assumptions, traditions, blind spots, and habits that shape our practice. We might not like what we hear. Reflective practice requires framing and reframing one's practice in light of this information. **Listen with curiosity to understand the perspectives being offered.** What's at the heart of the discussion? What can I learn about myself from what's being offered? How do I change based on what's revealed to me?

Be self-aware of how the information might make you react, despite the *good intentions.* Your emotional responses are valid. Acknowledge them and strive to understand them.

Human Social Experiences that can trigger emotional responses (*1*)	Questions for Reflection
Status My sense of importance or self-worth. My perceptions of position in the social hierarchy.	• Do I feel the information isn't true? Why? • What does it mean if there are aspects of truth? • Am I feeling concerned about what my colleagues think of me if this was true? • Do I feel my ways aren't valued?
Relatedness My feelings of safety and connectedness with others. My sense of belonging in the community.	• Am I feeling this way because of the person giving the information? How would I feel if the same information was given by someone else? • Am I feeling my colleagues think I don't belong?
Fairness My perception that exchanges between people are unbiased and based on shared standards. My sense that decisions are just.	• Do I feel I'm being treated unfairly? Why? • Do I feel I'm being singled out for something everyone does or that is out of my control to change? • Am I feeling that the standards for what's ineffective are unjustly applied to me?

Remember:

- The feedback is based on only 5-7 minutes of your entire practice. It's just a glimpse.

- Not all feedback needs to be acted upon. If you pass on taking action, be mindful of your reasons. Is it something: beyond your control to address? you're avoiding? you need help to be able to achieve? the community needs to work on together? that needs conversations with trusted allies?

---

1. D. Rock, SCARF: A brain-based model for collaborating with and influencing others. *NeuroLeadership Journal* 1, 44–52 (2008).

This page may be duplicated for educational use.

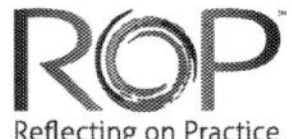

Educator Moves—Prior Knowledge

(Add your own examples in the space provided.)

How does the educator activate learners' prior knowledge?

Educator Moves	How the Educator Move looks in practice
<u>Prompt retrieval</u>. Educator prompts learners to think about their prior knowledge.	"What does this remind you of?" *or* "What does this look like to you?"
<u>Involve others</u>. Educator prompts learners to talk with one another about their prior knowledge.	"Tell the person next to you something you already know about this."
<u>Provide analogies</u>. Educator prompts learners to compare this to something similar or commonplace.	"Have you ever seen Patrick the sea star on Sponge Bob? What differences would you expect to see between the cartoon sea star and this live one?"

How does the educator connect learners' prior knowledge to ideas in the interaction?

Educator Moves	How the Educator Move looks in practice
<u>Harken back</u>. Educator prompts learners to recall what they originally thought about the topic.	"I remember you said _______ when you first examined this. What are you thinking about ___ now?"
<u>Reexamine ideas</u>. Educator prompts learners to reexamine and compare their ideas in light of what they've learned.	"How has your thinking on ___ changed? What did you do that helped you come up with a new explanation?"
<u>Make connections</u>. Educator prompts learners to connect the ideas.	"How do you think X is related to Y?" *or* "What do you think about what Tariq said?"

   This page may be duplicated for educational use.

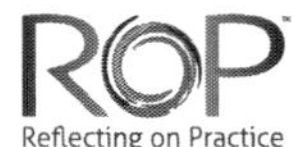

Educator Moves—Prior Knowledge *(continued)*

How does learners' prior knowledge affect the educator's moves?

Educator Moves	How the Educator Move looks in practice
<u>Use their language</u>. Educator uses examples & terms supplied by the learners.	"Sarah told us that tiny plants that live in the ocean are very important. Scientists call them plankton."
<u>Build on input</u>. Educator builds on/adds to what learners say and do.	"Mei showed us how to use a QR code app on her smart phone. You can do this with QR codes here in the museum."
<u>Add examples</u>. Educator expands on experiences & examples based on what learners say and do.	"Brian said ___; that makes me think about ______."

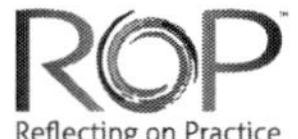 **Appendix A**

Educator Moves—Learning Conversations

(Add your own examples in the space provided.)

How does the educator encourage learners to make sense of the science through conversation?

Educator Moves	How the Educator Move looks in practice
Re-voice. Educator re-voices what learners say.	"So let me see if I've got your thinking right. You're saying…" [with space for student to follow up] *or* "Let me recap the different ideas. I heard: _______"
Request elaboration. Educator prompts learners to elaborate on and explain their reasoning and thinking.	"What makes you think that?" *or* "What evidence helped you arrive at that answer?" *or* "Say more about that."
Consider others. Educator prompts learners to consider other people's reasoning in relation to their own.	"How would you restate what Sophie just said in your own words?" *or* "How does her explanation connect with what you're thinking?" *or* "Turn to the person next to you and explain what you think."
Invite differences. Educator prompts learners to voice alternative views and disagreements.	"Do you agree or disagree, and why?" *or* "What do others think about that idea?" *or* "Find someone nearby who has a different view from yours and persuade them that you're right."
Push integration. Educator prompts learners to reconsider existing understanding and integrate new thinking, including the disciplinary (e.g., scientific or mathematical) view.	"Based on what you just read (or experienced), tell us how you're thinking about the concept now."
Apply thinking. Educator prompts learners to apply their thinking to a new context, encouraging use of the discipline's language.	"You've explained that density is causing the layering. How might this phenomenon relate to the melting glaciers and currents in the North Atlantic?"
Wait time. Educator provides wait time for learners to express themselves.	"Take your time…we'll wait."

    This page may be duplicated for educational use.

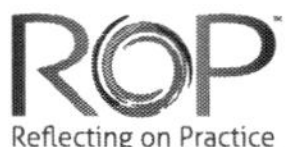

Educator Moves—Objects

(Add your own examples in the space provided.)

How does the educator encourage learners to engage with the objects and make their thinking visible to themselves and others?

Educator Moves	How the Educator Move looks in practice
<u>Set an example</u>. Educator models how to engage with the object.	"Hold your moon ball to the 'Sun' till you see a sliver of light; this is a crescent Moon. Now rotate to the left."
<u>Give it a go</u>. Educator prompts learners to give it a try and make the effort to use the object.	"Go ahead, put on these glasses and tell me what you see."
<u>Observe closely</u>. Educator invites learners to use their senses to study the object deeply.	"Check out all the fur. What do you notice? Describe how it feels."
<u>Evidence reasoning</u>. Educator prompts learners to use the object to gather evidence to support their reasoning, connections, and explanations.	"What additional information does the food coloring provide to help you explain what's happening in the two cups?"
<u>Learn together</u>. Educator prompts learners to work with each other to engage with the object and learn from each other.	"Try this out with your friend and see what you can figure out together."
<u>Wonder on</u>. Educator encourages learners to use object to ask and answer their own questions, or answer the educator's questions.	"Oh, so you're wondering whether we can see a full Moon during the day. Let's see if we can recreate that with your moon ball."
<u>Encourage free play</u>. Educator offers learners time to explore the object on their own.	"Keep working on this with your mom and I'll come back in a few minutes to see how you're doing."

© 2021 Regents of the University of California ■ **331**

Reflecting on Practice
Appendix B
Design and Teaching Tools

Tools

Learning Cycle Design Framework 334

Teaching Purposes 336

Thinking Moves & Thinking Routines 337

Discussion Map 338

Facilitation Approaches 339

Worksheets

Reflection Exercise Worksheet for Rapid Design Feedback 340

Reflection Questions (for Rapid Design Feedback) 341

Reflection Exercise Worksheet for Critiquing Objects in Learning Experiences 343

Learning Cycle Design Framework

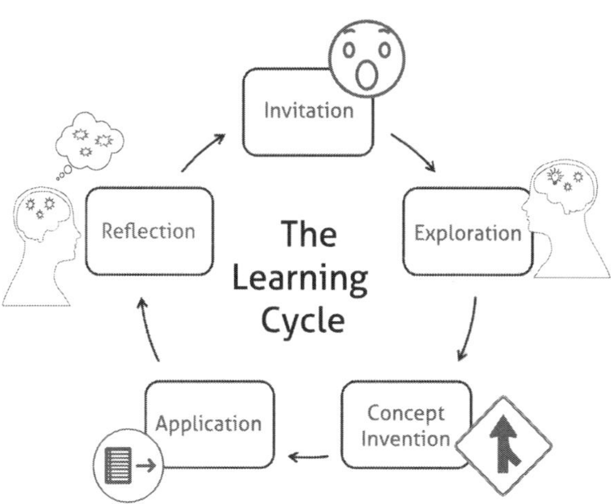

<u>Invitation</u>—Initiates the learning task and sets the context. The invitation phase sparks interest, and spurs learners to recall and retrieve past connections from their memories that may be relevant to present learning experiences. Generates anticipation, and captures learners' attention to focus on the topic of the upcoming activities.

> *Educator's Role:* Create interest and pique curiosity. Raise questions and problems to be explored. Elicit responses that uncover learners' current knowledge about the concept or topic.

- How will the task get learners interested in learning about the topic?

- How will the task provoke learners to access their prior knowledge?

<u>Exploration</u>—The exploration phase is driven mainly by learners' interest and questions, followed by discussion about observations, results, ideas, and more questions. It involves exploration of real phenomena, providing a common base of experiences for learners to develop new concepts and skills. Learners retrieve prior connections to examine alongside new information in an effort to determine the relevance to the new ideas. They begin to elaborate on memories related to the concept and use their everyday language to articulate their thinking.

> *Educator's Role:* Encourage learners to work together without direct instruction from you. Observe and listen to learners as they interact. Ask probing questions to redirect learners' investigations when necessary. Provide time for learners to puzzle through problems. Invite learners to listen to what peers are saying or noticing.

- How will the task provide learners with experiences that produce concrete observations and discoveries to help them make sense of the topic?

- How will the task urge learners to relate prior experiences with new ones?

   This page may be duplicated for educational use.

Learning Cycle Design Framework *(continued)*

Concept Invention—With their interest and attention captured, learners actively process the experience, review evidence and data they gathered through exploration, and try to make sense of it. As learners describe their observations and generate explanations, they make new and/or different connections to existing memories.

> *Educator's Role:* Encourage learners to explain concepts and definitions in their own words. Guide them to express out loud their evolving understandings with their peers. Reassure them that what they express doesn't need to be coherent just yet; the purpose is to test out connections. Ask for evidence and clarification from learners to help them make sense of their experience. Push for integrating new information with existing memory, including rearranging prior connections. Use and introduce technical language from the discipline. Connect to learners' direct experiences when explaining concepts.

- How will the task encourage learners to struggle with their understanding and negotiate their ideas with others?

Application—New connections are still tenuous. The application phase urges learners to apply new knowledge and skills to solving a problem or meeting a challenge, transferring their new knowledge to unfamiliar contexts in the world. Learners retrieve these fresh connections in their memories and elaborate on them further as they apply them to different contexts, which in turn strengthens the new connections.

> *Educator's Role:* Provide opportunities for learners to use vocabulary, definitions, and explanations, and transfer knowledge from familiar circumstances to novel ones in order to strengthen new connections.

- How will the task demand learners to apply what they've learned to a new situation or context?

- How will the task invite learners to make connections that are meaningful to them?

Reflection—Learners reflect on how their original notions have been or need to be modified. In this phase learners are consciously thinking about the new connections they've made to existing memories, and how their understanding is changing. Reflection cultivates metacognitive skills. It's time for learners to be conscious of what they learned and how they learned it, which becomes a resource for future experiences.

> *Educator's Role:* Encourage learners to confront their former ideas and evolve new ones, solidify conceptual framework connections, and build metacognitive skills.

- How will the task prompt learners to think back on the process for learning to help reinforce their understandings and make them better learners in the future?

Teaching Purposes

Teaching Purpose	Description
Activate prior connections	For <u>learner</u> to 1. trigger and access existing views and understandings of specific ideas and phenomena; 2. express (by talking, in writing, etc.) existing connections between ideas.
Give information	For <u>educator</u> to 1. summarize the explanation that unfolded from the community, to help learners follow how it developed; 2. provide the explanation and/or make the case for the disciplinary (e.g., scientific, mathematics) view; 3. bring together everyday and disciplinary views.
Check for understanding	For <u>educator</u> to 1. make the disciplinary view available for all learners in a social context, so that learners can compare their understanding with what is articulated by the group; 2. evaluate what learners understand about the ideas being taught so that educators *and* learners are made aware of flaws and accuracies in learners' understanding.
Express your thoughts	For <u>learner</u> to 1. verbalize/express potential connections between ideas as learner **integrates** new and existing knowledge and skills, and to make sense and meaning of the disciplinary ideas; 2. **listen to and consider** how others connect ideas; 3. reconsider and rearrange prior understanding in light of new knowledge and skills, and **internalize** it for oneself.
Transfer understanding	For <u>learner</u> to 1. apply the disciplinary ideas in a range of contexts; 2. take over responsibility from the educator for using those meanings.

    This page may be duplicated for educational use.

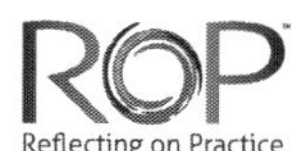

Thinking Moves & Thinking Routines

Thinking Moves

Actions that learners take when they think and work towards understanding (*1*).

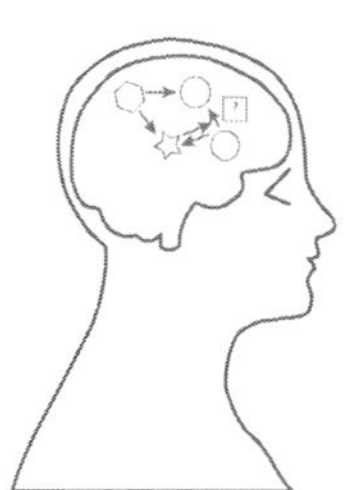

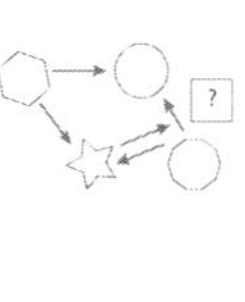

Thinking Routines

Sequences of steps that learners take to make their thinking actions visible to themselves and others. Educators integrate these sequences into the experiences they design and facilitate.

Observe closely and describe what's there	• Micro Lab • Notice/Think/Wonder
Build explanations and interpretations	• 3-2-1 Bridge • Jigsaw (Reading Partner, Expert Group, Case Studies) • Pictures & Words • Predict-Explore-Revise
Reason using evidence	• Jigsaw (Reading Partner, Expert Group, Case Studies) • Predict-Explore-Revise • Think-Pair-Share
Make connections	• 3-2-1 Bridge • Concept Map • Connect, Extend, Challenge, & Apply • Jigsaw (Reading Partner, Expert Group, Case Studies) • Pictures & Words • Think-Pair-Share • Walkabout
Consider different viewpoints and perspectives	• Jigsaw (Reading Partner, Expert Group, Case Studies) • Micro Lab • Predict-Explore-Revise • Sentence-Word-Phrase • Think-Pair-Share • Turn & Talk • Walkabout
Capture the heart (core) of a concept and form conclusions	• Pictures & Words • Sentence-Word-Phrase
Wonder, spark curiosity, and ask questions	• Notice/Think/Wonder • Predict-Explore-Revise
Uncover complexity and go below the surface of things	• Concept Map • Micro Lab • Pictures & Words

1. R. Ritchhart, M. Church, K. Morrison, *Making thinking visible: How to promote engagement, understanding, and independence for all learners.* (John Wiley & Sons, 2011).

This page may be duplicated for educational use.

Discussion Map

- **Pose a question that invites learners to retrieve and share their experiences, connections, and knowledge that may be relevant to the topic.**
  - Prepare your questions ahead of time so that they aren't too leading, but really draw out learners' thinking. In other words, be sure the answer you seek isn't embedded in your tone or in the wording of your query.
- **Listen with curiosity, probe for understanding, and encourage community to do the same.**
  - Use suggested questions and model the behaviors for the community.
  - Display the questions for everyone to use.

Questions to ponder while *listening with curiosity*	Questions to use and model to *probe for understanding*	Questions to pose to encourage *careful listening*
• What prior connections are learners recalling? • How are learners thinking those connections are relevant to the topic? • Are those connections relevant? • What information are the learners missing? Can they get this information from this exploration? • How are their connections flawed—gaps, inaccurate associations, inappropriate connections?	• What makes you think that? • Was there anything in the reading/task that supports the connections you're making? • What do you mean? • How does that work? • If that were the case, what would you expect to see? • What are you unsure about? • What additional information would be helpful to you?	• Can someone add to that idea? • If that were the case, what assumptions are we making? • X made a critical comment. Could someone restate what was said in a different way? • What do others think about that idea? • What is another way to consider that position? • How is that idea relevant to understanding the topic or answering the question?

- **Remember to stay neutral in your reaction.**
  - It helps to draw or write the contributions on the board while the conversation unfolds.
- **Connect back to the main topic.**
  - Reflect on ideas generated from the group conversation. Consider the connections and determine if more information or perspectives are needed.

Suggestions to connect ideas	Suggestions to seek more information
• Let's review the key ideas/observations we discovered. • What connections can we make between the ideas/observations and the main topic?	• Are there additional perspectives to consider? • What additional information would be helpful to understand the main topic/question?

- **Organize and summarize ideas.**
  - Display and review a summary of the ideas. Task learners to first revise write their initial thoughts, and then compare their revisions with your displayed summary.
  - Or invite learners to generate the summary together. *Can someone start us off to generate a summary of the ideas from this conversation? Who can add to these?*

 © 2021 Regents of the University of California   This page may be duplicated for educational use.

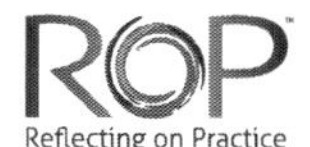

Facilitation Approaches

Educators have the task of providing guidance in learning. The educator is in a position of power, and learners need room to explore. Therefore, depending on the teaching purpose, educators can approach facilitating conversations in different ways (*1*).

There are two dimensions to facilitation approaches: <u>authoritative/dialogic</u> and <u>interactive/noninteractive</u>. These dimensions intersect to form four classes of facilitation approaches.

1. The **authoritative/dialogic dimension** refers to the extent to which the conversation <u>emphasizes one idea (authoritative) or invites many ideas (dialogic)</u>. It's important to understand that authoritative doesn't mean authority and dialogic doesn't mean dialogue.

2. The **interactive/noninteractive dimension** refers to the extent to which the conversation allows for <u>participation from more than one person (interactive)</u>, or <u>only one person, usually the educator (noninteractive)</u>.

Interactive-Noninteractive Dimension

<table>
<tr><th colspan="2">Dialogic-Authoritative Dimension</th></tr>
<tr>
<td>

Interactive-Dialogic

Purposes: Activate prior connections; Express your thoughts; Transfer understanding

Pattern*: Open chain
I-R-P-R-P-R-P-R-
I-R_1-P-R_2-R_3-P-R_1-R_4-...

</td>
<td>

Noninteractive-Dialogic

Purposes: Give information

Pattern: Educator monologue

</td>
</tr>
<tr>
<td>

Interactive-Authoritative

Purposes: Express your thoughts; Check for understanding

Pattern*: Closed chain
I-R-E
I-R_1-P-R_2-P-R_3-(E)

</td>
<td>

Noninteractive-Authoritative

Purposes: Give information

Pattern: Educator monologue

</td>
</tr>
</table>

*Key: I = Initiate (educator asks question); R_1= Respond (Learner 1), Rn= Respond (learner n); P = Probe (educator probes for evidence or elaboration); E = Evaluate (educator determines accuracy of learner response)

---

1. P. H. Scott, Teacher talk and meaning making in science classrooms: A Vygotskian analysis and review. *Studies in Science Education* **32**, 45–80 (1998).

This page may be duplicated for educational use.

Reflection Exercise Worksheet
for Rapid Design Feedback

This Reflection Exercise prepares the community to engage in Rapid Design Feedback by sparking inklings for what revisions to make. The purpose of the Rapid Design Feedback is to get input quickly for revising the design of a learning experience before committing further to these modifications.

(A) Choose, with your partner, the learning experience to revise.

(B) Be prepared to share details of the context during the Rapid Design Feedback.

- **Setting:** Where does this learning experience take place? (e.g., classroom, outreach)

- **Audience:** Who are the learners? (e.g., adults, school groups, families...)

- **Timeframe:** What is the (intended or actual) time duration for the experience?

- **Outcomes:** What are the desired learning outcomes?

- **Other:** What else is important for peers to know about this experience?

(C) Decide what aspect of the experience will be the focus of the revision.

- Learning fundamentals, use Five Foundational Ideas on Learning

- Sequence and flow, use Learning Cycle Design Framework

- Rationale for segments, use Teaching Purposes

- Strategies for making thinking visible, use Thinking Moves/Thinking Routines

- Leading inclusive conversations, use Discussion Map/Facilitation Approaches

(D) Discuss the set of reflection questions related to your chosen focus.

Each set of reflection questions invites partners to scrutinize the existing design with that aspect in mind. Overall, the questions push partners to contemplate: *Does the existing design make sense considering the community's conversations on learning and teaching? What changes do we want to make?*

(E) Be prepared to share your inklings for revisions to get feedback from peers.

- What changes do you want to make? It's possible there may be a combination of minor tweaks and massive rethinking.

- Why do you feel those changes are needed?

- What are possible options for the changes, including why you're suggesting these possibilities and not others?

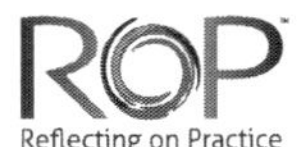

Reflection Questions
for Rapid Design Feedback

Learning fundamentals

How do the Five Foundational Ideas on Learning show up in this experience? What changes are needed?

Which of the foundational ideas on learning are addressed?

- In what ways are they addressed in this design?

- Are these ways effective for supporting the learning outcomes? How do you know?

- Which parts need minor tweaks? Which parts require deeper consideration?

Which of the foundational ideas on learning aren't addressed?

- How might they be included?

Sequence and flow

What tasks are included in the experience and how do they match up with the different phases of the Learning Cycle? How do these tasks work together to support the learning outcomes for this experience? What changes are needed?

Invitation. What task(s) gets learners interested in learning about the topic? How does the task(s) invite learners to retrieve their prior knowledge?

Exploration. What task(s) encourages learners to make observations and discoveries to help them make sense of the topic? How does the task(s) urge them to relate their prior experiences with new ones?

Concept Invention. What task(s) prompts learners to struggle with their understanding? How does the task(s) push them to negotiate their thinking with others?

Application. What task(s) asks learners to apply what they've learned to a new situation? How does the task(s) encourage them to examine their new connections in a different context?

Reflection. What task(s) has learners thinking back on the process for learning to strengthen their understandings? How does the task(s) help them be better learners in the future?

Rationale for segments

How do the Teaching Purposes show up within and across the phases of the Learning Cycle in this experience? What changes are needed?

Are there enough opportunities in the experience for learners to Express Their Thoughts in all the Learning Cycle phases?

- What are the different ways provided for learners to Express Their Thoughts?

(continues)

Reflection Questions *(continued)*

- How do these opportunities for expression build on each other to connect and deepen learners' understanding?

What are the different ways information is given?

- Learners shouldn't be expected to already know or construct in the moment all the information needed to participate in the learning experience. The educator telling learners the information shouldn't be the only way to give information.

- What information can learners get from the objects? How are learners tasked to get this information?

Strategies for making thinking visible

How is learners' thinking made visible to them and their peers? What changes are needed?

What kinds of Thinking Moves are helpful for understanding the ideas in this experience?

Which Thinking Routines offer opportunities for the kinds of thinking helpful for understanding?

- Do the Thinking Routines need to be customized for this experience? How?

- How do the different Thinking Routines included in the experience build towards the learning outcomes?

Leading inclusive conversations

How does the design make space for the educator to facilitate learning conversations? What changes are needed?

Where are the opportunities in this experience for the educator to use the Discussion Map?

- What prompts are used, or could be incorporated to encourage learners to elaborate on their thinking?

- What prompts invite learners to listen closely to what's shared?

Where are the opportunities for the different classes of Facilitation Approaches to be used?

- How does/could the educator solicit many ideas from many learners (interactive-dialogic)? What does/could the educator do with all the ideas solicited?

- What is the big idea (e.g., science concept) the educator should be attending to for the interactive-authoritative approach? How does the design move from many diverse, and/or everyday ideas offered by learners towards the one discipline-based idea for learners to negotiate and come to a deeper understanding? How might the design include an opportunity to expand back out for learners to propose many new ideas?

     This page may be duplicated for educational use.

Reflection Exercise Worksheet for Critiquing Objects in Learning Experiences

What learning experience with a focal object did your group decide to focus on?	Why did the group choose that focal object in that experience?
	What are the learning outcomes for the experience?

Below are questions for critical review of the focal object in the learning experience. Answer the questions individually, then discuss with your group.

Our Current usage.

1. How is the object currently used? What information can learners get from the object that is relevant to the experience? How are learners supported to get this information? What kinds of questions are learners asked? What are learners asked to do with the object?

2. Does our current use of the object support our understanding of how people learn? If not, how might we integrate that? What else/what more can we invite learners to do with the objects?

3. Are we providing learners the opportunity to examine/explore/mess with the object in order to answer the questions we ask them and their own questions? In what ways? If not, how could we introduce such opportunities?

Learners' current usage.

Recall how learners typically engage with the objects. What do they talk about, or want to talk about? What are the interesting conversations learners have about the objects?

Pushing ourselves.

What more might we do to encourage learners to explore the object to deepen their understanding? What are some other ways to use the object that we haven't yet considered?

This page may be duplicated for educational use.

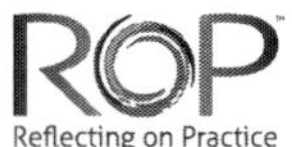

Index

Page numbers in *italic* indicate illustrations, **bold** for a table.

3-2-1 Bridge routine: ideas on learning 19–20; revisited 308–310; sequence for thinking actions 337

About the Program, handout 31–32

active reading 162, 222, 226, 246

analogy, definition 20, 309

Brahms, Lisa 275–276

Case Studies routine: directions 263–265; how learners interact with objects 272, 273-282; sequence for thinking actions 337

category (knowledge grain size) 138–139, 150–151

Checklist for Reflection Sessions: introduction 168; use to prepare for Video Reflections 168, 251, 310, 318–319

cognitive system and prior knowledge 116–117

collective recall: application 25; definition 25

Community Agreements: introduction, 51; use with Video Reflections 166, 249, 306, 317, 319

concept map, directions 105–109

conceptual change: cognitive reasoning 151, 199; knowledge condition factors 139; misconceptions, origins and adaption 152–153

Connect, Extend, Challenge & Apply routine: Learning Cycle Design Framework 109–114; objects and learner interaction 269; sequence for thinking actions 337; teaching purposes and Discussion Map 188–189; Tell a Friend 81–82

Continue the Learning assignments: manageability considerations 14

conversations: characteristics 179; family 237–238; power relationships 236

critical friendships: concept 48; in reflective practice 48–49, 316

Crossan, Mary 3–4

Crowley, Kevin 273–274

Crowl, Michele 277–278

Cup & Card, hands-on activity 22–23

culture of science 81, 83–84

Design & Teaching Tools Match Up hands-on activity: Card Sets 298; directions 292–293; Instruction Cards 299

Dewey, John 60, 62

Discussion Map: experience reflections 215, 224; first mention 129; formal introductin 188-189; tool 338

Donegan, Kathleen 289, 296–297

Eberbach, Catherine 273–274

Educator Moves: Learning Conversations 247–248, 330; Objects 304-305, 331; practice documentation 43, 165, 249, 306, 319; Prior Knowledge 164–165, 328–329; Thinking Moves, complementary use 290; Video Reflection exercises 159–160, 167, 247–248, 307–308, 322-323

engagement: behavioral, cognitive, emotional 118-119

Expert Jigsaw routine: learning conversations, facilitation 216–217; reflective practice exercise 45–47

externalize 117–118, 202; sequence for thinking actions 337

Facilitation Approaches: applied in design of Two Balloons activity 218–224; examples 231–232; introduction 218; relationship to other tools 224–225; tool 339

false belief 93, 133, 149, 150

families: interaction with objects studies 273–274, 277–278, 281–282; interactive engagement 265; learning facilitation 237–238; science aspirations 85–86; unit of learning 118

Feedback Chart introduction 55-57; use 57, 166, 249, 305; tool 319, 326-327

feedback, for learning: 187; talking to learn benefits 202–203

feedback, peer: giving 55–56; Rapid Design Feedback 294–295, 340–342; receiving responses and review 56–57; reflection exercise worksheet 320–321; with Educator Moves 167

fidget toys 12

Five Foundational Ideas on Learning: introduction 25, 33-34; Walkabout, discussion exercise 27–29

frame of reference: definition 47; problem presentation/resolution 53; reflection exercise worksheet 320; reflective practice influence 47, 63–64

From Learners' Perspective task: Table Talk 24–25, 77–79, 104–105; Turn & Talk 167, 224–226, 250

goal orientation, motivations 62

grain size of knowledge 132–133, 149–151

Halversen, Catherine 31

Halverson, Erica 275–276

Hands-on activities: Cup & Card 22–23; Design & Teaching Tools Match-Up 292–293; Ice Cubes Investigation 131–137; first practice, Video Reflection 51–58; key objectives 22; learning conversations, Video Reflection 249–250; Mystery Tube 75–76; objects, Video Reflection 305–306; Phases of the Moon 95–103; playing with objects 265–269; prior knowledge, Video Reflection 165–167; program design element 31; Role-Plays 179–185; Two Balloons 217–224

Hohenstein, Jill 279–280

How people learn, reading *107*, 116-121

Iaccarino, Maurizio 71, 87

Ice Cubes Investigation, hands-on activity: directions 131-137; explanation, for participants *148*, 148; Science Content Guide, for facilitator 142–147

ideas, expression: elaboration strategies 205; externalizing and articulating 117–118, 202

individual belief (knowledge) 133, 138–139, 149

Institutional Practices: documentation for community 43, 167, 243, 303; part of Video Reflections exercise 58, 159–160, 250, 306

Jacobs-Priebe, Phoebe 275–276

jigsaw, routines for Research Discussion: *see also* Expert Jigsaw; Reading Partner Jigsaw

Kisiel, James 281–282

knowledge, specialized 63

knowledge systems 80, 84

Kopczak, Charles 281–282

Lane, Henry 3–4

learners' perspective: being a learner 22; communication skills 64; expertise perceptions 147; facilitating conversations, tools and ideas 224–226; models, explanatory aids 133; motivation and engagement 118–120; perception of science, group evaluation 77–79; personality traits 117–118; sense making experiences 104–105; video reflection experience 167, 250

Learning Cycle Design Framework: introduction 109–113; example, Phases of the Moon 115, **115**; example, Two Balloons 231-232, **231-232**; relationship to other tools 225; tool 334–335

learning environment inequalities 5–6

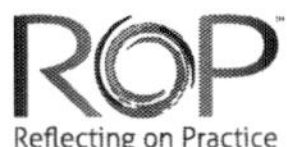

Learning Journal: Discussion Map experiences 215; how people learn activity 96, 103, 106, 109; ideas on learning 19–20, 308–310; individual practice review 59, 81, 168, 250; key objectives 16; learning experience design 288; objects in learning experience 264; personal goals review 30; personal records 14; Playing with Objects activity 266, 267–268, 269; program design element 32; teaching purpose 188–189; Two Balloons activity 218–219, 226

learning styles, arguments discussion 25–27

Let's Talk Practice task: Educator Moves, tool applications 165; individual practice inquiry 59, 81–82; learners' prior knowledge 139–141; Learning Cycle Design Framework 109–114; new tools and modifications 226–227; objects and learner interaction 269; objects, Educator Moves 304–305; objects in learning experiences 306–308; personal goals review 30; program design element 32; teaching purposes 188–189; videos, educational benefits 164; what thinking looks like 289–291

Littis, Breanne 275–276

Marshik, Tesia 13, 25–26

Mazzocchi, Fulvio 71

McClain, Lucy, R. 277–278

memory: long-term, working 116, 200-201; making and retrieval *199–200*, 199–201

mental model 116–117, 138–139, 149-150

metaphor 20, 309

Micro Lab routine: facilitating conversations 226–227; rapid design feedback 293–294; sequence for thinking actions 337

mindset characteristics 62

Minute Paper routine: individual practice inquiry 59; personal commitment to reflective practice 168, 250; sequence for thinking actions 337; setting personal goals 30

misconceptions 137, 139, 151-152

models, explanatory aids 133

motivation: learners' engagement 118–120; reflective practice 62

Mystery Tube, hands-on activity 71, 75–76

neurons, neural pathways *199*, 199

Noninteractive authoritative facilitation approach 339

Noninteractive dialogic facilitation approach 339

Notice/Think/Wonder routine: part of Observation Instrument 53, 324-325; sequence for thinking actions 337

objects: defined 261; observation chart 260; types and features 272

objects, Case Studies: explanatory conversations 279–280; making, positive responses 275–276; object type and engagement 273–274; observational tools and uses 277–278; scientific reasoning activities 281–282

Observation Instrument: introduction 53; tool 319, 324–325

online classroom platform 213

organizational learning framework 3–4, *4*

Owens, Trevor 275–276

Phases of the Moon, hands-on activity: designed using Learning Cycle Design Framework 109–114, 115; directions 95-103, *96, 97*

power relations, conversations 217, 225, 236

prior knowledge: assess vs retrieval 130; conditions and accuracy 132–133, 138, 149; conditions and misconceptions, discussion 137–139; facilitating learner experience 6, 95, 130; families as facilitators 237–238; grain size of knowledge 132–133, 138–139, 149–151; Learning Cycle phases 110, 112; learning foundation 33; mental models 116–117

Prior Knowledge and Conceptual Change, reading 149-154

privacy considerations, video recordings 43

problem setting 47, 320

professional learning: definition 31, 45; deeper discussion 162–164; knowledge development 60; reflective practice, integral role 61, 163; time and resource investment 44; *see also* reflective practice, effective

professional learning community: critical friendships 48; definition 16, 31, 45; reflective practice influence 65

professional recognition 19

Protocol for Video Reflection: introduction 52; use 166, 249, 306; tool 319, 322–323

Quick Write: Discussion Map experiences 215; learning experience design 288; What is and is NOT science? 73–75

Rapid Design Feedback: learning experience design 294–295; reflection exercise worksheet 340; reflection questions 341–342

readiness, concept 56, 62

Reading Partner Jigsaw routine: how people learn 105–106; prior knowledge and conceptual change 137–139; sequence for thinking actions 337; talking to learn 185–188

Reflecting on Practice program: design elements 31–32; educational equity guidance 5–7; format and content 32, 32; implementation guidance 7–8; institutional adoption 31; key objectives 1, 31; organizational learning framework 3–4, 4, 4; professional recognition 19; scholarship areas 2, 2–3; time and resource investment 14, 44

reflection sessions: community agreements 317; introduction 52; preparatory checklist 318; purpose and benefits 316; roles and responsibilities 317; tools, objective and usage 319

reflective practice: concept development 60; critical reflection and change 63–64; journey in *Reflecting on Practice* program *316*, 316; motivation and its influences 62; problem setting and resolution 63; professional learning 61; professional learning community 65

Reflective Practice and Professional Learning, reading 60-67

Research Discussion tasks: concept map, creative process 105–109; facilitating conversations, topic expert groups 216–217; Five Foundational Ideas on Learning 25–29; learning and talking revisited 245–247; literature as practice resource 27; objects and learner interaction 262–265; practical guidance 27; prior knowledge and conceptual change 137–139; program design element 32; reading partner jigsaw 105–106; reflective practice and professional learning 45–47, 162–164; science and knowing, viewpoint review 79–81; talking to learn topics 185–188; tension, power, and family 216–217

Retrieve & Connect tasks: 3-2-1 Bridge 19–20, 308–310; Minute Paper 168–169, 250; Quick Write 73, 215, 288; Table Talk 247–249; Think-Pair-Share 44–45, 129–130, 178–179, 261–262; Thought Swap 17–18; Turn & Talk 20–21

Role-Play, hands-on activity: checking for understanding (Script 2) 182–183, 194–195; Debrief Chart 175–176, 180, completed example 190; discussion points 185; educators give information (Script 1) 180–182, 191–193; learners' own thinking (Script 3) 183–184, 196–198; outline 179

Rosaen, Cheryl 65

rough design 266

Rowe, Shawn 281–282

scaffolding 236

Schön, Donald 60

Science and Knowing, reading 83-87

science capital 85–86

science, nature and practice: knowledge systems 83–84; science and engineering 83; scientific process misinterpretations 84–85; youth perceptions, related influences 85–86

sense making 101, 104–105

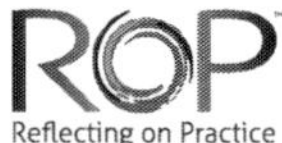

Sentence-Phrase-Word routine: learning and talking 245–247; reflective practice & professional learning 162–164

Sheridan, Kimberley 275–276

Sherin, Miriam 65

social cognitive system 117–118

staff development: counterproductive practice 2; frame of reference 47

Table Talk routine: Cup & Card activity 24–25; facilitation guidance 24–25; How Science Works flowchart 77–79; objects, critiquing 306–308; objects, Educator Moves 304–305; Phases of the Moon activity 104–105; prior knowledge, Educator Moves 164–165; Video Reflection experiences 247–248

Talking to Learn, reading 199-207

teaching purpose: informs facilitating conversation 203–205; reading partner discussions 185–188; role plays, discussion points 182, 183, 184–185, 224

Teaching Purposes, tool 336

Tension, Power, and Family, reading 235-239

Think-Pair-Share routine: learners' prior knowledge 129–130, 139–141; learning conversation characteristics 178–179; professional learning 44–45; sequence for thinking actions 337; teaching with objects 261–262

Thinking Moves: introduction 289–291; tool 337

Thinking Routines: connection to Thinking Moves 290–291; key objectives 17; tool 337

Thought Swap routine: 17–18; sequence for thinking actions 337

touch of silence 17

Tran, Lynn Uyen 31, 61, 279–280

Turn & Talk routine: facilitating conversations 219–220, 224–226; key objectives 20; learning beliefs and behaviors 20–21; sequence for thinking actions 337; video reflection experience 167, 250

Two Balloons, hands-on activity: design analysis **231–232**; directions 217-224; Ocean: A Giant Heat Reservoir handout 222, 233–234; prior conceptual knowledge 229; purpose and approach 217–218; Science Content Guide 228–230

value-laden language 55–56

Vartabedian, Melanie A. 281–282

Video Reflections, first practice: critical friendships 49; educator experiences 65; facilitator and topic selection 39–40; learning requirements 14–15; participant concerns 50; practice roles, responsibilities and agreements 51–52; privacy considerations 43; procedure elements and planning 40–43; program design element 32; protocol, guided practice 53-58; protocol, sections overview 52; purpose and shared benefits 49–50

Video Reflections, whole-community: critical-colleague group preparations 168–169, 251, 310; Educator Moves, first introduction 165; educator experiences 165, 247–248; preparatory tasks 159-160, 243–244; 303

Vygotsky, Lev 2, 33, 65, 201

Walkabout routine: facilitation guidance 25–26, 28–29; Five Foundational Ideas on Learning 25–29; routine's progression 27–28; sequence for thinking actions 337; science and knowing 79–81

White, Roderick E. 3–4

workplace learning: attitudes and approaches 1–2; individual to organization 3–4

youth, science career barriers 85–86

Zimmerman, Heather, T. 277–278

zone of proximal development 236